A GUIDE TO *Modern Biology*

A GUIDE TO
Modern Biology
GENETICS, CELLS AND SYSTEMS

ELEANOR LAWRENCE

Copublished in the United States with
John Wiley & Sons, Inc., New York

Longman Scientific & Technical
Longman Group UK Limited
Longman House, Burnt Mill, Harlow
Essex CM20 2JE, England
and Associated Companies throughout the world.

Copublished in the United States with
John Wiley & Sons Inc., 605 Third Avenue, New York, NY 10158

First published in 1989

British Library Cataloguing in Publication Data

Lawrence, Eleanor
A guide to modern biology.
1. Organisms. Cells 2. Molecular genetics
I. Title
574.87

ISBN 0-582-44272-9

Library of Congress Cataloguing in Publication Data
Data is available
ISBN 0-470-21401-5 (USA only)

Set in 9½ / 11½ Sabon Roman
Produced by Longman Group (FE) Limited
Printed in Hong Kong

PREFACE

This book was initially inspired by the explosion of new work in many areas of biology that has been fuelled directly or indirectly by the application of recombinant DNA techniques. On the one hand, the increasingly rapid pace of research has led to increasing specialization, jargonization and fragmentation in the scientific literature, making it more difficult for the student or the non-specialist to get their bearings. But behind this apparent fragmentation, a universal language of modern cell biology is emerging, and is beginning to make an impact on higher-level problems such as those posed by development or the workings of the immune system. This book is intended as a selective guide to the present state of certain key areas of biology, the framework of ideas in which new work is interpreted, and the connections that are being made between hitherto separate areas of research. The fruits of recombinant DNA research also affect us all more directly, in their applications to medicine, agriculture and industry, and these aspects are also covered.

The book is intended both as a quick reference source for individual topics, and as more general introductory reading for the student or non-specialist embarking on a new or unfamiliar subject. Some general knowledge of biology, basic biochemistry and chemistry has had to be assumed in most sections. Those sections of more general interest should, however, be accessible to anyone with an interest in biology. Each chapter starts with a general introduction, setting the subject in context. This is followed by a series of short sections on individual topics. These are arranged so that those unfamiliar with the subject area can work their way through from the beginning; the more knowledgeable reader can use the index to find individual sections of interest.

The scope of the book has been deliberately restricted to areas of biology that can be directly addressed in terms of molecular genetics and cell structure and function, with a strong bias towards animal cells and mammalian and human biology. It therefore does not cover subjects such as ecology, animal behaviour and classical evolutionary biology, or psychology, which are concerned with a different level of biological organization. With regret, also, much work specifically related to plants has also been omitted, as in a book of this size and level it is difficult to do equal justice to the very different modes of life and specializations of plants and animals.

A short, purely personal, selection of useful basic texts is given at the end of the book, along with guidance for the non-specialist on where to look for more information. In addition, a brief list of further reading (mainly review articles and recent papers) is provided at the end of each chapter. This will provide an entry to the current scientific literature. References in the further reading lists are arranged in the order in which topics are covered in the text. A deliberate decision has been made not to include the names of individual scientists in the text, except in the few cases where their name has virtually become synonymous with the theory or technique mentioned.

This book would never have been started, let alone completed, without the encouragement and patience of the staff at Longmans. My thanks also to Gillian Anlezark, David Pritchard, Helen Skaer, Howard Stebbings, Peter Sudbery, and Robin Weiss who very kindly read individual chapters and made many constructive suggestions for improvement and to Miranda Robertson for her helpful advice. Any remaining errors of fact or interpretation are my responsibility. Sarah Bunney's careful editing greatly improved the clarity of the text. Finally, grateful thanks to my husband and daughter who have also lived with this book and its tribulations for the past two years.

Eleanor Lawrence
London 1989

CONTENTS

ACKNOWLEDGEMENTS

We are grateful to the following for permission to reproduce copyright material:

Academic Press Inc for figs 10.12 (a) & (b) adapted from an article by J E Sulston et al pp 64–119 *Developmental Biology* Vol 100 (1983) and figs 2.10 (a) & (b) adapted from *In The Proteins* by R E Dickerson (Vol 2, 2nd Edition); The American Association For The Advancement of Science for fig. 6.1 from the article 'The Fluid Mosaic Model of the Structure of Cell Membranes' by S J Singer & G Nicholson pp 720–731 *Science* Vol 175 (18/2/72) Copyright 1972 by AAAS and fig 10.13 from the article 'Control Circuits for Determination and Transdetermination' by S A Kauffman pp 310–318 *Science* Vol 181 (12/7/73) Copyright 1973 by AAAS; Annual Reviews Inc for fig 3.4 from an article by N Maeda & D Smithies pp B1–108 *Annual Review of Genetics* Vol 20 (c) 1986 by Annual Reviews Inc and fig 10.17 from an article by P W Sternberg & H R Horvitz pp 489–524 *Annual Review of Genetics* Vol 18 (c) 1984 by Annual Reviews Inc; The Benjamin/Cummings Publishing Co for fig 3.1 adapted from *Molecular Biology of The Gene* by J D Watson et al (Vol II, 4th Edition) (c) The Benjamin/Cummings Publishing Co and fig 12.21 redrawn from, and table 12.2 adapted from, *Immunology* by L E Hood et al (2nd Edition) (c) 1984 The Benjamin/Cummings Publishing Co; Blackwell Scientific Publications Ltd for figs 10.6 & 10.14 adapted from *An Introduction to Development Biology* by J McKenzie (1976); Cambridge University Press for figs. 8.11 (b) & (c) & 10.4 adapted and redrawn from *The Invertebrates* by R McNeill Alexander (1979); Cell Press for fig 3.6 adapted from an article by L Patthy pp 657–663 *Cell* Vol 41 (1985); Garland Publishing Inc for fig 1.1 redrawn from *Molecular Biology of The Cell* by B Alberts et al

(1983); Macmillan Magazines Ltd for fig 6.9 redrawn from an article by M Noda et al in *Nature* Vol 320 (c) 1986 Macmillan Magazines Ltd, fig 4.4 redrawn from an article by J Marks, J-P Shaw and C-K Chen in *Nature* Vol 321 (c) 1986 Macmillan Magazines Ltd, fig 10.21 adapted from an article by P A Lawrence et al in *Nature* Vol 328 (c) 1987 Macmillan Magazines Ltd, fig 12.17 adapted from an article by J A Todd, J I Bell & H O Devitt in *Nature* Vol 329 (c) 1987 Macmillan Magazines Ltd, fig 10.16 adapted from an article by H Woodland & E Jones et al in *Nature* Vol 332 (c) 1988 Macmillan Magazines Ltd, fig 11.3 adapted from an article by R Doll pp 589–596 *Nature* Vol 265 (c) 1977 Macmillan Magazines Ltd; The Rockefeller University Press for fig 8.9 (b) adapted from an article by D L Taylor, J A Rhodes & S A Hammond in *The Journal of Cell Biology* Vol 67 (1975); Springer-Verlag New York Inc for fig 7.5 adapted from *Ultrastructure of The Mammalian Cell* by R V Krstic (1979).

CHAPTER 1

THE CELLULAR WORLD

This book is concerned largely with what cells do and how they do it. It deals almost exclusively with **eukaryotic cells**, the cells of plants and animals, whose vast range of structural specialization enables them to perform an equally wide range of tasks. This brief introductory section sets the eukaryotic cell in its wider context and introduces some basic terminology.

THE GREAT DIVIDE: PROKARYOTES AND EUKARYOTES

The cell is the basic building block of the living world. Despite a great variety of shape, size and internal structure there are only two basic designs of cell. These mark a deep divide within living organisms, an evolutionary division far deeper than the traditional distinction between animals and plants.

Prokaryotes

On one side are the **prokaryotes** – the bacteria and the blue-green algae (cyanobacteria). Their small cells are simple in internal structure, lacking a distinct nucleus. They also lack other organelles such as mitochondria and chloroplasts. The molecular apparatus associated with energy conversion in respiration and photosynthesis, which in more complex cells is enclosed in these organelles, is carried in the cell membrane or in simple membranous structures in the cytoplasm. Their 'chromosome' is made of a single circular molecule of doxyribonucleic acid (DNA), anchored to the cell membrane and not enclosed in a nucleus. They reproduce by binary fission after replicating their DNA. Prokaryotes are invariably unicellular. Some form filaments or loose colonies of cells but there is nothing approaching the cellular division of labour seen in truly multicellular organisms. The true bacteria and the cyanobacteria possess rigid proteoglycan cell walls which differ in

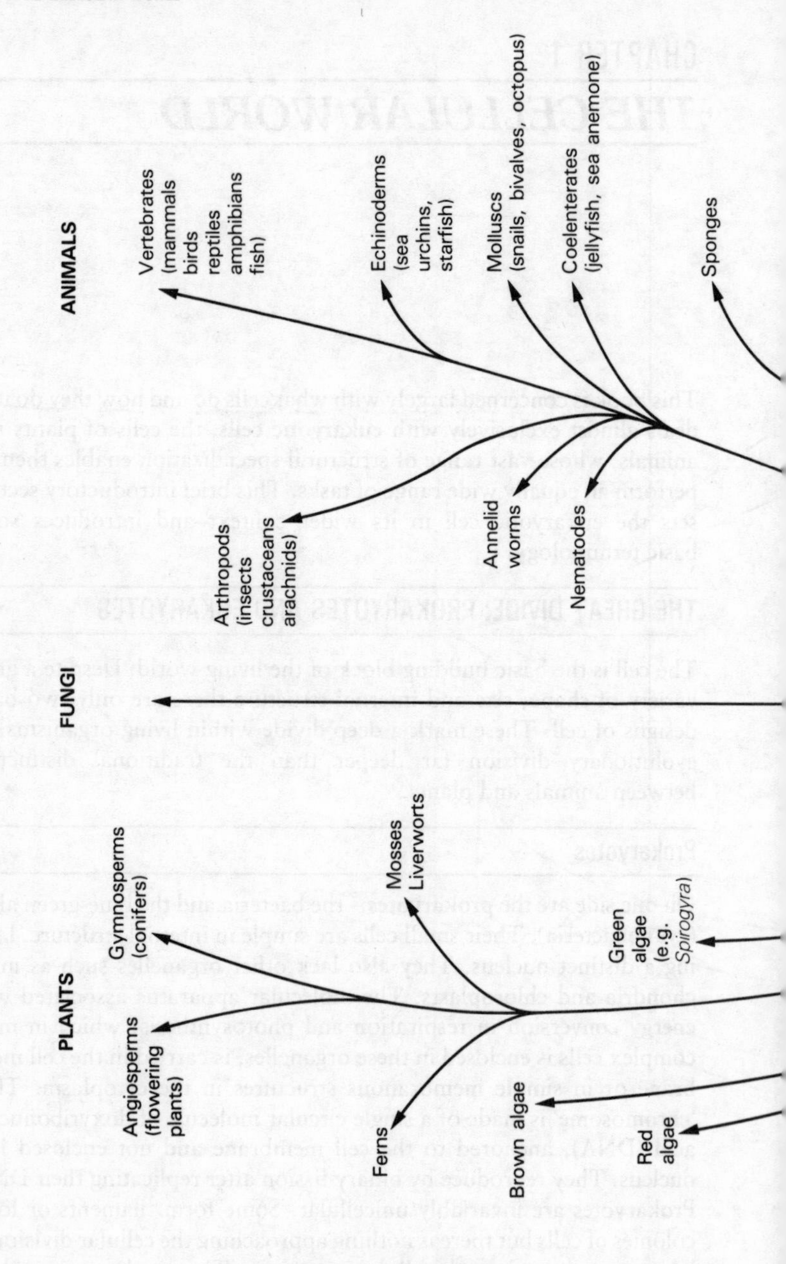
ANIMALS
Vertebrates (mammals birds reptiles amphibians fish)
Echinoderms (sea urchins, starfish)
Molluscs (snails, bivalves, octopus)
Coelenterates (jellyfish, sea anemone)
Sponges
Arthropods (insects crustaceans arachnids)
Annelid worms
Nematodes
FUNGI
PLANTS
Gymnosperms (conifers)
Angiosperms (flowering plants)
Ferns
Mosses Liverworts
Green algae (e.g. Spirogyra)
Brown algae
Red algae
MULTICELLULAR EUKARYOTES

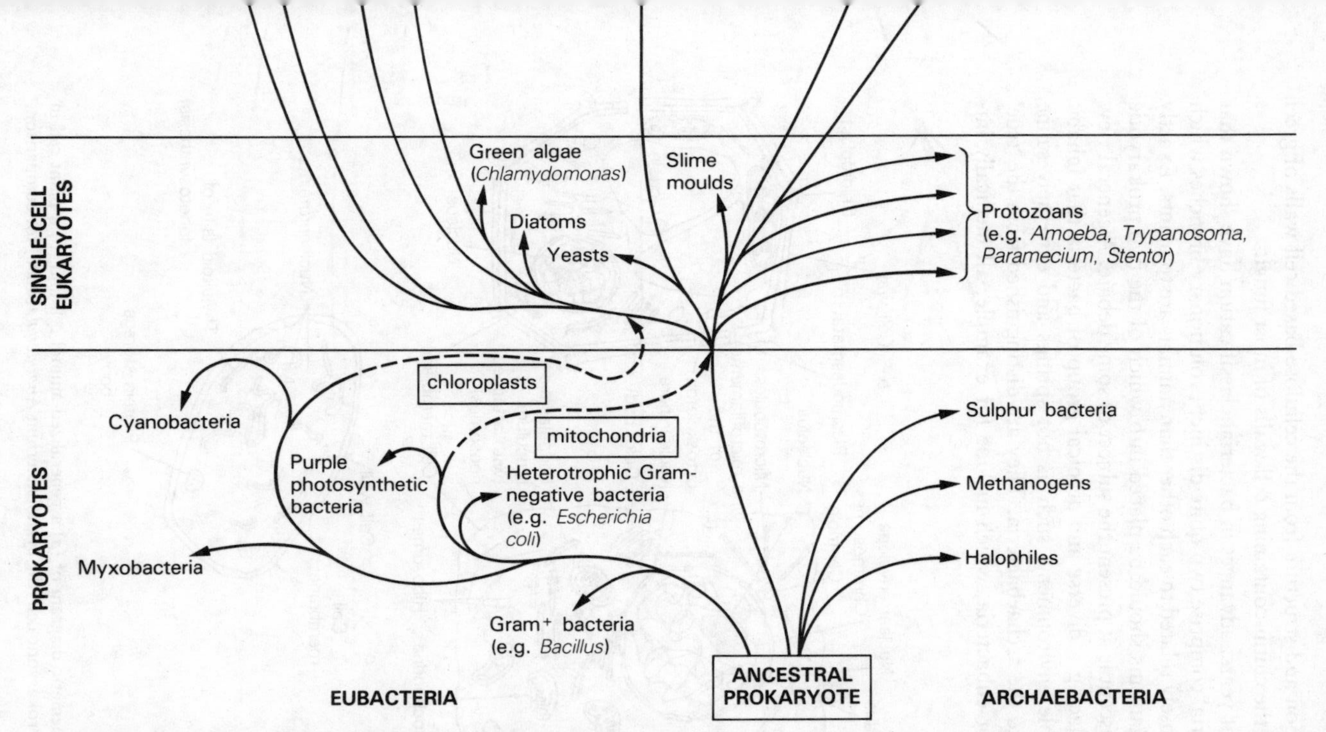

Fig. 1.1 The main groups of living organisms and their presumed descent from an ancestral prokaryote. The diagram is organized to show lines of descent and groups of organisms of similar levels of organizational complexity. It should not be read as a conventional evolutionary family tree where the vertical dimension indicates the passage of time. Redrawn with modifications from Fig. 1-38 of B. Alberts *et al.*, *Molecular Biology of the Cell* (Garland, New York, 1983).

composition and structure from the cellulose-based cell walls of green plants or the chitin-containing cell walls of most fungi.

In recent years, advances in bacterial classification have shown that the 'bacteria' comprise two quite distinct evolutionary branches which are less closely related to each other than animals are to plants. Exactly which organisms should be placed into which of the two prokaryotic lines of descent is at present the subject of some debate. A general view favours placing a diverse and atypical group of bacteria that inhabit inhospitable environments such as hot springs and extremely saline waters into the **archaebacteria**. They are distinguished from the 'true' bacteria or **eubacteria**, which include for example, all medically im-

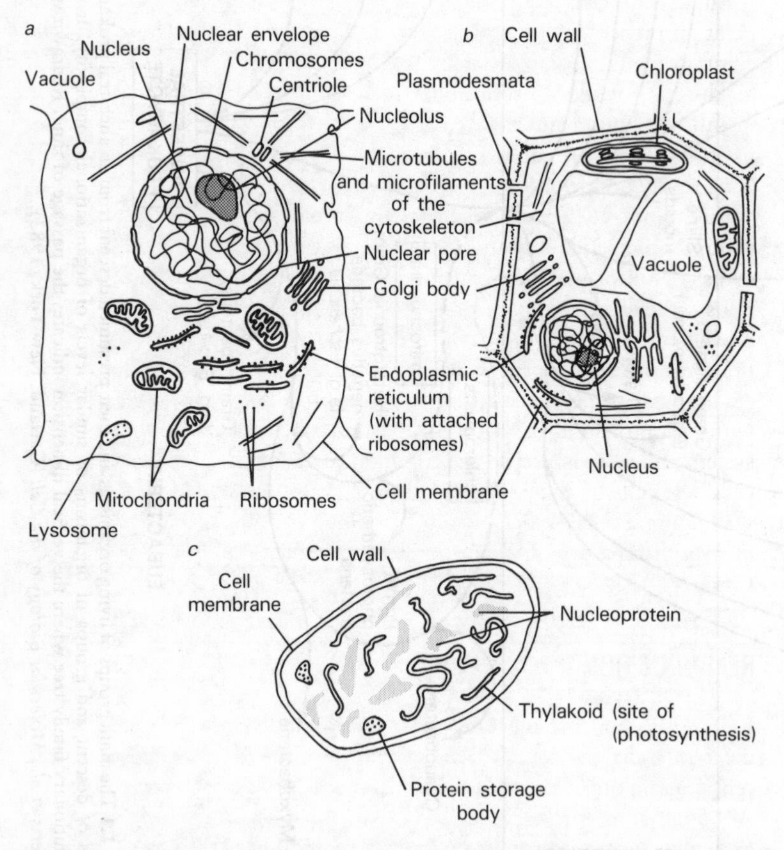

Fig. 1.2 Schematic diagram of (*a*) a generalized animal cell and (*b*) a plant cell in section. *c*, Prokaryotic cell (the cyanobacterium *Anabaena azollae*) for comparison.

portant bacteria, familiar soil bacteria and the cyanobacteria. The archaebacteria are believed to represent the oldest group of living organisms still in existence.

Eukaryotes

Eukaryotes comprise the remainder of the living world. All animals, plants, multicellular algae and fungi are eukaryotes, as are some micro-organisms – yeasts and other unicellular fungi, slime moulds, protozoans, and the unicellular green algae such as *Chlamydomonas* (Fig. 1.1). Protozoans, algae, slime moulds and some fungi are often grouped together as the Protista.

Eukaryotic cells are typically larger than prokaryotic cells. Despite great differences in size and structure all eukaryotic cells share a typical compartmentalized internal design (Fig. 1.2). Their DNA, and the machinery for its transcription into RNA is sequestered in a distinct membrane-bound **nucleus**. Aerobic respiration and, in green plants, photosynthesis, are carried out in specialized **organelles – mitochondria** (sing. mitochondrion) and **chloroplasts** respectively.

Eukaryotic cells possess extensive internal membrane systems – the **endoplasmic reticulum** and **Golgi apparatus** (see Chapter 7) – and a **cytoskeleton** of protein fibrils and tubules (see Chapter 8) that maintains the distinctive shapes of eukaryotic cells and is involved in cell movement. Other specialized internal organelles serve various functions in different types of cell. Not all organelles are present in all eukaryotes; some species (mostly protozoan parasites) even manage without mitochondria, but a nucleus, an extensive internal membrane system and a cytoskeleton are common to all and define the eukaryotic cell. The cells of algae, plants and fungi are surrounded by a rigid cell wall; animal cells lack cell walls. The eukaryotic cell is believed to have evolved from a prokaryotic ancestor some 1500 million years ago (see Chapter 5).

REPRODUCTION

Eukaryotes and prokaryotes differ considerably in their mode of cell division and reproduction. Prokaryotes are invariably **haploid** – their cells contain only one set of the genes characteristic of the given species. Although some prokaryotes have a form of sex, in which genes are exchanged or transferred from one cell to another of the same (or even a different) species, there is nothing resembling the sexual reproduction that is characteristic of most eukaryotes. A bacterial colony is in general made up of genetically identical individuals, the product of successive asexual divisions of a single cell.

Almost all eukaryotic species can reproduce sexually, even if, as with many microorganisms and plants, they often reproduce asexually. Most familiar plants and animals are therefore **diploid**, that is they contain two complete sets of genetic instructions in all their cells, in the form of two sets of chromosomes, one derived from each parent when the male and female single-celled **gametes** (e.g. ovum and sperm) fuse at fertilization.

Meiosis, or reduction in chromosome number, takes place in eukaryote cells during gamete formation. This ensures that each gamete only receives a single chromosome set, the full diploid complement being restored when two gametes fuse to form the **zygote** or fertilized egg, from which a new individual develops (Fig. 1.3).

The body (**somatic**) cells of this individual will all be diploid. When eukaryotic cells divide, their chromosomes, which have already replicated in readiness, go through a strict programme of **mitosis** which

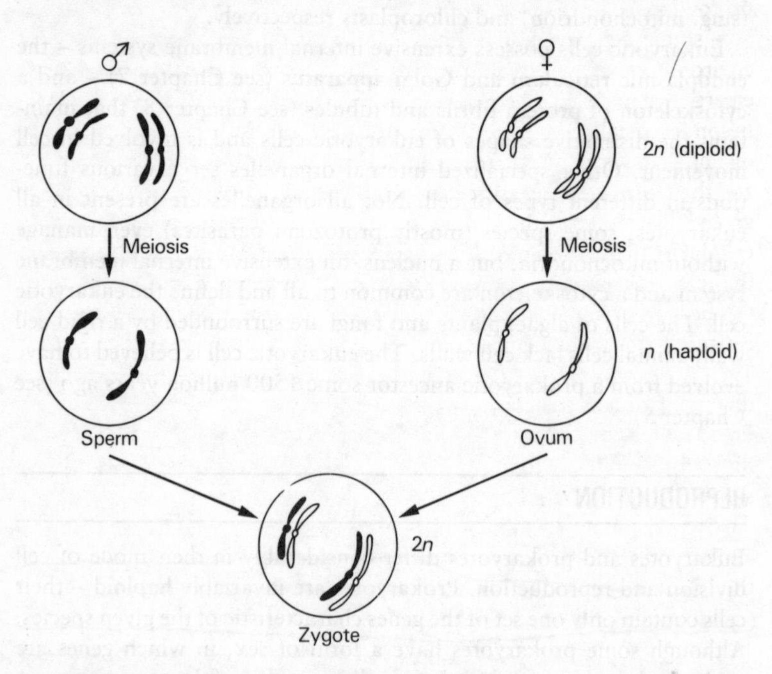

Fig. 1.3 Gamete formation and reconstitution of diploid zygote on fertilization. Only two pairs of chromosomes shown for convenience.

ensures correct separation and segregation into the two daughter cells so that each contains a full diploid complement (see Fig. 2.1).

Mosses and ferns, many algae and many fungi pass through a distinct 'haploid generation' after meiosis, producing an independent haploid organism. In vascular plants (e.g. flowering plants and gymnosperms (conifers)) and most animals the haploid generation is reduced to the gamete.

THE ADVANTAGES OF BEING MULTICELLULAR

Some protozoans are highly complex in structure, with different parts of the cell specialized to a high degree for such functions as feeding and locomotion. But there is a limit to how much can be achieved within a single independently living cell. In particular there seem to be finite limits on the size of free-living cells – the largest single-celled organisms are giant amoebae whose multinucleated cells can reach up to several centimetres in diameter, but most unicellular organisms fall within the range of a few micrometres (bacteria) to a few millimetres (the larger protozoans).

A single cell, even if mobile, can exploit only a tiny part of its environment. New possibilities for making use of available resources emerge when cells come together to form larger cooperative units and become specialized to serve different functions. This division of labour distinguishes true multicellular organisms from those algae and protozoa that form colonies of identical individuals. In a few cases (e.g. the green alga *Volvox*) the colonies foreshadow the evolution of multicellularity in having a primitive differentiation of function between cells that can reproduce to form a new colony and those that cannot.

The cells of multicellular organisms reveal the extraordinary versatility and evolutionary potential of the basic eukaryotic cell design (Fig. 1.4). The evolution of eukaryotes represents an elaboration and diversification of structure, in cells and in the organisms they make up. Prokaryotes, by contrast, have evolved in the direction of metabolic diversity, being able to make use of a much wider range of materials as sources of energy and building blocks for growth than eukaryotes can. They carry out a vast range of chemical conversions and syntheses with an ease envied by the chemist, who has yet, for example, to duplicate the efficiency of biological nitrogen fixation.

The range of specializations shown by eukaryotic cells is enormous. For example, in any vertebrate there are more than 200 distinct types of cell. Some, such as fat cells, are entirely given over to storing food reserves. Muscle cells are filled with contractile protein fibres,

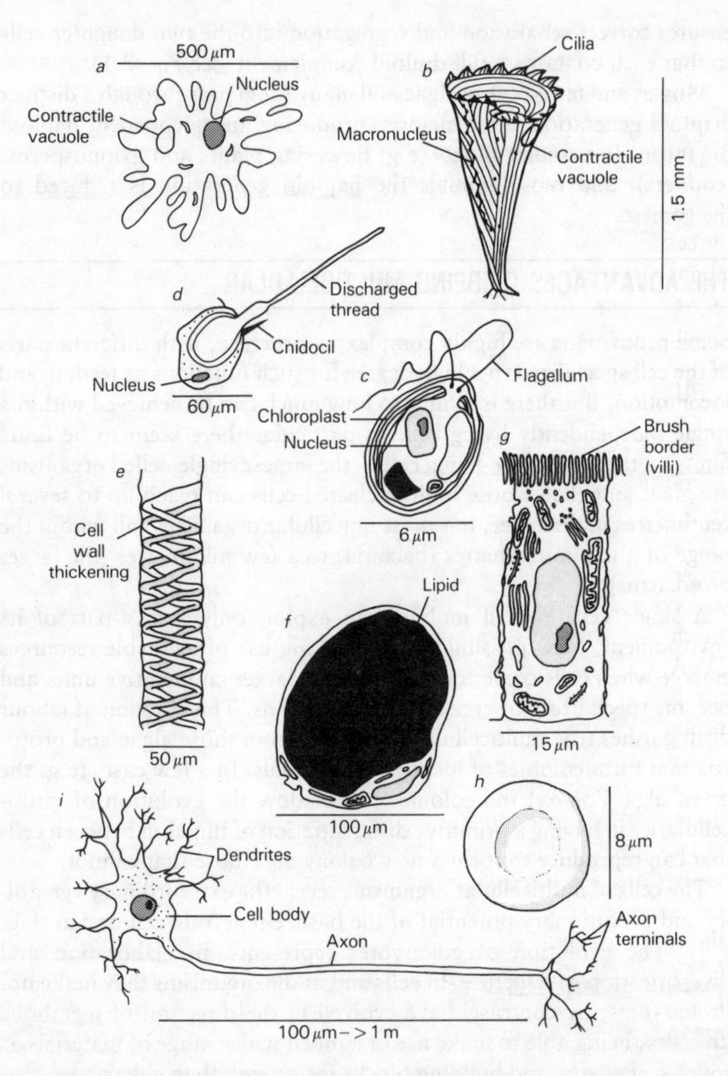

Fig. 1.4 Some eukaryotic cells: *a*,*b*,*c*, the free-living protists *Amoeba*, *Stentor* and *Chlamydomonas*; *d*, cnidoblast (stinging cell) from sea anemone; *e*, xylem cell from a woody plant; *f*, mammalian fat cell; *g*, epithelial cell from lining of gut; *h*, mammalian red blood cell; *i*, nerve cell.

specializing them for force generation. Connective tissue cells secrete materials to make supporting structures that range from elastic ligaments and tendons to load-bearing bone. Nerve cells generate and convey electrical signals.

Multicellular organisms are more than a simple sum of the independent properties of their cells. When cells are organized into tissues and organs new properties emerge. Connected in large arrays nerve cells generate a formidable capacity for information processing, and although the contraction of a single muscle cell has little effect, as part of a larger muscle controlled by nerves it can pump blood round the body or power the wingbeat of a fly. Nerve and muscle working together with highly developed sensory systems enable animals to move, to search for prey and escape predators. Plants have evolved in other directions. Stacked end to end, the xylem cells in the trunk of a tree can support a structure many metres high, that may live for hundreds of years, and that draws nourishment from air and soil and energy from sunlight.

VIRUSES

The other inhabitants of the living world are the **viruses** – minute particles visible only in the electron microscope, which can multiply only inside a living cell (Fig. 1.5). They are composed of a protein shell (the capsid) enclosing genetic instructions in the form of either of the two nucleic acids (DNA or RNA). These direct the manufacture of new virus particles. However viruses do not contain their own biochemical machinery for translating their genetic instructions and making new DNA, RNA or protein. They can only replicate by infecting a cell and using its enzymes and accessory proteins for DNA or RNA replication, transcription and protein synthesis. The newly synthesized virus particles are then released from the cell, sometimes destroying it in the process. Most known viruses have been identified because they cause disease; some are among the most dangerous human and animal pathogens known, others cause only a mild and limited infection. Some viruses are even carried semi-permanently from generation to generation as harmless passsengers in animal chromosomes. Viruses also infect plants and bacteria. Most plant viruses are RNA viruses but a few plant DNA viruses have been discovered. Both RNA viruses and DNA viruses infect animals and bacteria. Viruses that attack bacteria are known as **bacteriophages or phages**.

One class of animal viruses of particular interest to the molecular biologist are the **retroviruses**. These are small RNA viruses that instead of immediately replicating themselves after infection make a DNA copy of their RNA. This is then inserted into one of the host cell chromosomes and may be carried along for many cell generations. Retroviruses have been of great interest ever since their discovery, as they include viruses known to cause cancer in animals and in humans (see Chapter 10). The human immunodeficiency virus (HIV) that causes acquired immunodeficiency syndrome (AIDS) is also a retrovirus.

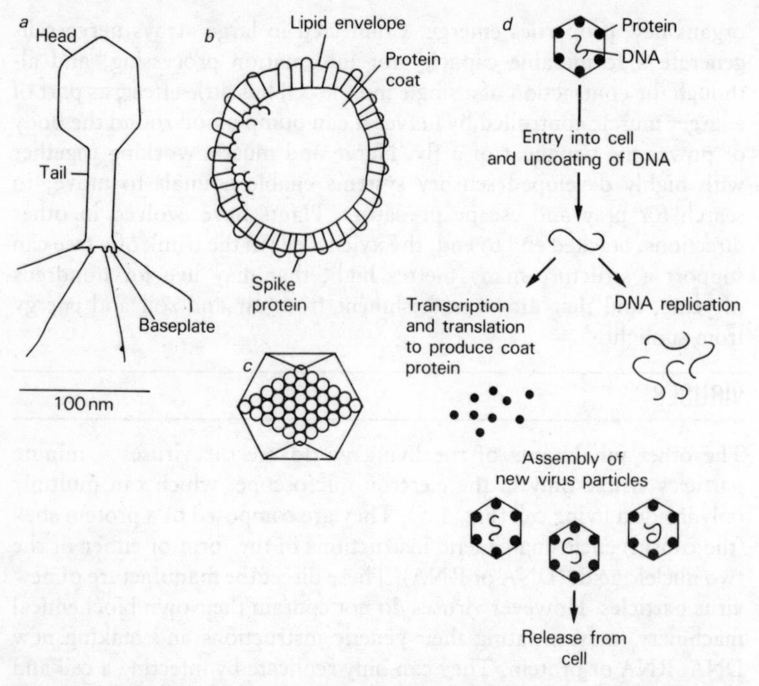

Fig. 1.5 *a*, T4 phage, a bacterial virus that attaches to the outer surface of a bacterium and injects its DNA into the cell. *b*, Influenza virus, a large RNA virus, has an outer envelope composed of a layer of host cell lipid membrane that it picks up as it leaves the cell. *c*, Model of an adenovirus particle showing its icosahedral symmetry and the regular arrangement of the protein subunits that make up its coat. *d*, Life cycle (simplified) of a DNA virus.

FURTHER READING

See general reading list (e.g. Raven et al.; Alberts et al.; Darnell et al.).

D.W. FAWCETT, *The cell*, W.B. Saunders, Philadelphia/London, 1981. Electron micrographs of mammalian cells.

T.D. BROCK, *Biology of microorganisms*, 3rd edn, Prentice-Hall, Englewood Cliffs, NJ, 1979.

S.E. LURIA, *General virology*, 3rd edn, Wiley, New York/Chichester, 1978.

G. R. SOUTH and A. WHITTICK, *Introduction to phycology*, Blackwell Scientific, Oxford, 1987. An up-to-date introduction to the algae.

PART 1 *The Genetic Revolution*

CHAPTER 2

BASIC GENETICS

Determining the material basis of the genes and how they act has been one of the chief preoccupations of biology in the latter half of the twentieth century. The first 50 years saw the rediscovery and vindication of Gregor Mendel's theory of inheritance and solutions to the immediate questions of inheritance – how genetically-based characteristics are passed down from one generation to the next. The properties and behaviour of the gene as the unit of inheritance were exhaustively explored, and the often dramatic effects of alterations (mutations) in individual genes were uncovered in their hundreds in experimental organisms such as the fruitfly *Drosophila*.

But the gene remained a somewhat abstract entity until the discovery of its chemical basis in DNA and the equally important demonstration that genes specify the synthesis of proteins. This opened the way to finding out how genes in general, and individual genes in particular, are acting to influence the development, physiology or behaviour of an organism at the biochemical level.

In attempting to answer the question of what genes do and how they do it, biochemical or molecular genetics has become an indissoluble part of every aspect of biology rather than a self-contained discipline. Its technical language and methodology are often impenetrable, even to other biologists, but the results it is now able to deliver, both in terms of a greater understanding of gene organization and control in higher organisms and also as a purely technical aid for getting to the heart of essentially physiological or developmental questions, are indispensable.

During the past decade genetics has taken a revolutionary and sometimes controversial turn with the development of **recombinant DNA technology** (see Chapter 4). This is the umbrella term for a variety of techniques that are used to isolate, analyse and manipulate individual genes, something that was not technically possible before the mid-1970s. This at last means that the molecular structure and mode of regulation of genes from multicellular animals and plants can be studied in great detail, and has revealed many unsuspected features of gene organization and action.

Molecular genetics based on recombinant DNA techniques is providing new approaches to many of the crucial questions posed by multicellular life. It provides developmental biologists with a way in to the intricate networks of gene activity that regulate embryonic development, and the associated question of how specialized cells differentiate from unspecialized precursors, and how they maintain and pass on their differentiated state to their descendants. Immunologists have been able to solve the longstanding problem of how the body can produce an apparently limitless number of different antibodies to fight infection and are using recombinant DNA techniques to develop new vaccines. Recombinant DNA techniques provide a short-cut to determining the molecular structure of proteins, which cells use to control and carry out their basic metabolism, to communicate with each other and to perform specialized tasks. Recombinant DNA techniques have broadened our knowledge of human genetics in particular, uncovering the genetic defects responsible for many heritable conditions and providing new and powerful methods for detecting an affected foetus or carriers at risk of producing an affected child.

The controversial aspects of recombinant DNA technology spring from its capacity to alter the genetic makeup of plants, animals and bacteria as never before, chiefly by introducing genes from one species into another. Plant and animal genes are isolated and multiplied for further study by transplantation into rapidly multiplying micro-organisms (see Chapter 4: DNA cloning) and bacterial and yeast cells carrying human, animal and viral genes are now commonplace. **Genetically engineered** cells are used commercially to mass-produce hormones, vaccine components, and other useful proteins, and a multimillion pound biotechnology industry has sprung up in the last decade to exploit the potential of recombinant DNA. Plants are being engineered for genetic resistance to insect pests or herbicides by the introduction of genes from other species, and genetically engineered bacteria and viruses are now being tested for release as biological pest control agents or in live vaccines. Genetic engineering is now also poised for application to farm animals (see Chapter 4: The genetic modification of plants and animals). Replacement of a person's own genetically corrected bone marrow cells to alleviate some genetic defects may soon be a practical option (see Chapter 4: Human gene therapy), but the most controversial possibility of all – the genetic modification of human embryos – is not being pursued, and as a means of correcting genetic defects would be a quite uneccessary option given advances in other medical technologies (see Chapter 4: Research on human embryos).

This chapter outlines some basic genetic principles and introduces DNA, RNA and proteins and the way in which information encoded in DNA is used to make proteins. Recombinant DNA techniques have opened up the subject of gene organization and the control of gene expression in multicellular organisms and some of the most important new discoveries are outlined in Chapter 3. The techniques that underlie recombinant DNA technology and some of its medical, commercial and agricultural applications have been placed together for convenience in Chapter 4.

INHERITANCE AND GENETIC VARIATION

Most of the genetics in this book is concerned with gene organization in terms of DNA, and the general principles of gene action and regulation at the biochemical level. This is, however, only one aspect of the broader study of heredity, the latest chapter in the long history of modern genetics, which is generally held to have started with Mendel's revolutionary insight into the nature of inheritance in the mid-nineteenth century. The molecular biologist's gene often seems far removed from the everyday experience of inheritance – the seemingly hit and miss way in which characteristics are handed down from parents to offspring in plants and animals. Our genes unambiguously define us as human, dog or daffodil but at the same time appear to be capable of generating endless variations on the theme.

It has been assumed throughout this book that most readers will be acquainted with the basic principles of inheritance in sexually reproducing organisms and the reason for genetic variation, but for those with only hazy memories of monastery gardens and peas, some basic background and terminology are given here.

The mechanism of heredity

The complex nature of inheritance in most eukaryotes arises from sexual reproduction and the consequent random partitioning of a set of each parent's genes into the gametes (e.g. ovum and sperm in animals and pollen and ovum in flowering plants) and the coming together, again at random, of two gametes to produce the fertilized ovum from which a new organism develops.

Genes and Chromosomes

Every living thing inherits a set of genes that determine what sort of organism it will develop into and its individual appearance, and which also mastermind much of its day-to-day existence. In eukaryotic cells the genetic instructions are stored in the nucleus. They are carried in the **chromosomes**, which only become visible under the light microscope as short, thick, densely staining rods just before cell division. Chromosomes are composed of protein and DNA (deoxyribonucleic acid) and it is the DNA that encodes the genetic instructions (see later sections on DNA, etc.). The information encoded in DNA chiefly directs the synthesis of the many different kinds of proteins, which as structural components of cells and as essential biological catalysts (enzymes) are the biochemical basis of life. The fundamental unit of inheritance – **the gene** – is in material terms a stretch of DNA that specifies the manufacture of a single type of protein (or for some genes, certain

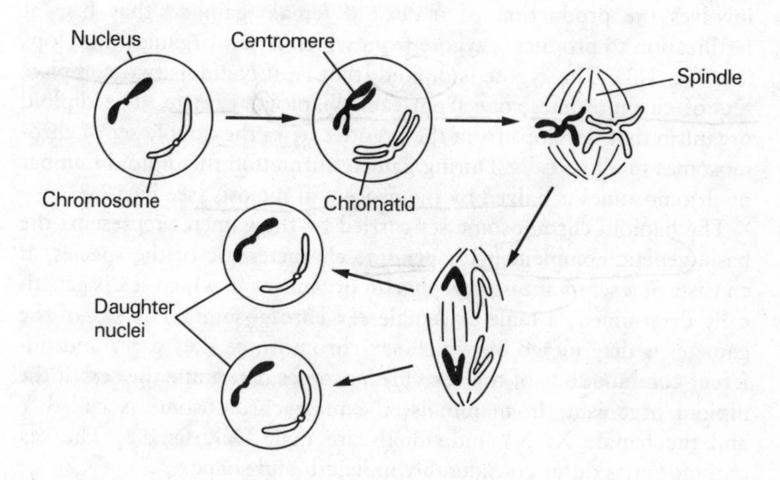

Fig. 2.1 Mitosis. Only two chromosomes shown for convenience. Each chromosome is first duplicated resulting in two sister chromatids which remain attached at the centromere. The duplicated chromosome is called a bivalent. The nuclear envelope breaks down and the chromosomes become aligned on the mitotic spindle (see Chapter 8). Sister chromatids then separate, each moving in an opposite direction. After separation, each chromatid is called a chromosome. Nuclear membranes re-form around each new set of chromosomes to give two daughter nuclei. Usually the cell itself then divides (cytokinesis).

RNAs). Genes are disposed along the chromosomes in a fixed linear order characteristic of the species. The number of genes for any species is not yet known accurately but probably varies from several thousand (for a simple bacterium) to many thousands for a mammal.

When a eukaryotic cell divides into two daughter cells the chromosomes, which have already been duplicated in readiness for division, go through a series of movements called **mitosis** (Fig. 2.1) which separate and segregate the duplicate sets so that each daughter cell receives a full complement.

Sexual and Asexual Reproduction

Bacteria and many simple eukaryotes generally reproduce asexually by cell division or budding off a new cell. The offspring of asexual reproduction are genetically identical to their parents as they simply receive a duplicated copy of the chromosome or chromosomes of the parent cell.

Most eukaryotes, however, can also reproduce sexually, and sexual reproduction is the only mode of reproduction in most animals. It involves the production of male and female gametes that fuse at fertilization to produce a zygote from which a new organism develops (see Fig. 1.3). The zygote is diploid, that is, it contains two complete sets of chromosomes, one from each (haploid) gamete. The diploid organism that develops from the zygote carries the double set of chromosomes in all its cells. During gamete formation the diploid number of chromosomes is halved by the process of **meiosis** (see Fig. 2.3).

The haploid chromosome set carried by the gametes represents the basic genetic complement or **genome** characteristic of the species. It consists of a set of **autosomes** plus (in organisms in which sex is genetically determined) a male or female **sex chromosome**. The sex of the gametes is determined by which sex chromosome they carry and different combinations of the sex chromosomes determine the sex of the diploid organism. In mammals, the male chromosome is called Y and the female X. XY individuals are male, XX female. The sex chromosomes differ considerably in length and shape.

The number of chromosomes in a haploid set varies widely from species to species. For example, the fruitfly *Drosophila* has 4, humans have 23. Every diploid human body cell (or **somatic cell**) therefore carries 46 chromosomes – 2 sex chromosomes and 44 autosomes which comprise 22 pairs of corresponding autosomes originating from the male and female parent respectively. The maternal and paternal chromosomes of a pair are said to be **homologous**, and in normal cells are identical in shape and size.

Genetic variation: why everyone is different

Heterozygosity

Diploid cells contain two copies of each autosomal gene, one inherited from each parent, but these copies are often not identical. Within each species many genes naturally exist in two or more alternative forms (**alleles**), any of which can occupy the appropriate site on the chromosome (the **locus**). Diploid organisms usually carry different alleles at some, at least, of their loci, and different combinations of alleles can lead to considerable differences in outward characteristics (this is discussed further in the section on genotype and phenotype below). As just one example, in snapdragons (*Antirrhinum*) different combinations of alleles of a gene involved in specifying flower colour variously give white, pink or red flowers (Fig. 2.2). An individual carrying different alleles of a gene is said to be **heterozygous** for that gene, one carrying identical alleles **homozygous**. In most cases, the alleles on both chromosomes of a homologous pair are active (the mammalian X chromosome and similar sex chromosomes in other animals are the exception, see below).

The existence of variant forms of a gene occurring naturally within a population is termed **genetic polymorphism**. Many genes are known to

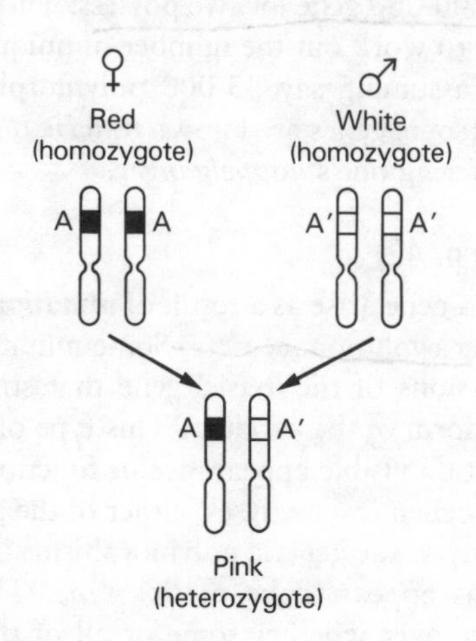

Fig. 2.2 Inheritance of flower colour in *Antirrhinum* showing the effects of different combinations of alleles. *A* and *A′* are alternative forms of a gene involved in specifying flower colour.

be polymorphic in both vertebrates and invertebrates, and some, such as those that specify the histocompatibility antigens that determine tissue type in mammals, are highly polymorphic, with more than a hundred different forms of some histocompatibility genes – hence the difficulty in tissue matching for organ transplantation (see Chapter 12).

Most work on genetic variation has so far (for obvious technical reasons) looked at genes encoding easily identified enzymes, and in large-scale studies about 30 per cent of these turn out to be polymorphic in humans. The average heterozygosity for these genes amongst humans was around 7 per cent – that is, a typical individual will be heterozygous at around 7 per cent of these loci. We do not yet know whether these levels of variability are typical of other sorts of genes but DNA sequence studies confirm a considerable level of variability in other types of gene. Similar or even greater levels of variation are found in other organisms. Because two unrelated individuals are more likely to carry different alleles of any given gene, whereas related individuals have a much greater chance of carrying the same ones, outbreeding (mating between unrelated individuals) tends to increase heterozygosity, inbreeding to reduce it. Inbred and genetically uniform strains of laboratory animals, crop plants, etc. are produced by repeated inbreeding.

The unique genetic makeup of each human individual (identical twins excepted) is therefore the result of various permutations of alleles at the (estimated) 100 000 gene loci we possess. Those mathematically minded might care to work out the number of unique individuals that can be produced, assuming say 33 000 polymorphic loci with four alleles at each one (some genes are known to have many more) and the chances of encountering one's *doppelgänger*.

Mutation (see also p. 42)

Different alleles of a gene arise as a result of **mutations** that occur in an 'ancestral' gene over evolutionary time. Some mutations produce very slightly altered versions of the 'basic' gene that still produce a fully functional variant form of the product. This type of allele usually has no detectable effect on visible appearance or function and can only be distinguished by biochemical analyses, either of the protein product or the gene itself. Many of the genetic polymorphisms uncovered so far in natural populations appear to be of this type. There is at present considerable debate over whether some or all of them are in fact as 'neutral' as they appear or whether this type of variation is maintained by selective forces, each variant form having (or having had in the past) some subtle advantage in certain circumstances that caused it to be positively selected for and maintained in the species gene pool.

In other cases, mutation causes changes in a gene such that it no longer works properly and either does not produce any gene product, or produces one that is altered so that it is inactive or acts in a different way. Such alleles often give rise to quite distinct changes in visible appearance or physiological function (see next section for more details). Alleles of this type produce differences in eye colour, flower colour, etc. Grossly defective alleles of genes vital for survival are usually sooner or later lost from the population, but in sexually reproducing diploid organisms even potentially harmful alleles can persist, if their effects are masked in heterozygotes by the presence of the good gene (see Chapter 4: Genetic disease).

The terms *mutation* and *mutant* as used in general biology usually describe changes that produce organisms showing some altered and usually deleterious abnormal characteristic not seen in the natural population. However, molecular geneticists sometimes term any change in DNA, even those that produce no effect, as a mutation.

Recombination

The effect of any particular gene is usually greatly influenced by the actions of many other genes, and therefore by the particular combination of alleles of other genes that an individual has inherited. Sexually reproducing organisms generate new combinations of alleles on each chromosome during gamete formation by the process of **genetic recombination**. Pieces of homologous chromosomes are exchanged in reciprocal fashion during meiosis by a process known as **crossing-over**, which involves breakage and reunion of chromatids from different homologous chromosomes in a complex series of molecular manoeuvres. The chromosomes passed on to a gamete therefore each contain contributions from both its grandparents (Fig. 2.3). Recombination reconstructs a chromosome containing grandparental alleles in new combinations and is an important source of genetic variation between offspring of the same parents.

Environment

The form and function of an organism are the result of the actions of thousands of genes, their protein products all interacting in the biochemical networks that underlie development and physiology. At all stages the final outcome of gene action may be considerably influenced by external 'environmental' factors – nutrition, climatic conditions, stress or trauma, competition from other animals and plants, etc. – which can lead to considerable differences in final size and appearance

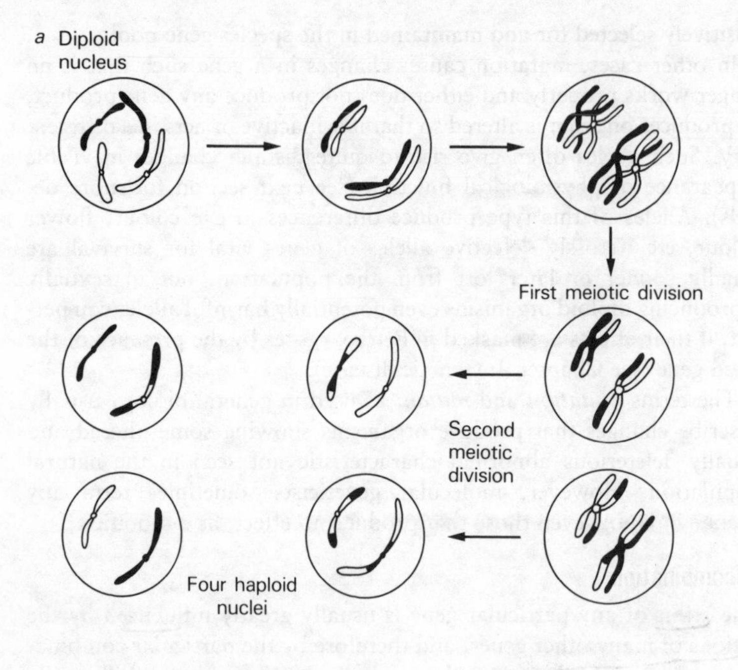

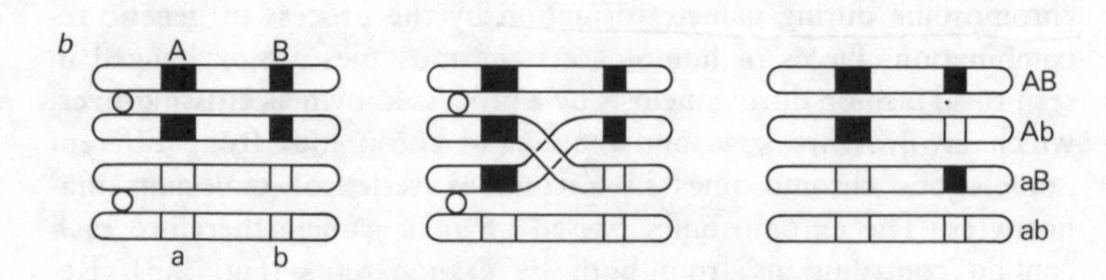

Fig. 2.3 Meiosis and genetic recombination. Only two pairs of chromosomes are shown for convenience. *a*, The stages of meiosis. Homologous chromosomes pair up along their length (synapsis) and are duplicated without separating. Recombination (crossing over) between chromatids of the different chromosomes occurs. At the first meiotic division the two chromosomes separate and are segregated into different nuclei (the reduction division of meiosis). Each new nucleus now contains half the diploid number of chromosomes. These nuclei then divide once more by mitosis (the second meiotic division), segregating each chromatid of the bivalent into a different nucleus. Four haploid nuclei are therefore formed from the original diploid nucleus. *b*, Reciprocal recombination and its outcome. Uppercase and lowercase letters represent different alleles of the same gene.

between two organisms of the same species and similar genetic constitution but living in different conditions. However, differences due to environmental factors are *not* heritable.

PHENOTYPE AND GENOTYPE

Until the nineteenth century the apparently haphazard nature of inheritance in plants and animals bedevilled all attempts to provide coherent and universal rules of heredity. Plant or animal hybrids usually resemble one or other of their parents in some respects but show quite considerable differences from either in others. A distinctive feature may disappear for a generation or two and then just as mysteriously reappear in a succeeding generation. We now know that these baffling phenomena are the direct result of diploidy, meiosis, past mutation and resulting heterozygosity, and genetic recombination.

The 'classical' genetic analysis of diploid organisms seeks to describe in formal terms the complex relation between outward appearance and underlying genetic constitution that arises from the mechanism of inheritance in sexual reproduction. After the demonstration in the 1940s that genes encode proteins, it also became possible to try and explain the effects of different alleles in biochemical terms – that is, in terms of which proteins they produce (or fail to produce) and to relate this to visible or otherwise detectable effects.

A crucial distinction in genetics is that between **phenotype** and **genotype**. Phenotypic characters are outward characteristics, such as eye colour, hair colour, flower colour and height, as well as less obvious attributes, such as whether an organism is able to produce a particular enzyme. The genotype on the other hand is the actual genetic makeup of the individual. In broad terms this means the gene or genes that underlie and determine the phenotype, and in the more precise terminology of classical genetics the exact combination of alleles that underlie alternative states of a particular phenotypic character (see Fig. 2.4).

Phenotype is ultimately determined by genotype but is also greatly influenced by environmental factors. Height in humans is a good example. One's height potential is genetically determined but the adult height reached depends on a variety of external factors such as good nutrition in childhood. The fact that a mammal's extremities (paws, and tips of tail, ears and nose) are habitually colder than the rest of the body is the reason for the markings of Siamese cats – in this case the environmental factor of temperature is interacting with the genetically determined process of pigment production to produce a distinctive phenotype.

Some characteristics are obviously influenced to a much greater degree than others by environmental factors. For example, the relative intrinsic genetic contribution to human personality traits, behaviour and measured intelligence, which are moulded to such a great extent by an individual's personal experience, and are moreover difficult to measure objectively, is extremely difficult to determine and has long been the subject of heated 'nature versus nurture' debates.

Relation of genotype to phenotype

Multifactorial (Polygenic) Inheritance

Many obvious phenotypic features in higher organisms, such as body size, height, the shape of a hand or grandmother's nose, are the result of the combined effects of many different genes acting independently of each other. In many cases there is also considerable environmental input. When the number of genes involved is very large or the environmental contribution very great it is impossible to identify the effects of individual genes and determine the underlying genotype. Phenotypic characters determined by large numbers of genes are said to show **multifactorial** or **polygenic inheritance**. In most cases, especially where there is considerable environmental influence as well, there is a smoothly graded spectrum of variation rather than a number of discrete phenotypes; such traits are said to be **continuously variable**. Although they do not show a predictable pattern of inheritance – for example, a child's exact height cannot be predicted from that of its parents – each of the underlying genes will, of course, be inherited according to the normal rules. The study of variation in a population arising from multifactorial inheritance is largely the province of quantitative genetics and is not covered further here. An understanding of the nature and inheritance of continuously variable characteristics is, however, of great importance, especially in ecological and evolutionary studies.

Single-gene Traits

In contrast to multifactorial traits, some phenotypic characteristics show up in two or more quite distinct contrasting states, either as naturally occurring alternatives or as 'mutant' versus 'normal', which are inherited in various predictable ways. These represent characters that are essentially determined by the effects of different alleles at a single locus and can therefore be used to dissect the inheritance and function of single genes. They are often called simple mendelian traits after Mendel who first discovered a pattern to heredity and outlined a *a* Purple (♂) *CC* White (♀) *cc* Parental generation

All *Cc* (purple) allowed to self F1 generation

Male gametes

Female gametes	*C*	*c*
C	*CC*	*Cc*
C	*Cc*	*cc*

CC (purple) (25%) *Cc* (purple) (50%) *cc* (white) (25%) F2 generation

b

Male gametes

Female gametes	*C*	*c*
c	*Cc*	*cc*
c	*Cc*	*cc*

Offspring are 50% heterozygous purple and 50% homozygous white

Fig. 2.4 *a*, Cross between two pure-breeding lines of garden pea, one with purple flowers, the other with white. The hybrid progeny of the first generation (F1) are all purple flowered, but when these are allowed to self-fertilize or are intercrossed, white-flowered plants reappear in their progeny in a ratio of approximately 1:3. This ratio emerges however many times the cross is repeated and is unchanged if the male parent is white flowered and the female purple flowered. Purple is said to be a dominant trait, white recessive.

These phenotypes are described in terms of the following genotypes. *C* and *c* are alternative alleles of a gene determining pigment production. The pure-breeding parents are homozygous – *CC* (purple) and *cc* (white) – and produce respectively all *C* or all *c* gametes. *C* pollen combining with *c* ova (or vice versa) gives *Cc* plants in the F1 generation. Since these are always purple *C* is said to be dominant over *c*, which is termed the recessive allele.

Each *Cc* parent will produce *C* and *c* gametes in approximately equal quantitites. The offspring of the second generation cross – the F2 generation – therefore consist of genotypes *CC*, *Cc*, *cC* and *cc* in approximately equal amounts. These give only two phenotypes – purple and white – in the proportions of approximately 3:1.

b, Backcrossing the heterozygote to the homozygous recessive parent gives another typical pattern of inheritance which confirms that the characteristic being considered is determined by a single locus.

correct theory of inheritance long before the actual chromosomal mechanism was discovered.

Figure 2.4 illustrates the relationship between genotype and phenotype for one such character, and explains the genetic terms **dominant** and **recessive**. The example is one of Mendel's original crosses between two colour strains of the common garden pea. The phenotypic character being analysed is pigmentation – one true-breeding parental strain is pigmented (i.e. purple flowered, brown seeded) and the other is unpigmented (white flowered, green seeded).

In humans, the tendency to early baldness in males, the presence of hairs on the middle segment of the fingers and the ability to taste the compound phenolphthiocarbamide (it tastes disagreeably bitter to some people but not to others) are some of a very small number of easily detectable and common simple mendelian traits. Those mentioned all happen to be caused by dominant alleles at single loci (whose biochemical functions are unknown) and are inherited along the lines of 'purple' in the example above. Many inherited diseases also show simple mendelian inheritance, indicating that they are caused by a (usually recessive) defect in a single gene. Recessive genetic defects are inherited along the lines of 'white' in the above example. The effect of a recessive allele is only revealed when it is in the homozygous state, whereas a dominant allele shows an effect even when present in only a single copy (i.e. in the heterozygous state).

Mutations that produce serious dominant abnormalities, even if not immediately lethal during embryonic development or early life, do not usually survive in a wild population and are eliminated by natural selection. Recessive mutations, even if equally harmful in the homozygous state, can, however, persist at a low frequency, sheltered by the masking effects of the normal allele in heterozygotes. Even when two heterozygotes mate, only a proportion of their offspring will be homozygous (see Fig. 2.4) and in a relatively large population heterozygotes for the locus will always far outnumber homozygotes. The persistence of recessive alleles has evolutionary advantages, as it retains a pool of potentially useful variation. A trait that is deleterious in one environmental situation may confer a positive selective advantage if circumstances change.

Not all pairs of alleles at a locus show a dominant–recessive relationship. Many show various degrees of co-dominance, contributing more-or-less equally to the phenotype. The heterozygote sometimes shows up as a phenotype intermediate between the two parental types. The effects of different combinations of alleles in snapdragons shown in Fig. 2.2 is an example of co-dominance.

Sex-Linked Traits

The patterns of inheritance described above are seen only for genes carried on the autosomes. Genes caried on the sex chromosomes show a different pattern. This is termed **sex-linked inheritance** and, in humans, is most evident in the inheritance of such genetic diseases as haemophilia A, which is caused by a recessive defect in the gene for the blood-clotting factor VIII located on the X chromosome.

Females always inherit two X chromosomes, one from their father and one from their mother, but one is inactivated at random in all their cells early in development. Females are therefore genetic mosaics for all their X chromosome genes, some cells carrying the paternal X chromosome and some the maternal X chromosome. Most body tissues contain some of each type of cell and therefore will effectively express both alleles. In the case of haemophilia, if a woman inherits a defective X chromosome from her mother and a normal one from her father the normal one is able to compensate.

Males, on the other hand, only inherit one X chromosome – from their mother – and one Y chromosome (which carries very few genes, and almost none in common with the X chromosome) from their father. If they inherit a defective X chromosome there is no second X chromosome to compensate and they show the disease.

X-linked recessive traits appear only extremely rarely in females, as this requires the coming together of two defective X chromosomes. X-linked traits are therefore usually transmitted through heterozygous female carriers and appear only in males (Fig. 2.5).

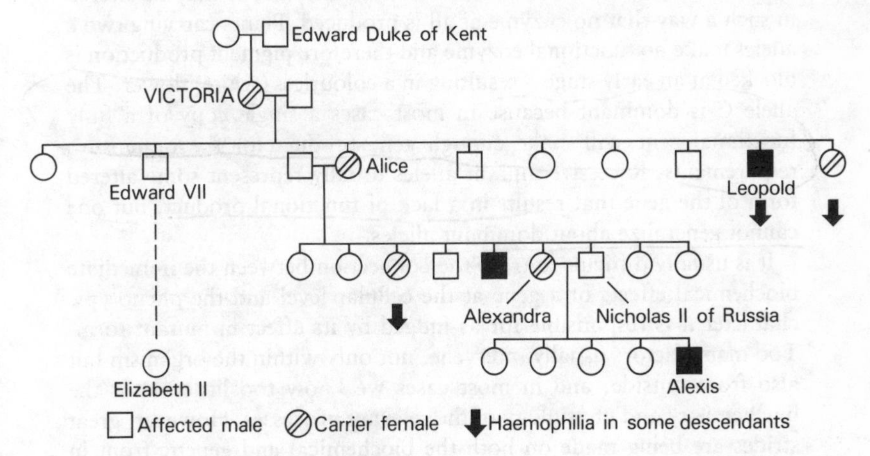

Fig. 2.5 The inheritance of haemophilia A in the descendants of Queen Victoria.

Linkage

The examples given above deal with inheritance patterns of alleles at a single locus. When the inheritance of alleles at two separate genetic loci is tracked simultaneously several different patterns of inheritance may be seen. If the two loci are on different chromosomes their alleles will segregate independently into the gametes and will therefore behave in a cross quite independently of each other. But alleles carried on the same chromosome will, in the absence of recombination, be inherited together. They are said to display **linkage**. The further apart the two loci are, the more likely it is that recombination will occur between them during meiosis (see Fig. 2.3). The degree to which recombination disrupts linkage is a way of determining the positions of genes relative to each other (see Chapter 4: Gene mapping).

The biochemical effect of genotype

Why a particular pair of alleles has a given effect depends on what the product specified by that locus is, and how the two alleles differ in their ability to make it. For example, in the case of pigment production we might find that the *C* gene (see Fig. 2.4) specifies a key enzyme early in a metabolic pathway involved in pigment synthesis. The *C* allele represents a 'normal' gene that directs the synthesis of fully functional enzyme. The allele *c*, on the other hand is a defective variant which may either direct the manufacture of an inactive enzyme, or may be altered in such a way that no enzyme at all is produced. Plants carrying two *c* alleles make no functional enzyme and therefore pigment production is blocked at an early stage – resulting in a colourless (white) flower. The allele *C* is dominant because in most cases a single copy of a fully functional gene will make enough gene product for the organism's requirements. Recessive mutant alleles usually represent some altered form of the gene that results in a lack of functional product, but one cannot generalize about dominant alleles.

It is usually difficult to trace the connection between the immediate biochemical effects of a gene at the cellular level and the phenotypic character it is responsible for as judged by its effect in mutant form. Too many factors usually intervene, not only within the organism but also from outside, and in most cases we know too little about the biochemistry and physiology of the relevant processes. However, great strides are being made on both the biochemical and genetic front in understanding, for example, how a single defective gene produces the complicated pathology of such conditions as cystic fibrosis and muscular dystrophy (see Chapter 4: Genetic disease).

THE PLAYERS IN THE GAME

How genetic information is embodied in DNA and is converted into the executive machinery of the cell – the proteins – is described briefly in the next sections.

The players in the genetic game are the two nucleic acids, DNA and RNA, and the proteins. The synthesis of other constituents of living matter – lipids, carbohydrates, and the many other compounds made by a cell – is not specified directly by DNA but indirectly through the enzymes and other proteins involved in their manufacture.

The **central dogma** of molecular biology emphasizes the universal and irreversible one-way nature of the information flow from nucleic acid to protein. Information can pass back and forth between RNA and DNA, as in the life cycle of some RNA viruses, but there is no way a modification made to protein after it is synthesized can be transmitted back up the pathway to DNA.

DNA (deoxyribonucleic acid)

Two years after Mendel published his first results in 1865, a German physiologist, Friedrich Miescher, reported the isolation of a substance from the nuclei of pus cells from discarded bandages which he named nuclein and which we would now call **nucleoprotein**. A few years later, nuclein was found to be composed of a protein component and an organic acid – **nucleic acid**. DNA had been isolated. From this somewhat unsavoury beginning it was another 80 years before DNA's role in heredity was demonstrated and genetics and biochemistry were united.

A DNA molecule consists of two long, paired chains of **nucleotide** subunits, wound around each other in a double helix (Fig 2.6). Each nucleotide consists of a nitrogenous base – **adenine**, **guanine** (the **purines**), **cytosine** or **thymine** (the **pyrimidines**) – linked to a molecule of the pentose sugar deoxyribose, which also carries a phosphate group.

Individual nucleotides are joined together by regularly repeating ester bonds between the phosphate group of one and the sugar group of another to form a **polynucleotide chain**. The bases project inwards, stacked up on each other and the helix is held together by specific pairing by relatively weak hydrogen bonding between adenine (A) and thymine (T) and between guanine (G) and cytosine (C). Adenine always pairs with thymine and guanine with cytosine. The two strands of a DNA helix are therefore *complementary* to each other, not identical.

Because of its regular repeating linkage, a polynucleotide chain has directionality, with one end being chemically quite distinct from the other. The end with the free phosphate group is known conventionally as the 5′-end because the phosphate is attached to carbon atom 5 of the deoxyribose moiety; the other end carries a free hydroxyl group on carbon 3 of deoxyribose and is called the 3′-end. In DNA the two chains run in opposite directions (antiparallel).

Genetic information is encoded in the sequence of the bases along the strands of the helix (see below, The genetic code) and represents instructions for making proteins and RNAs. A gene is defined at the DNA level as a stretch of DNA that directs the synthesis of a particular

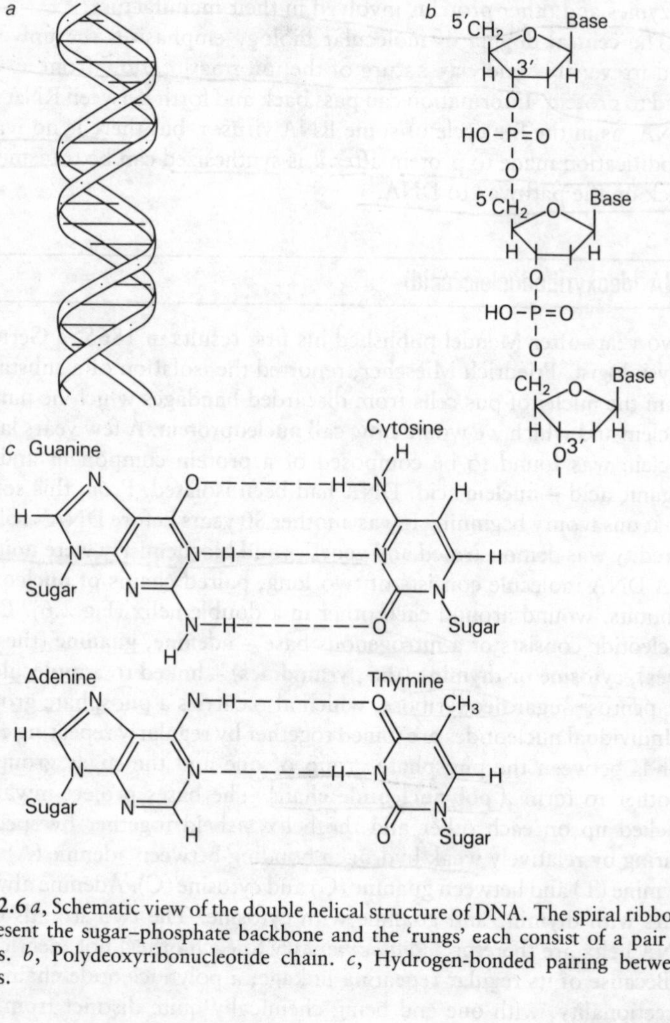

Fig. 2.6 *a*, Schematic view of the double helical structure of DNA. The spiral ribbons represent the sugar–phosphate backbone and the 'rungs' each consist of a pair of bases. *b*, Polydeoxyribonucleotide chain. *c*, Hydrogen-bonded pairing between bases.

protein (or RNA). The order of nucleotides along DNA is called variously its **base sequence, nucleotide sequence** or **DNA sequence**. There is no chemical constraint on the order in which bases may be linked together so that DNA's coding potential is virtually unlimited. DNA molecules are typically very long – a typical eukaryotic chromosome consists of a DNA molecule many millions of nucleotide pairs long. The length of a DNA molecule is usually given in base pairs (bp) or kilobases (equal to 1000 bases or base pairs and abbreviated kb or kbp).

DNA molecules are enzymatically degraded by **deoxyribonucleases** (DNases) that cut the sugar–phosphate backbone between nucleotides. There are many different DNases each with a particular biological role. Some are involved in DNA repair (see below) cutting out mismatched bases or portions of damaged DNA. Others are involved in the reshaping of DNA during genetic recombination (see below). Yet others recycle DNA by degrading it to its constituent nucleotides. Some DNases act only on double-stranded DNA, others cut only across a single strand. DNases divide into two main groups: **exonucleases**, which degrade DNA by clipping successive nucleotides off one end, and **endonucleases** which cleave bonds within DNA.

DNA Replication

Any molecule that is to serve as a repository of genetic information must be able to act as a template for the synthesis of an exact copy of itself to pass on to daughter cells at cell division. Amongst biological molecules only the nucleic acids (DNA and RNA) can do this. DNA replicates by **semiconservative replication**. The strands of the double helix unwind and separate, and complementary strands are progressively synthesized along each parental strand following the base-pairing rules – A with T and C with G (Fig. 2.7). The enzymes that catalyse this process are the **DNA polymerases**. The result is two identical DNA molecules each containing one original and one new strand. In reality the process is complicated, involving a large number of accessory proteins (e.g. helicases, unwindases and gyrases) that help to unwind the helix and separate the strands.

Replication of DNA always starts at a special site (or sites) termed **replication origins**. The very long DNAs of eukaryotic chromosomes have multiple replication origins, allowing their DNA to be replicated in a reasonable length of time; smaller DNAs, such as those of mitochondria, chloroplasts and animal and plant DNA viruses, have a single origin of replication. A DNA that forms a complete replicating unit is often termed a **replicon**. DNA replication in a eukaryotic

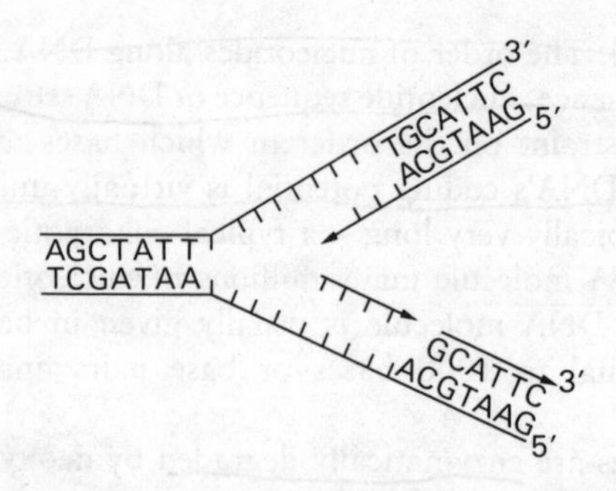

Fig. 2.7 DNA replication. Arrows indicate the direction of new DNA synthesis. At the replication fork, the two strands unwind and DNA polymerase synthesizes a complementary strand along each one. DNA polymerase can proceed only in a 3′ to 5′ direction and DNA is synthesized discontinuously in short stretches (Okazaki fragments) along one of the strands (termed the lagging strand) which join up later.

chromosome proceeds at a rate of around 50 nucleotides per second, in bacteria at around 500 nucleotides per second.

Reverse Transcription

DNA can also be copied from RNA by the enzyme **reverse transcriptase**, which is produced by some small RNA viruses (the retroviruses). (For the biological significance of this synthesis see Chapter 3: Repetitive DNA; and Chapter 10: Cancer Research)

DNA Repair

DNA is the permanent information store of a cell, and there are many safeguards to protect it from damage and maintain its nucleotide sequence unchanged through many rounds of replication. Very rarely, a wrong nucleotide may be incorporated during replication, but DNA polymerases have an inbuilt 'proofreading' capacity that allows them to recognize the mistake and remove the incorrectly paired nucleotide. Other enzyme systems repair more extensive damage that can occur – for example, on exposure to ultraviolet light or certain chemicals. Repair processes, while restoring the integrity of the DNA molecule and thus ensuring the cell's survival, sometimes introduce alterations (mutations) in the nucleotide sequence. The double-stranded nature of DNA is essential to its ability to repair itself as damage on one strand can only be correctly repaired by reference to the undamaged strand.

Recombination

The third property of DNA that has important implications for its biological role is the ability of two DNA molecules to undergo **recombination**, that is, to exchange pieces with each other or to combine with each other.

Reciprocal general recombination occurs between paired homologous chromosomes at meiosis. The two sets of chromosomes (one derived from each parent) pair up and each pair undergoes more or less extensive random reciprocal exchange of DNA in the process known as **crossing over**. This brings together genes from both parents in the resulting recombinant chromosomes, which are then distributed to the gametes (see Fig. 2.3).

As well as the reciprocal type of recombination, which requires long stretches of DNA of similar sequence, some DNAs, most notably small viral and plasmid DNAs, **integrate** themselves into another DNA molecule – usually a host cell chromosome – by a different type of recombination process. This requires little or no homology between the two DNAs. Various types of **integration** are accomplished by different molecular mechanisms but they all result in a continuous piece of DNA into which the new DNA has been incorporated.

RNA (ribonucleic acid)

The second nucleic acid in living cells is **RNA**. Like DNA it is composed of a sequence of nucleotide units strung together in a long chain. Unlike DNA, however, it is usually single-stranded. It is present in all cells as **messenger RNA (mRNA)**, **transfer RNA (tRNA)** and **ribosomal RNA (rRNA)**, which are involved in realizing the information stored in DNA and translating it into protein (see below). There are also many small RNAs in eukaryotic cells whose functions are largely unknown (but see Chapter 3: RNA Splicing). Some viruses, termed the **RNA viruses**, have single or double-stranded RNA instead of DNA as their genetic material. All RNA is made by **transcription from DNA**, or by replication of a pre-existing RNA template as in some RNA virus replication.

RNA is made of nucleotide components (**ribonucleotides**) similar to those in DNA except that thymine (T) is replaced by the structurally very similar base uracil (U), which also pairs with adenine (A), and the sugar in RNA is ribose rather than deoxyribose. In single-stranded RNA, regions with complementary base sequences often fold back on themselves and form limited stretches of double helix. tRNAs and rRNAs have extensive **secondary structure** of this type (Fig. 2.8).

In present-day cells DNA constitutes the permanent information

Uracil

Ribose

store and RNA the executive arm of the genetic apparatus. Much of a cell's RNA is short-lived, being produced for a particular purpose and then broken down. However, various aspects of cell biochemistry suggest that RNA, not DNA, was the original genetic material in the first living cells, DNA having evolved much later as a more stable form of information storage. Ribozymes – RNAs with enzymatic activity – may represent relics of this original 'RNA world' and their recent discovery has provided a new impetus to the study of RNA (see Chapter 5).

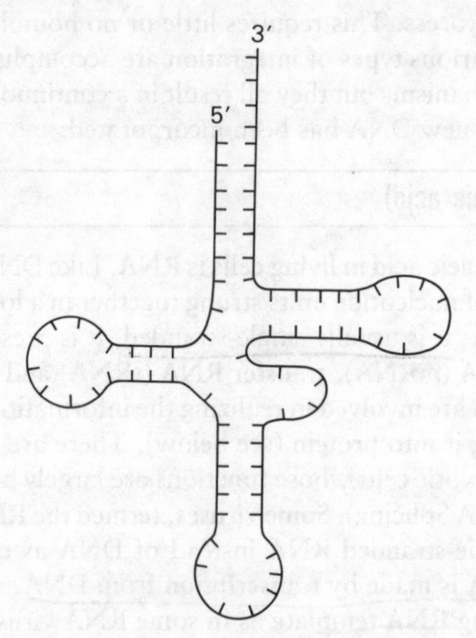

Fig. 2.8 Schematic diagram of tRNA molecule showing secondary structure due to base pairing.

Proteins

Proteins are essential constituents of living matter, and their infinite structural variety and chemical versatility underlie life itself. Cells extract energy from food and (in green plants) light and rebuild the constituent atoms of food, air and water into cell components by chemical reactions catalysed by a multitude of protein **enzymes**. Proteins form the basic structure of a cell (see Chapters 6, 7 and 8) and the machinery by which cells respond to signals from the environment and from other cells (see Chapter 9). DNA cannot replicate or be transcribed without proteins to unwind it and catalyse DNA or RNA synthesis. The selection of particular genes for expression is regulated largely by proteins that interact with control sites on DNA and RNA (see Chapter 3: Control of Gene Expression). RNA cannot be translated into proteins without the aid of protein, an evolutionary paradox that has plagued discussions on the origin of life (for a possible resolution of this issue see Chapter 5).

Protein Structure

The great functional versatility of proteins is the result of their chemical structure. Proteins are large molecules made up of one or more unbranched chains of amino acid units strung end to end (a **polypeptide chain**), linked through the amino group ($-NH_2$) of one unit and the carboxyl group ($-COO^-$) of the next (Fig. 2.9). Some 20 different types of amino acids are found in proteins (Table 2.1) and there can be up to several thousand amino acid residues in each chain. A short chain of

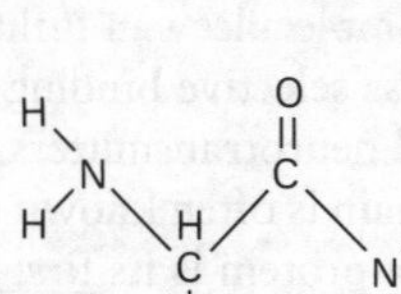

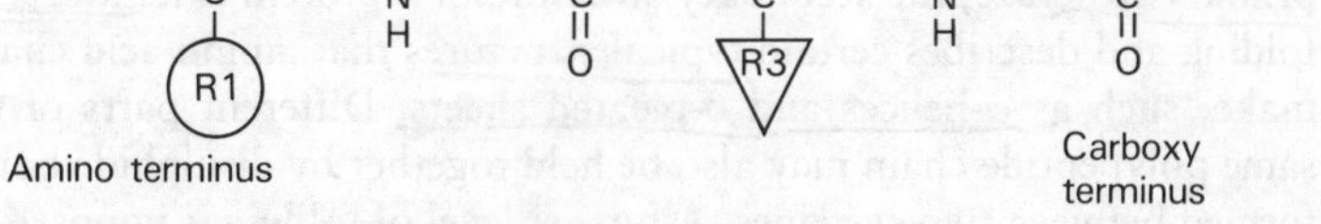

Fig. 2.9 *a*, Amino acid. R may be any of a wide range of side chains (see Table 2.1). *b*, Linkage of amino acids to form a polypeptide chain.

amino acids, up to 20 or 30, is generally called a **peptide**. Proteins are often characterized by their molecular weight (in daltons) (now often given as molecular mass relative to a hydrogen atom, M_r) which is invariably in the thousands – hence the commonly used terminology, K (e.g., a 50K protein is one of M_r 50 000).

Protein synthesis is under the direct control of the genes. The amino acid sequences of the many thousands of different polypeptide chains are each specified by a different gene. As soon as it is synthesized a polypeptide chain folds up into a three-dimensional structure that appears to be largely determined by its amino acid sequence. Different amino acid sequences, therefore, can produce proteins differing both in shape and in the chemical and physical properties of their surface. Proteins can be roughly globular (as are many enzymes), elongated and thread-like (fibrous proteins such as the keratins of hair, and the elastin and collagen of connective tissues) and many shapes in between. Some proteins (e.g. many enzymes, haemoglobin and cytochromes) contain a metal ion or other inorganic group (e.g. haem), which is essential for their function, and which is bound to the protein chain after it is synthesized. A large and important class of proteins – the **glycoproteins** – contain carbohydrate in the form of chains of sugars covalently linked to certain amino acid residues.

The particular chemical, physical and biological properties of a protein depend on its surface conformation. The folding of the polypeptide chain creates sites on the surface with a particular 'shape', chemical reactivity and electrostatic profile, which are determined by the nature and arrangement of the amino acid side groups. At these sites other molecules can be selectively bound and interact with the protein. Some proteins recognize other proteins, self-assembling into regular structures such as the protein coat of a virus or the internal cytoskeleton of a eukaryotic cell. Enzymes possess **active sites** at which a particular substrate is temporarily bound and a chemical reaction catalysed, and also in many cases regulatory sites at which other molecules can influence the enzyme's activity. Receptor proteins possess selective binding sites for signalling molecules such as hormones and neurotransmitters.

The amino acid sequence of a polypeptide chain is often known as its **primary structure**; the **secondary structure** of a protein is its first-level folding and describes certain typical structures that amino acid chains make, such as α-helices and ß-pleated sheets. Different parts of the same polypeptide chain may also be held together by disulphide bonds formed between two cysteines. A further level of folding is imposed on this to give the final three-dimensional conformation (the **tertiary structure**) of the protein (Fig. 2.10*a*). The folded polypeptide chain often comprises several distinct regions of compact structure (**domains**), which are separated by more flexible regions. Different functions are often located in different domains.

Although the 3-D folding of a polypeptide chain is thought to be determined by its amino acid sequence, the rules governing its folding are not yet known. It is therefore not yet possible to predict the full 3-D

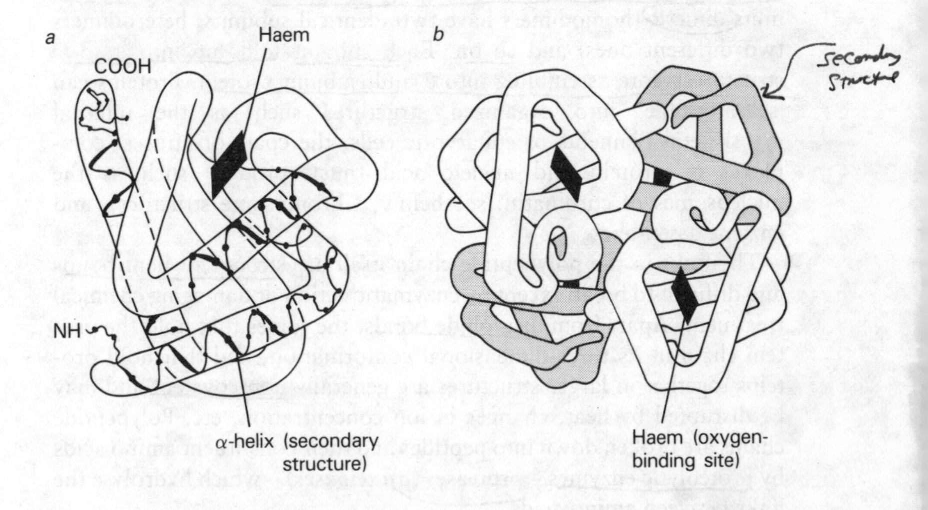

Fig. 2.10 *a*, The structure of the oxygen-carrying protein myoglobin, showing parts of the path of the polypeptide chain. Myoglobin contains a single haem group with which oxygen combines. (After R.E. Dickerson, in *The Proteins*, 2nd ed., vol. 2, Academic Press, New York, 1964.) *b*, Three-dimensional structure of the haemoglobin molecule. It consists of two each of two different subunits – α and ß globin – each of which contains a single haem group. Note the resemblance of the globin subunits and myoglobin.

structure of a protein from its amino acid sequence, although secondary structure can be predicted to a limited extent. The emerging art of **protein engineering**, which aims to change a protein's amino acid sequence by tinkering with the DNA that encodes it, is a powerful new tool for investigating the relationship of amino acid sequence to 3-D structure.

The amino acid sequence of a protein is now a fairly routine matter to determine, either by direct sequencing of the protein itself or from the nucleotide sequence of its corresponding DNA. Determining its tertiary structure is more difficult. **X-ray crystallography** can reveal 3-D structure down to the atomic level in proteins that will form suitable crystals. But many proteins, especially those associated with cell membranes, are difficult to purify in crystalline form, and lower-resolution techniques, such as image-enhanced electron microscopy, have been used to gain some idea of their structure.

Some proteins are composed of several polypeptide chains (subunits) either of the same or different kinds (Fig. 2.10*b*). Proteins composed of a single polypeptide chain are termed monomeric, those with two subunits dimeric (homodimers have two identical subunits, heterodimers two different ones) and so on. Each subunit folds up into its 3-D structure before assembling into a multisubunit protein. Proteins can 'self-assemble' into organized structures such as the internal cytoskeletal filaments of eukaryotic cells, the coats of viruses, complexes of protein and nucleic acid (nucleoprotein) such as the nucleosomes of chromatin (see below, Chromosome structure), and enzyme assemblies.

The links in the polypeptide chain itself are strong covalent bonds and difficult to break except by enzymatic action or damaging chemical treatment. Apart from disulphide bonds, the forces that hold the protein chain in its three-dimensional conformation, and that hold proteins together in larger structures are generally non-covalent and may be disrupted by heat, changes in ion concentration, etc. Polypeptide chains are broken down into peptides and their constituent amino acids by proteolytic enzymes – **proteases** (**proteinases**) – which hydrolyse the links between amino acids.

Allostery

Many proteins are **allosteric**: a molecule binding at one site on their surface disturbs their three-dimensional structure (a conformational change), which allows (or prevents) them interacting with another molecule at another site, or, more subtly, makes it easier or more difficult. Allosteric effects are an important and widespread mechanism in many biochemical pathways. Many multisubunit enzymes show allosteric interactions between their constituent subunits. Binding of substrate at one active site makes it easier for substrate to bind at the other active sites. Binding of a regulatory molecule such as cyclic AMP or calcium to one site on a protein may cause a conformational change that either activates or inactivates the active site. Receptor proteins are also believed to act through conformational changes induced when a signalling molecule binds. These structural perturbations transmit the information that a signal has been received to other parts of the receptor protein, which then pass on the information to the biochemical machinery within the cell through which it makes its response. Allosteric effects are also seen in such proteins as haemoglobin, where its affinity for oxygen increases when one of its four oxygen-binding sites is filled.

DECODING DNA

Transcription and messenger RNA

DNA does not participate directly in protein synthesis. When a gene is **expressed** – that is when the information encoded in its DNA is used to synthesize protein – it is first **transcribed** into a *complementary* RNA copy (a **transcript**).

Transcription is catalysed by the enzyme **RNA polymerase,** which moves along the DNA in a 3′ to 5′ direction, locally separating the strands of the helix and breaking the weak bonds between the base-paired nucleotides. The nucleotides in DNA are successively matched to their complementary ribonucleotides (C to G, G to C, A to U and T to A) and at each step the RNA polymerase adds the correct nucleotide to the growing RNA chain. (RNA is synthesized from 5′ to 3′ so that the 5′ end of an RNA is the 'beginning'.) As the RNA lengthens, it separates from the DNA, allowing several RNA polymerase molecules to progress along the same stretch of DNA at the same time. Only one of the DNA strands in any given gene is transcribed. The DNA strand carrying the same base sequence as the RNA is called the **coding strand**; the opposite DNA strand, which acts as template for transcription, is the **anticoding strand**. Signals encoded in the DNA sequence indicate points at which RNA polymerase binds to DNA and starts and ends transcription (see Chapter 3: Fig. 3.1).

In eukaryotes, transcription takes place in the nucleus and a primary RNA transcript is extensively modified (processed) by enzymatic action to form mature RNAs, which are exported from the nucleus. Transcripts of protein-coding genes are processed and exported as **messenger RNA** to be translated in the cytoplasm into protein (see Chapter 3: From gene to protein). In prokaryotes, the primary RNA transcript acts as messenger RNA without further processing, and translation starts at the free 5′ end of the transcript even before transcription is complete.

The genetic code

Proteins and nucleic acids are both linear permutations of a limited number of different subunits. The formal coding problem is how a four-'letter' code (A,G,C, and T or U) in DNA and RNA can specify the protein code of some 20 letters corresponding to the 20 amino acids that are the basic building blocks of proteins (see Table 2.1). The answer turns out to be a non-overlapping **triplet** code in which a group of three consecutive nucleotides specifies a single amino acid (see Table 2.2). Because proteins are synthesized from an RNA template the triplets or **codons** are given in the RNA form in Table 2.2 (to get the DNA form of the codon replace U with T). The genetic code is redundant, in that almost all amino acids are specified by more than one triplet. This redundancy means that although the amino acid sequence of a protein can be unambiguously deduced from its corresponding DNA or mRNA (provided one can identify the correct starting point), the exact nucleotide sequence of DNA cannot be similarly deduced from a protein sequence.

A small number of codons do not code for amino acids but have the effect of stopping translation, marking the end of the protein. These are the 'nonsense' or termination codons. The code is not punctuated internally and correct readout depends on starting at a particular place and reading the message in the correct **reading frame** (Fig. 2.11). There appears to be no chemical or structural basis for the correspondence between codons and particular amino acids and the origin and evolution of the code is still much of a mystery (but see next section on tRNA). Because there seems no reason why different codes could not have been equally effective and have been produced in the earliest stages of the evolution of life, the universality of the genetic code amongst living organisms provides a strong argument for the descent of all life on Earth in the form we know it from a single cell lineage.

Recently, some exceptions to the universality of the code have been discovered in mitochondria and in ciliate protozoans, but the differences are small and are generally thought to be later divergences from the universal code rather than relics of some alternative form. In

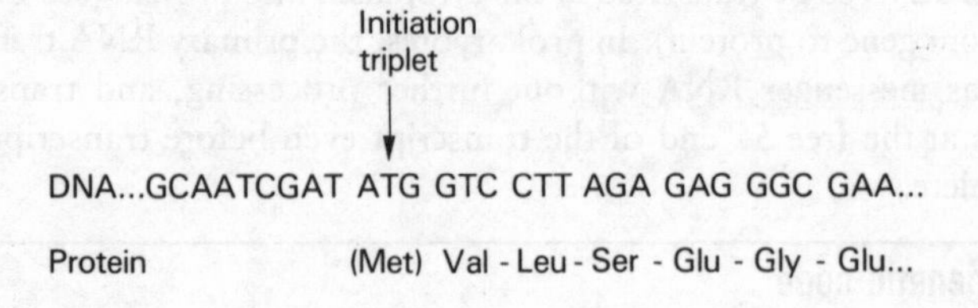

Fig. 2.11 Setting the reading frame. Protein-coding sequences invariably start with a methionine triplet.

Table 2.1 The amino acids commonly found in proteins

Amino acid	*Abbreviations*		*R group (side chain)*
Glycine	Gly	G	H
Alanine	Ala	A	CH_3
Valine	Val	V	$CH(CH_3)_2$
Leucine	Leu	L	$CH_2CH(CH_3)_2$
Isoleucine	Ile	I	$CH(CH_3)CH_2CH_3$
Serine	Ser	S	CH_2OH
Threonine	Thr	T	$CH(OH)CH_3$
Lysine	Lys	K	$(CH_2)_4NH_2$
Arginine	Arg	R	(CH_2) S $NHCNHCH_2$
Histidine	His	H	
Aspartic acid	Asp	D	CH_2COOH
Asparagine	Asn	N	CH_2CONH_2
Glutamic acid	Glu	E	$(CH_2)_2COOH$
Glutamine	Gln	Q	$(CH_2)_2CONH_2$
Proline	Pro	P	
Tryptophan	Trp	W	
Phenylalanine	Phe	F	
Tyrosine	Tyr	Y	
Methionine	Met	M	$(CH_2)_2SCH_3$
Cysteine	Cys	C	CH_2SH

Two cysteine side groups can form a covalent disulphide bond S = S. Disulphide bonds between cysteines are widespread in proteins and are important in determining three-dimensional conformation and in holding together multisubunit proteins (see text).

Table 2.2 The genetic code

First base	Second base: *U*		*C*		*A*		*G*	
U	UUU	Phe	UCU	Ser	UAU	Tyr	UGU	Cys
	UUC		UCC		UAC		UGC	
	UUA	Leu	UCA		UAA	Stop	UGA	Stop
	UUG		UCG		UAG	Stop	UGG	Trp
C	CUU	Leu	CCU	Pro	CAU	His	CGU	Arg
	CUC		CCC		CAC		CGC	
	CUA		CCA		CAA	Gln	CGA	
	CUG		CCG		CAG		CGG	
A	AUU	Ile	ACU	Thr	AAU	Asn	AGU	Ser
	AUC		ACC		AAC		AGC	
	AUA		ACA		AAA		AGA	Arg
	AUG	Met	ACG		AAG		AGG	
G	GUU	Val	GCU	Ala	GAU	Asp	GGU	Gly
	GUC		GCC		GAC		GGC	
	GUA		GCA		GAA	Glu	GGA	
	GUG		GCG		GAG		GGG	

mitochondria one difference is that the 'stop' codon UGA is used to encode tryptophan; in ciliate protozoans (e.g. *Tetrahymena*) the 'stop' codons UAA and UAG are used to encode glutamine.

Translation: deciphering the genetic code

Proteins are synthesized in the cytoplasm using an mRNA template which specifies the order in which amino acids are successively linked to form a polypeptide chain. Amino acids do not recognize the codons in mRNA directly, and their addition in the correct order is mediated by adaptor RNAs – **transfer RNAs** (**tRNAs**) – each of which carries a three-base **anticodon** which recognizes and temporarily pairs with the complementary codon in mRNA. Each tRNA picks up only the amino acid corresponding to the codon it recognizes. This attachment is catalysed by a battery of **aminoacyltransferases**. There are 20 of these enzymes, each specific for a given amino acid and the tRNAs corresponding to it. It is believed that each tRNA carries features

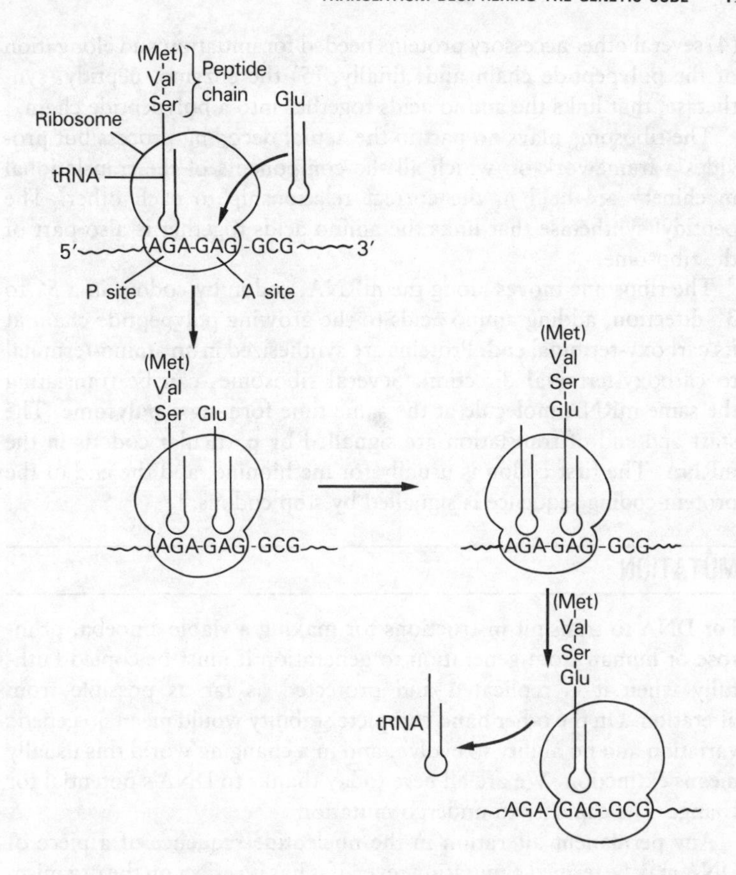

Fig. 2.12 Steps in the addition of an amino acid to a peptide chain at the ribosome. A tRNA bearing the peptide chain that has already been synthesized occupies the P site on the ribosome. The appropriate aminoacyl-tRNA lines up beside it at the A site. The peptide chain is transferred to the new charged tRNA in a reaction that requires elongation factors, GTP and the enzymatic action of peptidyl transferase. The ribosome moves along the mRNA, releasing the uncharged tRNA, and leaving the A site empty for a new aminoacyl-tRNA.

recognized by a specific aminoacyltransferase but the structural basis of this second genetic code is not yet well understood.

Translation is an exceptionally complex affair (Fig. 2.12) requiring (as well as mRNA and amino acids) (1) tRNAs, (2) ribosomes, which are composed of several types of **ribosomal RNA (rRNA)** and around 20 different ribosomal proteins, (3) the tRNA aminoacyltransferases,

(4) several other accessory proteins needed for initiation and elongation of the polypeptide chain and, finally, (5) the enzyme, peptidyl synthetase, that links the amino acids together into a polypeptide chain.

The ribosome plays no part in the actual decoding process but provides a framework on which all the components of the translational machinery are held in the correct relationship to each other. The peptidyl synthetase that links the amino acids together is also part of the ribosome.

The ribosome moves along the mRNA, codon by codon, in a 5′ to 3′ direction, adding amino acids to the growing polypeptide chain at its carboxy-terminal end. Proteins are synthesized in an amino-terminal to carboxy-terminal direction. Several ribosomes can be translating the same mRNA molecule at the same time forming a **polysome**. The start and end of translation are signalled by particular codons in the mRNA. The first codon is usually for methionine, and the end of the protein-coding sequence is signalled by stop codons.

MUTATION

For DNA to transmit instructions for making a viable amoeba, primrose or human from generation to generation it must be copied faithfully when it is replicated and protected as far as possible from alteration. On the other hand complete stability would mean no genetic variation and no ability to evolve, and in a changing world this usually means extinction. We are all here today thanks to DNA's potential for change – its capacity to undergo **mutation**.

Any permanent alteration in the nucleotide sequence of a piece of DNA may be termed a mutation, even if it has no effect on the organism that carries it. In general biology and genetics however, the term mutation is usually used to describe a change that causes a detectable, and in practice usually adverse, change in an organism's appearance or physiology. The difference in terminology between molecular and general biology is a potential source of confusion for the unwary.

Once a permanent change occurs in DNA it is passed on from the affected cell to all its descendants. When a mutation occurs in a germ cell (such as an ovum or sperm or the cells that give rise to them) it can therefore be passed on to that organism's offspring, affected offspring carrying the mutation in all their cells. Such **mutant** offspring can in their turn pass the mutation on to succeeding generations. If the mutation occurs in a body cell (somatic cell) on the other hand, it will only be transmitted to the immediate progeny of that cell, forming a group of mutant cells against a background of otherwise normal tissues as, for example, the abnormal cells of a malignant tumour.

When mutations occur in DNA sequences that encode proteins or RNAs or in their control regions they often alter the product of the affected gene in some way. They may either destroy the ability of the gene to make anything at all, or change the amino acid sequence of the protein specified. Mutations that result in non-production of a gene product or its synthesis in a much-altered or inactive form usually produce changes in the appearance or properties (the phenotype) of the cell or organism they affect. (In diploid organisms a mutant gene often has to be present in the homozygous state to produce a detectable effect on phenotype. See earlier sections on phenotype and genotype.) The discovery of the phenomenon of spontaneous mutation early this century, followed soon by the discovery that mutations could be induced by physical and chemical agents provided the impetus for the developing science of experimental genetics. A mutation with an obvious phenotypic effect that is inherited in a mendelian fashion marks out a single gene, whose inheritance and properties can then be further analysed.

Mutant organisms showing specific alterations in behaviour, morphological development, immunological competence or physiology play an important part in many areas of biology other than formal genetics, as they can pinpoint a biochemical pathway or piece of molecular machinery that is crucial to the normal working of a complex biological process. The large numbers of mutant strains that have been generated over the years in organisms such as the fruitfly *Drosophila* have identified hundreds of gene loci, some of which show fascinating developmental and behavioural effects when mutant. The advent of recombinant DNA technology now holds the promise of being able to identify the normal gene products of these loci, and possibly their normal role in the organism.

Types of mutation (Fig. 2.13)

Mutations affecting only a single base pair are termed **point mutations**. They arise chiefly as the result of the incorporation of an 'incorrect' nucleotide during DNA replication. This can occur for example either as a result of a (rare) mistake in the replication process itself or if a resident nucleotide undergoes a chemical transformation into another nucleotide which then directs the incorporation of its complementary nucleotide at the next round of replication.

Mutations can also be the result of **insertions** or **deletions** in the DNA sequence. Both can arise through certain types of genetic recombina-

...ATG GTG CTC TCT ATA GCT...
Met Val Leu Ser Ile Ala

a Point mutation

...ATG GTG **G**TC TCT ATA GCT...
Met Val Phe Ser Ile Ala...

b Frameshift

...ATG GTG CTC **AT**T CTA TAG...
Met Val Leu Ile Leu Stop

c Insertion (leading to frameshift)

...ATG GTT **CGA TAT CTC TGT G**CT CTC TAT AGC T...
Met Val Arg Tyr Leu Cys Ala Leu Tyr Ser...

d Deletion (– 11 nucleotides)

...ATG GTG CCT...
Met Val Pro... (the rest of the sequence will be frameshifted)

e Silent mutation

...ATG GTG CT**A** TCT ATA GCT...
Met Val Leu Ser Ile Ala... (no change)

Fig. 2.13 Types of mutation.

tion and transposition (see Chapter 3: Transposons) or after chemical treatments. Large deletions or insertions usually completely destroy the function of the gene in which they occur. Small deletions or insertions of one or two bases (or more – except, for obvious reasons, multiples of three) result in a change of reading frame and are known as **frameshift mutations**. The frameshift usually throws up a nonsense codon in the new reading frame, which stops translation prematurely.

Physical rearrangement of large pieces of DNA can also occur. Chromosomes undergo visible **translocations** and **transversions**, in which pieces break off and are rejoined to a different chromosome, or **inversions** where a piece becomes inserted in a different position or orientation in the same chromosome. In many cases, this leads to a change in gene expression, with genes that are normally silent becoming activated or genes that are normally active becoming switched off.

Mutations may occur spontaneously, as a result of rare errors in normal DNA replication and recombination. The changes in genes that occur over evolutionary time do not involve only simple nucleotide substitutions, deletions and insertions, but also extensive gene duplication and exchange of DNA between different genes.

Mutations may also be caused by external agents. Ultraviolet light, X-rays, acridine dyes and many other chemicals are **mutagens**, reacting with DNA in various ways to produce different kinds of mutations. Mutations caused by mutagens arise from attempts of the cell to replicate the damaged DNA, or sometimes from its attempts to repair the damage. Cells have enzymes that recognize many types of DNA damage and repair it, maintaining the integrity of the DNA and thus ensuring the cell's survival but sometimes causing a mutation in the process.

Because many mutagens are also carcinogenic, screening chemicals for their mutagenicity is one way of identifying potential carcinogens. Simple tests (e.g. the Ames test) that make use of a chemical's ability to cause mutation in bacteria or cultured cells are now widely used as a first screen for carcinogens.

Silent and Neutral Mutation

Because of the redundancy of the genetic code, mutations affecting the third position of a codon do not generally lead to any change in the amino acid specified and are 'silent' (see Fig. 2.13*e*). In other cases the amino acid specification may be changed but have little apparent effect on the normal function of the protein. This means that a considerable proportion of the difference seen at the DNA level between, say, corresponding genes from different species or alternative alleles of the same gene is neutral with respect to phenotype and is therefore not liable to be under any selective pressure from environmental or biological factors.

'Neutral' or 'silent' nucleotide positions, and indeed whole DNA sequences unconstrained by having to maintain a particular coding or control function, tend to change more rapidly over evolutionary time than those that are under selective pressure to remain unaltered in order to produce a functional protein or RNA. With the availability of extensive DNA sequences this phenomenon can now be put to practical use in molecular evolutionary studies to calculate 'basic' mutation rates for DNA in different organisms, unbiased by selection which throws out all mutations that lead to a non-functional protein. (See Chapter 5 for more on the use of DNA sequences in molecular evolutionary studies.)

CHROMOSOME STRUCTURE

One of the fundamental differences between eukaryotic and prokaryotic cells is in the way their DNA is packaged. Prokaryotic cells have no nucleus and their 'chromosome' is a single circular molecule of DNA anchored to the cell membrane. The prokaryotic chromosome is a complex of about 20 per cent protein of several different types and 80 per cent DNA and forms a compact mass – the **nucleoid**. Surprisingly little is known about the way in which protein is associated with DNA in the bacterial chromosome.

The chromosomes of eukaryotic cells on the other hand are highly organized structures containing about twice as much protein as DNA, and in which the DNA thread is wound up into a compact nucleoprotein fibre and further coiled and looped on itself. The material of eukaryotic chromosomes is known as **chromatin**. The structure of eukaryotic chromatin has important implications for operations such as replication and transcription, which must involve some local unpacking of chromosome structure, and for the control of gene expression. (For more on the relationship of chromatin structure to the control of gene expression see Chapter 3.) During chromosome duplication, not only must the DNA be replicated, but the complex protein composition and structure of the chromosome must be reconstituted on each new DNA.

Even the simplest eukaryotic unicell contains considerably more DNA than any prokaryote. The DNA is contained in separate chromosomes, each containing a single continuous linear DNA molecule. In a human diploid cell for example, nearly 2 m of DNA is packed into 46 chromosomes which, at their most compact, have a total length of no more than 200 μm. Some degree of ordered packing is essential to fit the total complement of DNA into a nucleus that on average measures about 5 μm in diameter. It also prevents the long DNA molecules (the average length of DNA in a human chromosome is of the order of 10^8 nucleotides or around 4 cm) becoming entangled, and ensures that they can be replicated and transcribed in an orderly fashion. The ends of chromosomes are sealed by regions known as **telomeres** which prevent the different chromosomes sticking to each other.

Heterochromatin and euchromatin

At times other than mitosis, chromatin appears to be present in at least two different states, as judged from its varying staining properties. **Heterochromatin** represents a compact form of chromatin in which genes are not transcribed. A familiar example of heterochromatin is the inactivated X chromosome in the cells of female mammals. This condenses to form a distinct 'heterochromatin body' which is visible throughout the cell cycle even when the rest of the chromosomes have become less compact. It replicates, but is not normally transcribed. (A recent finding that genes on the inactivated X chromosome are reactivated in 'aging' cultured cells may help to discover what keeps the X chromosome in its inactive state, as well as throwing light on what happens as cells, and organisms, age.) Some highly repetitive DNA sequences that are never transcribed also exist as regions of permanent heterochromatin. **Euchromatin**, on the other hand, represents a more relaxed form of chromatin and includes the genes that are available for transcription, although not necessarily being actively transcribed.

Nucleosomes

In chromatin, DNA is complexed in equal amounts (by mass) with small, basic, positively charged proteins, the **histones**. A continuous thread of DNA is wound round a succession of histone cores each

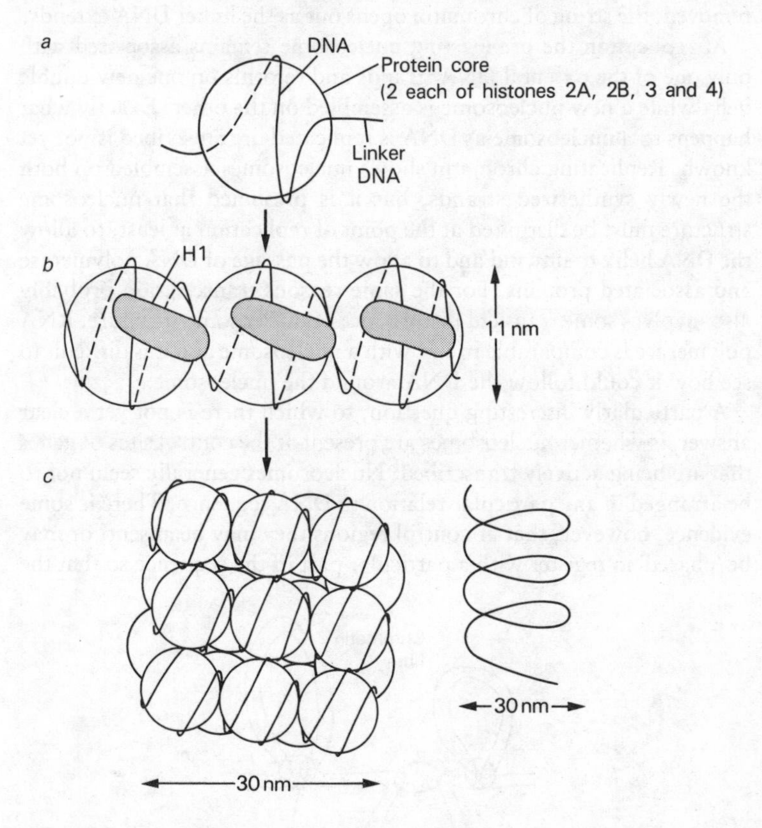

Fig. 2.14 *a*, Nucleosome. Nucleosome size is much the same in all species. A length of 140 base pairs of DNA is wound 1¾ times round the outside of the histone octamer in a left-handed spiral. The length of linker DNA between one nucleosome and the next varies between species and in different cell types from around 20 to 100 base pairs, but on average there will be one nucleosome for every 200 bases of DNA. *b*, String of nucleosomes. *c*, Possible arrangement of the nucleosome string in the 30-nm chromatin fibre.

consisting of two molecules each of four different histones (H2A, H2B, H3 and H4) to form **nucleosomes** (Fig. 2.14*a*). Nucleosomes occur at regular intervals along the DNA to give the 'beads on a string' effect that is clearly visible in extended fibres of chromatin in the electron microscope. The nucleosome structure is characteristic of all eukaryotic chromatin from yeast to human cells.

In mammalian cells (and those of many other eukaryotes) another histone, H1, binds to the short 'linker' DNA between each nucleosome, and keeps the nucleosomes packed up together (Fig. 2. 14*b*). If H1 is removed, the string of chromatin opens out as the linker DNA extends.

At replication the pre-existing nucleosome remains associated with only one of the parental DNA strands and reforms on one new double helix while a new nucleosome is assembled on the other. Exactly what happens to a nucleosome as DNA is replicated or transcribed is not yet known. Replicating chromatin shows nucleosomes assembled on both the newly synthesized strands, but it is presumed that nucleosome structure must be disrupted at the point of replication at least, to allow the DNA helix to unwind and to allow the passage of DNA polymerase and associated proteins. For the same reasons transcription probably also involves some transient disturbance of nucleosome structure. RNA polymerase is comparable in size with a nucleosome and it is difficult to see how it could follow the DNA around the nucleosome.

A particularly interesting question, to which there is not yet a clear answer, is whether nucleosomes are present at the control sites of genes that are being actively transcribed. Nucleosomes generally seem not to be arranged in any particular relation to DNA sequence. There is some evidence, however, that at control regions they may be absent, or may be 'phased' in register with a particular part of the sequence so that the

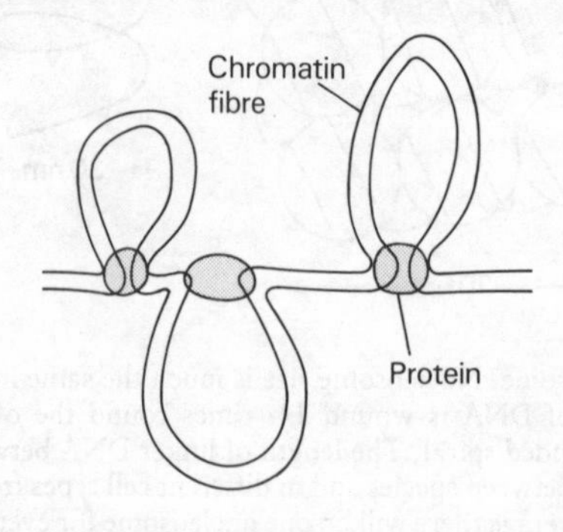

Fig. 2.15 Schematic view of the 30-nm chromatin fibre looped into domains.

DNA containing control sites falls in the linker regions where it may be more accessible to regulatory proteins.

Packing into nucleosomes already reduces the effective length of DNA by several orders of magnitude. The string of nucleosomes then coils on itself (Fig. 2.14*c*) to give a thicker fibre of 30 nm diameter (the **30-nm fibre**) which is the basic structure of native chromatin.

In the chromosome the chromatin fibre is believed to fold up into a series of loops (individual loops are often termed **domains**), held together by interactions between DNA-binding proteins at the base of the loops (Fig. 2.15).

Other chromosomal proteins

As well as DNA and histones, eukaryotic chromosomes contain hundreds, possibly thousands, of different types of **non-histone chromosomal proteins**. The functions of few of these proteins are known. Some probably have purely structural roles, providing the chromosome's permanent 'skeleton'; others certainly represent gene-regulatory proteins, involved either in chromatin packing and unpacking or as more selective regulators of transcription.

Mitotic chromosomes

For most of the life cycle of a cell the chromosomes are not visible under the light microscope, appearing only at mitosis and meiosis, when they condense dramatically to form short rod-like structures. At this stage they are completely inactive genetically. The DNA has already replicated and transcription has ceased. Gene expression and DNA replication take place in the **interphase** period between one mitosis and the next, when the chromosomes assume a more extended and relaxed structure.

The mitotic chromosomes represent the highest orders of chromatin packing. Here several centimetres of DNA are packed into a nucleoprotein rod a few micrometres long. Each mitotic chromosome consists of two newly copied identical sister chromatids lying side by side and still attached to each other at a central region called the **centromere**. The centromere is a constriction that divides chromosomes into two arms, often of unequal length – the **short arm** and the **long arm**. As chromosomes condense at mitosis, special staining procedures show typical **banding** patterns along their lengths. The arrangement of the bands on each chromosome is so invariant that it is used to identify individual chromosomes in a **karyotype** – the total set of chromosomes displayed

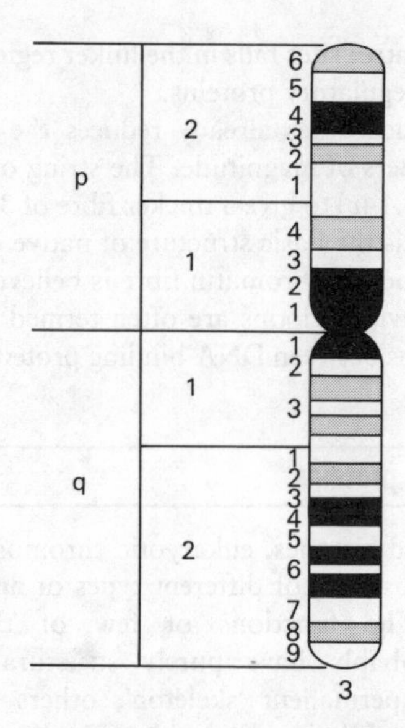

Fig. 2.16 Banding pattern on human chromosome 3 with Giemsa staining.

at mitosis. These banding patterns form the basis for the standard nomenclature for mammalian chromosomal regions (Fig. 2.16). The location 3p21 in a human karyotype, for example, indicates a position in band 1 of region 2 on the short arm (p) of chromosome 3 (the long arm is designated 'q').

Interphase chromatin

DNA is transcribed and replicated during the interphase period. The structure of interphase chromatin is therefore of particular interest. In general, this is difficult to determine visually, as interphase chromosomes appear in the electron microscope as tangled and dispersed threads of chromatin throughout the nucleus. Fortunately, an unusual set of interphase chromosomes – **polytene chromosomes** – are clearly visible even under the light microscope. They were discovered many years ago and have provided much of the evidence on which present models of chromosome structure and its influence on gene expression are based.

Polytene Chromosomes

Giant polytene chromosomes are found in the salivary glands of some insect larvae (e.g. *Drosophila* and the midge *Chironomus*). Polytene chromosomes arise through the repeated duplication of chromosomal DNA without separation of the copies. The DNA copies remain in register, forming ribbon-like giant chromosomes bearing a typical invariant pattern of many transverse dark bands separated by light interband regions. Banding is presumed to reflect differences in the packing of the DNA, but its significance is not known. The banding pattern is very much finer in scale than that of the mitotic chromosomes, with one band probably corresponding – very roughly – to one gene or part of a gene.

Some types of mutation (e.g. extensive deletions, additions and DNA rearrangements) cause visible disturbances in the banding pattern. This type of cytogenetic evidence was crucial in the early part of this century in proving that genes were indeed located on chromosomes, and in determining the chromosomal locations of genetically identified gene loci.

Polytene chromosomes also show that considerable changes in chromosome structure occur when genes are transcribed. 'Puffs' corresponding to actively transcribing genes can be seen even under the light microscope. Puffs represent a small region of the chromosome that swells up and becomes less dense. They have been known for many years and are a direct demonstration that transcription involves an unfolding of the chromosome structure.

FURTHER READING

See general reading list (e.g. Strickberger; Watson et al.; Hartl; Russell).

D. Hartl, *A primer of population genetics*, Sinauer/Blackwell, Sunderland, MA/Oxford, 1987.

L.L. Cavalli-Sforza and W.F. Bodmer, *The genetics of human populations*, W.H. Freeman, San Francisco, 1971.

F. Portugal and J.S. Cohen, *A century of DNA: a history of the structure and function of the genetic substance*, MIT Press, Cambridge, MA, 1977.

CHAPTER 3

GENE ORGANIZATION AND EXPRESSION

THE GENE DISPLAYED

A gene is usually defined at the molecular level as a stretch of DNA that specifies a single type of polypeptide chain, tRNA, rRNA or other small structural RNA. Depending on the context, this definition sometimes includes not only the coding region but also the adjacent control sites required for its correct expression. The discovery of split genes, overlapping genes and stretches of DNA that can encode two forms of a protein (see RNA splicing) has required some rewriting of the small print in this traditional definition, but for most purposes it is still valid.

Genes that specify the everyday products of all cells, such as rRNAs and tRNAs for translation, enzymes of basic metabolic pathways, structural protein components, proteins that ferry nutrients and ions across the cell membrane etc. are often known as **housekeeping genes. Cell-type specific** genes in multicellular organisms encode the **luxury proteins** produced only by specific cell types, such as hormones, haemoglobin, antibodies, antigen receptors, etc.

There is typically only one or at most a very small number of copies of each protein-coding gene in the haploid genome. Such genes are also called **single-copy genes** to distinguish them from sequences that are repeated, often many hundreds or thousands of times, in the genome. In eukaryotes the tRNA, rRNA and histone genes are present in multiple copies.

Housekeeping and 'luxury' genes are often known as **structural genes**. Genes that are involved in controlling the activity of other genes, by producing gene regulatory proteins or in some other way, are often known as **regulatory genes**. Regulatory genes distinguished by purely genetic studies may turn out to be control regions, not encoding any protein or RNA.

As well as specifying amino acids, DNA also encodes control signals that mark out sites at which transcription is regulated, and the beginning and end of transcription and translation. The nature of the control sequences and the ways in which genes are switched on and off are now well known in bacteria, but only recently has it become possible to study this in any detail in eukaryotes.

Transcription units

A stretch of DNA that is transcribed into a continuous RNA is called a **transcription unit**. In eukaryotes, each protein-coding gene is transcribed as a separate transcription unit, but in bacteria, groups of genes (operons) are typically 'co-transcribed' into a single mRNA. At the molecular level, there are considerable differences between the fine structure of bacterial and eukaryotic genes (see below, Split genes) both in internal structure and in their arrangement in relation to each other and mode of control. Schematic representations of a bacterial transcription unit, a typical eukaryotic split gene, and various other types of gene arrangement are given on the next few pages for easy comparison, showing the relation between gene and product (Fig. 3.1a–g). The caption briefly lists the main features encoded in the DNA. The nature of the control regions is explained in more detail in the section on regulation of gene expression later in this chapter. The production of proteins from split genes and overlapping genes is dealt with in more detail in the next few sections. The production of several different peptides from the pro-opiocortin gene (Fig. 3.1d) is an example of a quite widespread means of synthesizing small biologically active peptides. The pro-opiocortin gene is particularly versatile, encoding at least five peptides, but both the enkephalins (5 amino acids) and epidermal growth factor (53 amino acids) are also produced from a larger polypeptide encoding several repeated copies of the peptide.

The biological literature now abounds in DNA sequences of genes and their control regions. A full DNA sequence of both strands is conventionally written with the 'coding' strand uppermost, starting with the 'beginning' of the gene (the 5′ end of the coding strand and the 3′ end of the anticoding strand) to the left. The triplets can then be read off directly left to right as an amino acid sequence. It is of course the 'anticoding' strand that is actually transcribed. Likening transcription to the flow of a river, sequences to the 'left' of a given point are often described as **upstream** and sequences to the 'right' as **downstream** (see Fig. 3.1*a*).

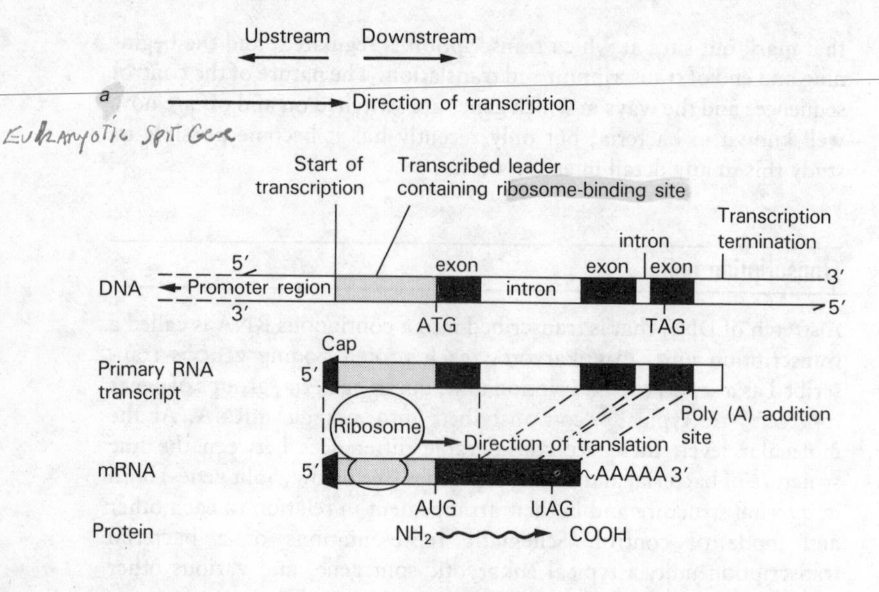

Fig. 3.1 From gene to product (not all the same scale). *a*, Typical eukaryotic split gene; *b*, bacterial operon; *c*, overlapping genes in bacteriophage ΦX174; *d*, production of peptide hormones from a single polyprotein encoded by the pro-opiocortin gene; *e*, 18S and 28S rRNA synthesis from tandemly repeated gene clusters; *f*, histone genes and histone synthesis; *g*, production of different mRNAs from the same DNA by differential DNA splicing. (*g* after Fig. 24-23, J.D. Watson *et al.*, *Molecular biology of the gene*, Vol II, Benjamin Cummings, Menlo Park, CA, 1987.)

Control signals in DNA

5′ control regions

Promoter, enhancer and other 'upstream' control sequences. See below, Control of gene expression for details.

The start of transcription

In both bacterial and eukaryotic genes the first nucleotide transcribed is usually a purine nucleotide, often, but not always, A, flanked by two pyrimidines. The triplet CAT commonly, but by no means invariably, marks the start of transcription in bacterial genes. Some genes have variable startpoints, allowing different gene products to be made from the same stretch of DNA.

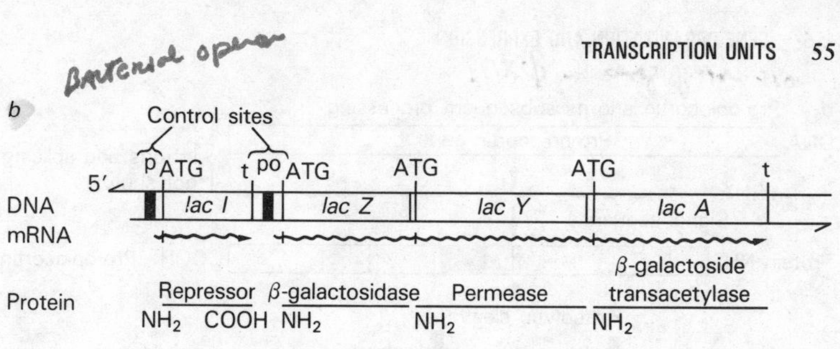

p, promoter; o, operator; t, terminator.

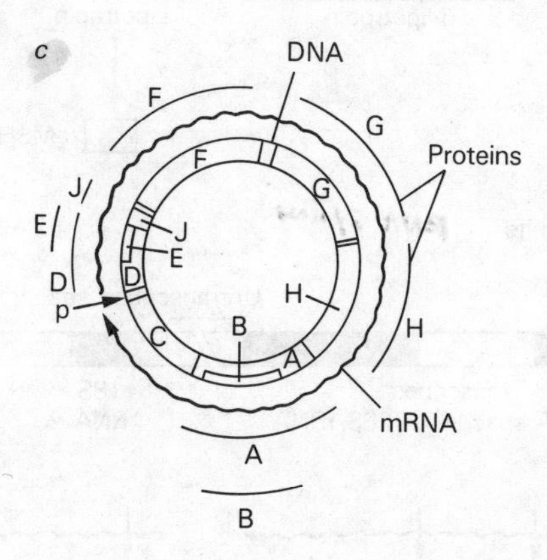

Transcribed leader sequence and ribosome-binding site
Between the start of transcription and the start of translation in both bacterial and eukaryotic protein-coding genes is a short sequence that is retained in the messenger RNA but not translated. It contains the site at which ribosomes initially bind to mRNA to start translation.

The start of translation
The start of translation is almost universally represented by the triplet ATG (AUG in mRNA) encoding the amino acid methionine. In bacterial genes a protein-coding sequence now continues uninterrupted until a terminal 'stop' triplet. An 'open reading frame' (ORF) is an appreciable stretch of DNA between a 'start' codon and a 'stop' codon that does (or could in principle) specify a protein.

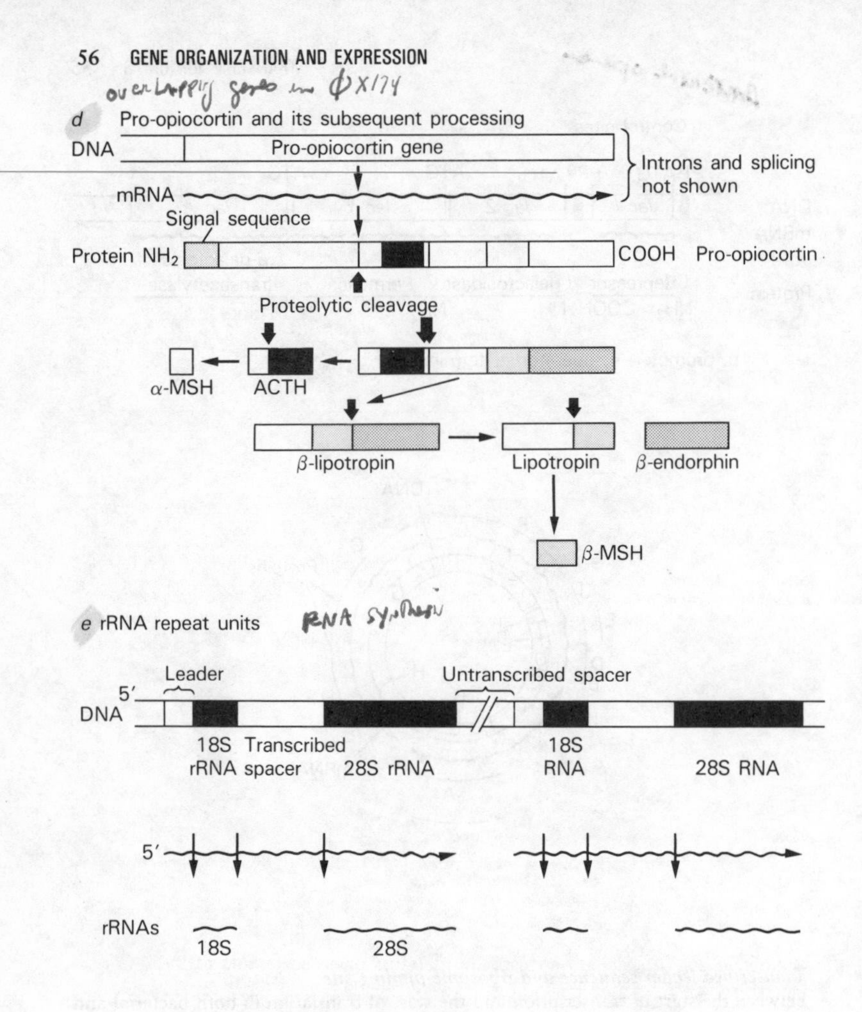

Introns and exons

The transcribed sequence of most eukaryotic genes is interrupted by non-coding regions – introns – which separate the coding region into individual exons. The whole sequence, exons and introns, is transcribed into RNA from which the introns are excised to give a continuous coding mRNA for translation, or an rRNA or tRNA.

Translated leader sequence

Many proteins destined to be incorporated into cell membranes or to be secreted from the cell contain a short sequence – the signal sequence – at the N-terminal end of the protein. This serves to direct the protein to its correct cellular location and is usually finally removed (see Chapter 7).

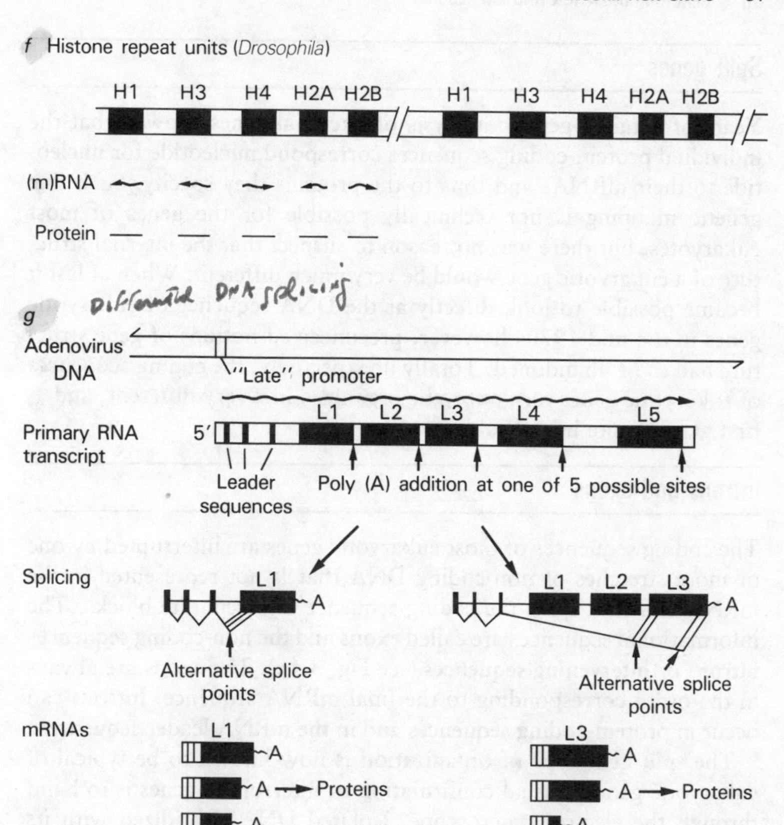

Alternative poly(A) addition and splicing patterns give rise to sets of mRNAs encoding the L2, L4 and L5 families of viral proteins

End of translation
This is signified by 'stop' or termination triplets – TAA, TAG or TGG (UAA, UAG or UGG in mRNA).

End of transcription
In bacterial transcription units, a terminator sequence often occurs just before the point of termination. In some cases termination also requires the action of an ancillary protein, rho, which interacts with the RNA polymerase. The point of transcription termination in eukaryotic genes is more difficult to determine. Many eukaryotic mRNAs have their 3′ ends trimmed off and a 'tail' of As added, which makes it difficult to determine the last base actually transcribed.

Split genes

Years of detailed genetic analysis of bacterial genes showed that the individual protein-coding sequences correspond nucleotide for nucleotide to their mRNAs and thus to the proteins they specify. Such fine genetic mapping is not technically possible for the genes of most eukaryotes, but there was no reason to suspect that the internal structure of a eukaryotic gene would be very much different. When at last it became possible to look directly at the DNA sequence of eukaryotic genes in the mid-1970s however, preconceived notions of gene structure had to be abandoned. Totally unexpectedly, the coding sequences of eukaryotic genes turn out to be organized in a very different, and at first glance quite inexplicable, way.

Introns and exons

The coding sequences of most eukaryotic genes are interrupted by one or more stretches of non-coding DNA that is not represented in the mRNA and that splits the coding sequence into separate blocks. The informational sequences are called **exons** and the non-coding sequences **introns** or **intervening sequences** (see Fig. 3.1*a*). The exons are always in the order corresponding to the final mRNA sequence. Introns can occur in protein-coding sequences and in the mRNA leader sequences.

The **split gene** type of organization is now known to be typical of eukaryotic genes. Visual confirmation of interrupted genes is to hand through the electron microscope. Isolated DNA hybridized with its corresponding mRNA clearly shows loops of unpaired DNA corresponding to the non-coding regions. Introns occur not only in genes coding for proteins but also in those encoding tRNAs and rRNAs.

Split genes immediately pose the problem of how a continuous mRNA (or tRNA or rRNA) is to be assembled from the separate chunks of information. The whole gene, exons and introns, is first transcribed into a long RNA molecule – the **primary transcript** or **pre-mRNA**. The introns are cut out one after another (not necessarily in the order in which they occur) and the exons joined up to each other to form a functional RNA. This is known as **RNA splicing** and has generated some exciting and novel biochemistry, including the discovery that RNA can act as an enzyme (see Chapter 5: Ribozymes). Different splicing patterns can also generate alternative mRNAs from the same DNA, a strategy used by some animal viruses to make the most of their limited genome (Fig. 3.1*g*) and by some cellular genes.

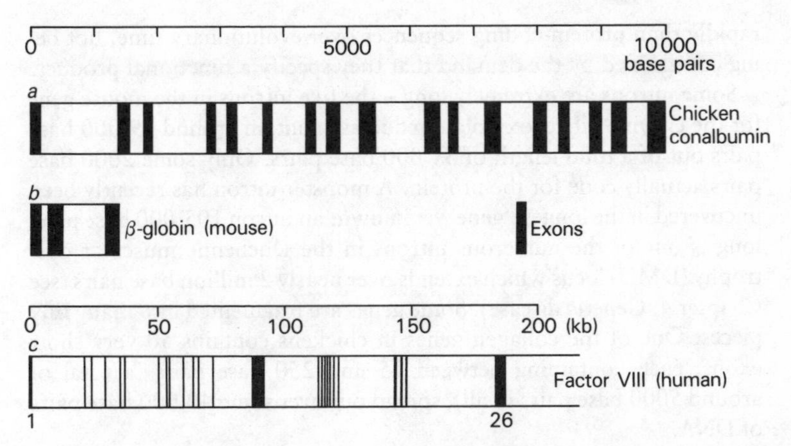

Fig. 3.2 *a*, Chicken conalbumin gene, total length 10 000 base pairs (bp), 17 exons; *b*, mouse ß-globin gene, 1400 bp, 3 exons; *c*, human clotting factor VIII gene, 180 000 bp, 26 exons.

Even before the discovery of split genes it was known that the information flow from DNA to messenger RNA in eukaryotic cells was more complicated than in bacteria. Only a small fraction of the RNA transcribed in the nucleus (**heterogeneous nuclear RNA, hnRNA**) ever reaches the cytoplasm as mRNA, and much of this nuclear RNA is considerably longer than a typical mRNA. The discovery of introns provides (at least part of) the explanation for the phenomenon of hnRNA.

The split gene type of organization has been found throughout the eukaryotic world from yeast to humans, and in most genes looked at. Introns occur in the nucleolar genes for rRNA in many organisms, in tRNA genes, in most protein-coding nuclear genes and in chloroplast genes and mitochondrial genes from some species (e.g. yeast mtDNA has introns, human mtDNA has not). An exception is the total lack of introns in any of the histone genes from a wide range of species. The few mammalian genes so far discovered to lack introns include the interferon genes and the gene for the ß-adrenergic receptor (see Chapter 9).

The number and lengths of introns vary enormously from gene to gene (Fig. 3.2). The gene for ß-globin, one of the first analysed, has two introns (as have all the members of the ß-globin family). They occur in exactly the same position in all mammalian ß-globin genes, but differ slightly in those of birds and amphibians, as might be expected from their more distant evolutionary relationship. Introns differ considerably from one another in DNA sequence, and they also change more rapidly than protein-coding sequences over evolutionary time, not being constrained by the demand that they specify a functional product.

Some introns are extremely long – the five introns in the mouse gene for the enzyme dihydroxyfolate reductase contain around 29 000 base pairs out of a total length of 31 000 base pairs. Only some 2000 base pairs actually code for the protein. A monster intron has recently been uncovered in the longest 'gene' yet known: an intron 105 000 base pairs long is one of the numerous introns in the Duchenne muscular dystrophy (DMD) locus which extends over nearly 2 million base pairs (see Chapter 4: Genetic disease). Some genes are fragmented into many tiny pieces. One of the collagen genes in chickens contains 50 very short exons, each containing between 25 and 250 base pairs (a total of around 5000 base pairs in all), spread out over some 40 000 base pairs of DNA.

The split gene type of organization is virtually absent from prokaryotic organisms (bacteria) and bacterial viruses (phages), providing yet another example of the deep divide between eukaryotes and prokaryotes. Two introns have been found in a gene in a bacteriophage (T4) that infects *Escherichia coli*. Split genes have also been found in several members of the 'archaebacteria', an unusual and in many ways atypical group of unicellular microorganisms only distantly related to the 'true' bacteria.

There has been much debate over whether the split gene organization is of ancient origin, and has been lost by most bacteria as they evolved ever more streamlined and energetically efficient genomes, or whether it arose in a direct ancestor of the eukaryotes after they diverged from the prokaryotic line.

A popular current view is that it is indeed very ancient, even predating the appearance of DNA, and that many introns represent the bits in between the earliest primitive coding sequences (which would probably have been RNA – see Chapter 5). As coding sequences became combined and rearranged to form more complex genes, the pieces in between were retained as introns.

This idea builds on evidence that the exons of many present-day genes correspond to distinct functional domains in the structure of the proteins they encode. The immunoglobulin antibody molecule provides a particularly clear example. An immunoglobulin molecule is made of two identical pairs of different but related polypeptide chains (see Chapter 12). Each chain consists of a number of repeated modules – modifications of a basic 'immunoglobulin domain', and each domain is neatly encoded within a different exon in an immunoglobulin gene. In the larger chain, one domain contributes to the antigen-binding site and confers its particular specificity as an antibody on the molecule, others determine the 'class' of the antibody – how it interacts with other cells of the immune system, for example, whether it is a circulating antibody, clearing antigen from the blood by interacting with phagogytes, or whether it causes allergic reactions by binding to mast cells in tissues to release histamine, etc.

The basic immunoglobulin unit turns up in a wide range of other proteins that are involved in cell–cell interactions, such as the histocompatibility antigens, several proteins present on the surfaces of immune system cells, and the cell adhesion molecule (CAM) that binds neurones together to form nerves. All these members of the immunoglobulin 'superfamily' are believed to have evolved either directly from each other by gene duplication and divergence or by **exon shuffling**, combining immunoglobulin exons with other exons to create proteins with novel properties.

It seems very likely that many eukaryotic genes have evolved by individual exons being duplicated and combined with others to create new genes and thus new proteins (see later, Gene families and superfamilies). This view regards the lack of introns in bacteria as an evolutionary streamlining. In eukaryotes, there is ample evidence from molecular evolutionary studies that a gene can both lose and acquire introns during evolution. Some introns of more recent acquisition may represent chance insertions of transposon-like DNA (see below, Transposons) or be relics of viral hitchhikers.

Production of eukaryotic messenger RNAs

In eukaryotes, the primary RNA transcript – the heterogeneous nuclear RNA or pre-messenger RNA – undergoes considerable modification before it leaves the nucleus. It is first modified at its 5′ end and immediately packaged into ribonucleoprotein particles. In this state it is spliced, the 3′ end trimmed and a tail of adenine nucleotides (the poly (A) tail) often added. The function of the poly (A) tail is unknown as some mRNAs seem to function perfectly well without it.

The production of tRNAs and rRNAs also involves extensive processing of a primary transcript.

Capping

All eukaryotic prospective mRNA is modified at its 5′ end by the process of capping. A guanosine residue is added to the 5′ end of the primary transcript. It is linked back to front compared with the normal chemical linkage of a nucleotide chain, to make a characteristic **cap** structure that is then methylated at the added guanosine and in some cases, at other nucleotides nearby. Capping seems to be necessary for the RNA to undergo further processing and be translated.

RNA splicing

Introns are removed from the primary RNA transcript before it leaves the nucleus.

Introns are delimited at either end by a characteristic short nucleotide sequence. This varies slightly from gene to gene and intron to intron but the first two bases of an intron in most nuclear genes of higher eukaryotes are always GT (U in the RNA) and the last two always AG. The GT site is the 5′ splice site and is often also termed the *donor*, as it is cleaved first, and the AG site is the 3′ splice site or *acceptor*.

Introns are not removed from primary transcripts in a fixed sequential order starting at the beginning. In genes with multiple introns, therefore, there must be some way of recognizing each intron and its splice points as an independent 'unit' and not missplicing a donor site to the wrong acceptor site. During splicing the pre-mRNA is held in a large ribonucleoprotein complex – the 'splicing-site complex' or **spliceosome**, and this may bring correct sites together (Fig. 3.3).

Spliceosomes contain several ribonucleoprotein (and possibly other protein) components only some of which have yet been identified. One component of the spliceosome is the so-called **snurp** – small nuclear ribonucleoprotein (snRNP). At least 10 different snRNPs have been identified in eukaryotic cells, where they appear to be involved in various aspects of RNA processing. Each type contains a single RNA

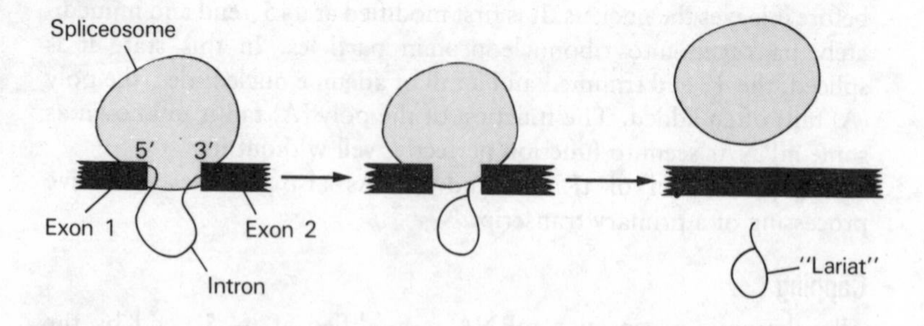

Fig. 3.3 Schematic diagram of nuclear pre-mRNA splicing. The 5′ end of the intron is cut and bonds to a 'branchpoint' nucleotide within the intron. The 3′ end of the intron is then cut and the two exons joined, releasing the intron as a 'lariat' structure.

molecule complexed with several proteins. In mammalian cells, U1, U2 and U5 snRNPs are involved in intron removal from nuclear RNAs. U1 binds specifically to the 5′ splice site, U5 to the 3′ site and U2 to the branchpoint (see Fig. 3.3). U1 and U2 can also bind to each other. How their interaction with the pre-mRNA leads to intron removal is not yet known.

Intron removal may not necessarily require the participation of conventional protein enzymes in the form of RNases and RNA ligases. The revolutionary discovery of introns that can, *in vitro* at least, splice themselves out of a primary transcript without the aid of proteins, suggests that proteins may be required simply to hold the RNA in the correct configuration for self-splicing to occur (see Chapter 5: Ribozymes). It may be the RNA in the snurps that is involved in pairing with the pre-mRNA and splicing it.

The 'spliceosome' model for intron removal refers mainly to nuclear protein-coding genes. Intron removal and RNA splicing in genes for rRNAs, tRNAs and in mitochondrial and chloroplast genes have their own distinctive features.

The phenomenon of *trans* splicing has recently been discovered in the genes for the surface proteins of trypanosomes. In this instance two separately transcribed RNAs are joined together to form the mRNA.

In many cases, two or more different mRNAs can be produced from a primary RNA transcript by **differential or alternative splicing**. These genes contain a number of potential splice sites, only some of which are used at any one time. Differential splicing was first discovered, along with split genes, in the adenovirus, where a long transcript can be spliced in several different ways to give different mRNAs and different protein products mRNAs during infection (see Fig. 3.1g).

It has since turned up in mammalian and other eukaryotic genes where it seems to be one means of producing different forms of a protein from the same gene in different tissues or at different stages of cell differentiation. Differential splicing, for example, underlies B cells' switch from producing membrane-bound immunoglobulin M (IgM) to secreted IgM antibodies (see Chapter 12).

Overlapping genes

Most genes from both prokaryotes and eukaryotes conform to the general rule that a gene is a discrete stretch of DNA specifying one, and only one, kind of polypeptide chain, but several exceptions have now been discovered.

In eukaryotic genes, differential RNA splicing can result in two quite distinct proteins being produced by the same stretch of DNA (see previous section). Another strategy for stretching the coding capacity of DNA to its limits is used by some tiny bacterial viruses. Bacteriophages such as ΦX174 and G4 pack the maximum amount of information into their limited amount of DNA by possessing truly **overlapping genes**.

A DNA/RNA sequence can in principle be 'read' in one of three **reading frames**. In most genes however, only one reading frame is **open**, the others having no appropriate start signal or being blocked by stop codons. ΦX174 and G4 produce a long mRNA transcript of their DNA. Parts of the mRNA transcript can be translated in different reading frames to give in one case two, and in another three separate proteins (see Fig. 3.1*c*). The 'overlap' usually consists of a short portion of the end of one gene and the beginnning of the next, but in G4 there are parts of the genome that are translated in all three reading frames, to specify part of three entirely different proteins. So far, this type of overlapping gene has been found only in phages and even so is rare.

In eukaryotic cells, however, several cases have been found of introns that also comprise separate genes in their own right. Some encode proteins required for their removal, but other large introns have been found that contain known tRNA and protein-coding genes. These genes are, as far as we know, under their own independent transcriptional control. One example is in the *dunce* gene of *Drosophila*. This contains a large intron that encodes a glue protein (proteins that form part of the insect's egg shell) and another, as yet unidentified protein. There seems no functional relation between the *dunce* gene, which is involved in some as yet unexplained way in determining the insect's capacity for memory, and the other genes that are embedded in it. The origins of such gene arrangements may lie in the random DNA movements that appear to be relatively frequent (on an evolutionary timescale) in eukaryotes. Such a complex arrangement, however, poses a considerable challenge to known mechanisms of, for example, RNA splicing and gene control.

Another type of gene overlap is found in both eukaryotes and prokaryotes, and this is where two genes are transcribed from opposite strands of the same piece of DNA.

THE EUKARYOTIC GENOME

The remarkable thing about most higher plant or animal DNA is how little of it is involved in specifying protein or RNAs. Around a third of the human genome (and much more in many other species) consists of DNA sequences that have no apparent role at all in specifying the final form and function of the organism. The remaining 60 per cent or so, which includes the genes and their control regions, also contains a substantial portion of non-coding DNA in the form of introns. This is in marked contrast to bacteria, which use all their DNA to encode information or control its expression.

The C-value paradox

The vagaries of the eukaryotic genome have been known to genetics for many years as the **C-value paradox** (C representing the amount of DNA in a haploid genome). Most eukaryotes have more DNA in their cells than they would appear to need just to code for all the necessary proteins and RNAs. The total amount of DNA in a genome also varies considerably throughout the eukaryotic world, often between closely related species, and bears little relation to degree of morphological complexity and position on the evolutionary ladder (Table 3.1). Flowering plants have the widest range of nuclear DNA content amongst the eukaryotes, some posessing genomes many times larger than those of humans. It seems unlikely that a lily or a newt needs ten times as many genes as a human – hence the paradox.

Table 3.1 Total DNA content of some representative haploid genomes (in base pairs)

Prokaryote	
Gram-negative bacteria (e.g. *E. coli*)	$3–5 \times 10^6$
Eukaryote	
Fungi	*c.* 2.5×10^7
Insects	$10^8–5 \times 10^9$
Amphibians	$10^9–10^{11}$
Reptiles	$2–3 \times 10^9$
Mammals	*c.* 3×10^9
Flowering plants	$3 \times 10^8–10^{11}$

Unique sequences

Some 20 years ago, analytical methods based on the rates at which single strands of denatured nuclear DNA reformed (rehybridized) into double-stranded molecules, first showed that the eukaryotic genome was made up of several quite distinct classes of DNA sequences. The first are the 'unique' sequences present in only one or a very few copies per genome. This class includes the genes that specify proteins – the so-called **single-copy genes** – but we do not yet know whether all unique sequence DNA represents such genes. Unique sequence DNA accounts for some 70 per cent or less of the genome.

Within the unique sequence DNA, a substantial proportion is non-coding in that it represents the introns that interrupt most eukaryotic genes, control regions and, possibly, some of the DNA between genes, and it has been estimated that actual coding sequences may represent as little as 10 per cent of the total genome. The eukaryotic genome has indeed sometimes been likened to a sea of introns out of which emerge tiny islands of information.

Multicopy genes

Genes for the DNA-packing proteins (the histones) and those for tRNAs and rRNAs are present in multiple identical copies – of the order of hundreds (16/28S RNA) to thousands (tRNAs, 5S RNA) in human DNA. This reflects a cell's need for large quantities of their products during DNA replication and protein synthesis at various stages in the cell cycle, and also the fact that in genes for structural RNAs the initial transcripts are the final product and are not amplified by subsequent rounds of translation. The rRNA genes are clustered in tandem repeats in regions in the chromosomes that come together to form the **nucleolar organizer**. The histone genes are also organized in tandem repeats of a five-gene unit comprising one each of the five types of histone gene (H1, H2A, H2B, H3 and H4) (see Fig. 3.1*f*). Multicopy genes of this sort, however, only account for a small fraction of the 'excess' DNA.

Repetitive DNA

The remainder of the genome typically consists of long and short DNA sequences repeated anything from tens to millions of times.

The function, if any, of most repetitive DNA in present-day genomes and its evolutionary origin is much debated (see below, Selfish DNA). Little repetitive DNA has any apparent protein-coding function, and when it has, appears simply to specify proteins necessary for its own duplication and dispersal around the chromosomes. Apart from these active mobile elements (see below, Transposons), which are present in relatively few copies, the remainder of the repetitive DNA falls into two classes.

Interspersed Repetitive Sequences

One is a rather diverse group of sequences and 'families' of related sequences typically a few hundred or thousand base pairs long and present in tens, hundreds, thousands or hundreds of thousands of copies dispersed around the genome. These are the **interspersed repeated DNAs**, and comprise around 20 per cent of a mammalian genome. Many are transcribed into RNA, either as short molecules or parts of longer transcripts, but do not appear to encode proteins.

These sequences belong to the class termed **moderately repetitive DNA** in the older DNA hybridization work. Although no function can be ascribed to these interspersed sequences as a class, it is likely that as they become better characterized, some may turn out to have useful roles as, for example, DNA replication origins (of which there could be around 30 000 in the human genome), and other control sites.

The origins of some of this repetitive DNA are now being uncovered and add an intriguing twist to the evolutionary story. A substantial proportion of the moderately repetitive DNA in the human genome (and that of other primates) is composed of one particular family of DNA sequences – the *Alu* family. There are around 300 000 copies of the various types of *Alu* repeat unit (of around 300 base pairs) dispersed throughout the human haploid genome, roughly one for every 6000 base pairs of DNA. Transcribed *Alu*-type sequences are found either as small RNAs or as parts of longer transcripts. Related sequences have been found in other mammals. *Alu* bears a suspicious resemblance to part of a small RNA (7SL RNA) encoded by a 'legitimate' gene and which is a component of the ribonucleoprotein signal recognition particle that directs proteins out of the cytosol to their final destinations (see Chapter 7).

The *Alu* family is believed to be derived from this 7SL RNA at some time in the past by a subversion of the **reverse transcription** process. DNA can be synthesized on an RNA template by the enzyme reverse transcriptase, and *Alu* is believed to represent reverse-transcribed copies of 7SL RNA that became inserted into the chromosomes (the integration of a reverse-transcribed DNA of this sort into a chromosome is called **retrotransposition**). Reverse transcriptase is a normal product of some RNA viruses (the retroviruses) that are relatively common inhabitants of many mammalian species and which use this method to hitch a lift in host-cell DNA; there may be cellular reverse transcriptases as well. 7SL RNA is found throughout the eukaryotes, but the repetitive *Alu* sequences have only been found in mammals so far.

Reverse transcription of other legitimate RNAs also accounts for some other repetitive sequences and reverse transcription of mRNA may also be responsible for several other features of eukaryotic genomes, such as the existence of 'processed' pseudogenes. These are inactive copies of known genes, but lack introns and contain features such as 'poly (A) tails' characteristic of mRNAs rather than 'genomic' genes. Other repetitive sequences may be derived, either directly or indirectly from retroviruses that infested the genome at some time in the past and have now lost the capacity to make virus particles, and yet others may represent inactive transposons.

Highly Repetitive DNA: Satellite DNA

The remainder of the genome is comprised of short (around 150 – 250 base pairs) and very short (fewer than 10 base pairs) DNA sequences that are serially repeated thousands or millions of times. They make up around 10 per cent of a typical mammalian genome. They are usually clustered in the inactive heterochromatin (see Chapter 2: Chromosome structure) at the centromeres and at the ends of chromosomes and are apparently never transcribed. In some species, this DNA forms a separate band from the bulk of nuclear DNA on density gradient ultracentrifugation because of its distinctive base composition, and is therefore also called **satellite DNA**. The very short repeated sequences are termed **minisatellites** and have been pressed into use as the basis of the novel technique of DNA fingerprinting (see Chapter 4).

Selfish DNA

Why we carry around so much excess baggage in our nuclei is still a mystery. We do not yet know enough about the long-range organization of the eukaryotic genome to say whether repetitive DNA has some important indirect function in overall gene regulation – for example, spacing out genes at some required interval.

Much repetitive DNA is therefore dismissed (perhaps prematurely) as 'junk' with, at most, a purely passive structural function in the chromosomes. This does not of course mean that it could be eliminated from present-day cells without any effect. Organisms have naturally evolved to accommodate the size of their own genome and differences in genome size can have some effect. For example, plants with large genomes tend to mature more slowly, presumably because of the longer time it takes to replicate the DNA, and are often perennials, plants with smaller genomes are often annuals or ephemerals. Ephemeral plants in particular never have genomes above a certain size. In this case the 'junk' DNA has a definite phenotypic effect, which will be subject to the normal process of natural selection. But in many species it does appear to be totally unnecessary. Some closely related species of amphibians with similar morphology and life history have genomes differing by almost an order of magnitude.

The question is why repetitive DNA arose in the first place and has persisted in eukaryotic cells. This is not a trivial consideration. Every time a cell divides this DNA has to be replicated at considerable metabolic cost. Bacteria, which multiply much more frequently and are under greater metabolic stress than most eukaryotic cells, possess streamlined genomes. There are few repeated genes of any sort in prokaryotes and introns are virtually unknown. The single-celled eukaryotic yeasts, whose lifestyle resembles that of bacteria, also have little repetitive DNA and although some of their genes do have introns they are much rarer than in multicellular organisms.

One proposed answer is that much repetitive DNA is **selfish DNA**, the ultimate parasite, which takes advantage of the DNA replication machinery of the cell to multiply itself in a quite arbitrary fashion, outside the control of the forces of natural selection. DNA sequences can be serially duplicated by DNA copying mechanisms that operate during normal genetic recombination, and transposition and retrotransposition would also provide possible mechanisms for its spread.

If selfish DNA has no direct effect on the phenotype, the only limit to its spread comes when it puts too much strain on the cell's metabolic resources and a balance is struck between its further multiplication and a selective disadvantage to the cell. The cells of multicellular organisms inhabit a relatively sheltered environment compared with those of unicellular microorganisms. They usually receive a fairly steady supply of nutrients and are not under such stringent pressure to economize on basic cell processes such as DNA replication. They also replicate their DNA much less frequently than bacteria and are therefore more tolerant of hitchhiking DNA.

Whether they have subsequently subverted it to their own ends has yet to be discovered.

How many genes?

If repetitive DNA is discounted the distribution of genome size throughout the living world begins to look more rational. The number of single-copy genes seems to increase up the ladder of morphological complexity, although we have realistic estimates of gene numbers for a very few species. At the lower end of the scale the bacterium *Escherichia coli* has around 3500 genes which are encoded in a genome of around 4.6×10^6 base pairs. The number of genes is estimated from the number of different proteins *E. coli* can produce as detected by sophisticated gel electrophoretic analysis. Yeast is thought to have around 4000 genes but has a larger total genome (around 2.3×10^7 base pairs). The fruitfly *Drosophila*, with a complex morphology and developmental programme has an estimated 5000 genes (on genetic and cytogenetic evidence). Mammals in general seem to require around 30 000–40 000 genes. Any given differentiated cell type produces around 10 000 different mRNAs, most of which are common to the majority of cell types, and it is thought that the total gene number is unlikely to be much more than two to four times this value. The number of 100 000 is often suggested for human genes (probably because it is a good round figure). The haploid human genome (including repetitive and non-coding sequences) is around 3×10^9 base pairs. Gene numbers in higher animals are as yet no more than guesstimates based on dividing the amount of 'single-copy DNA' by the number of base pairs in an 'average' gene (around 5000–6000 including introns). They are wide open to correction in the future, especially if many more genes of the length of the human DMD locus are found (around 2 million base pairs). Estimates of gene numbers in eukaryotes made before the discovery of introns and extensive repetitive DNA are usually far too high.

There are also ways of arriving at rough estimates of the number of genes in a haploid genome (or at least the number that are essential to life) indirectly from mutational evidence. In simple organisms that can be readily mutated and give large numbers of progeny, the average rate of deleterious mutation per gamete and the average rate of (deleterious) mutation per locus can be determined fairly accurately. Dividing the rate of mutation per gamete by the rate of mutation per locus gives an estimate of the number of genes required for survival.

Gene families and superfamilies

The cellular mechanisms for replicating and duplicating DNA that generate the large and apparently unnecessary amounts of repetitive DNA in eukaryotic genomes also have more productive evolutionary results.

Increases in levels of morphological and physiological complexity must be accompanied by the creation of at least some new genes that encode proteins with novel and useful properties. Evolutionary changes may also be driven by sets of genes coming under new control.

As more protein-coding genes are sequenced it is becoming obvious that most are members of large families and superfamilies of related genes (Table 3.2). Each of these **multigene families** is believed to have

Table 3.2 Some protein families

Globin superfamily (oxygen carriage)
α and β-globins (haemoglobin) (vertebrates)
Invertebrate haemoglobins
Myoglobin (vertebrates)
Leghaemoglobin and other phytohaemoglobins (flowering plants)
?Bacterial 'haemoglobin'
Immunoglobulin superfamily (cell recognition)
Immunoglobulins (*C* and *V* genes) (several hundred variant copies of the *V* genes)
T-cell receptors (C and *V* genes) (many variant copies of *V* genes)
Major histocompatibility antigens (MHC) (many variants of each type)
Thy1, CD4, CD8 (T-cell surface antigens)
NCAM (cell-adhesion molecule in nervous system)
Serine proteinases (see Fig. 3.6)
Transforming growth factor-β
TGF-β (see Table 9.9)
Inhibin A, inhibin B (polypeptides inhibiting secretion of follicle-stimulating hormone)
dpp gene product (protein required during development in *Drosophila*)
The '7 domain' receptors (see Table 9.8)
Rhodopsin
β-adrenergic receptor
Receptor for substance K
Muscarinic acetylcholine receptors

arisen by successive rounds of duplication of an 'ancestral' gene followed by divergence of the copies as a result of the accumulation of chance mutations over long periods of time. As well as duplication (Fig. 3.4) and divergence by point mutation there is increasing evidence for considerable exchange between the different members of sets of duplicated genes, and for new genes to be assembled from parts of others.

Multigene families provide strong evidence for the long-held theory of gene duplication as the chief means of generating novel genes during evolution. The idea was first proposed many years ago as the logical way round the problem of how a gene can evolve without destroying its original function and upsetting the delicately balanced cell economy or the developmental programme. This is a particularly important consideration in multicellular organisms. Over time, the various members of a multigene family can acquire different control sites, so that they become expressed in different tissues or at different developmental stages.

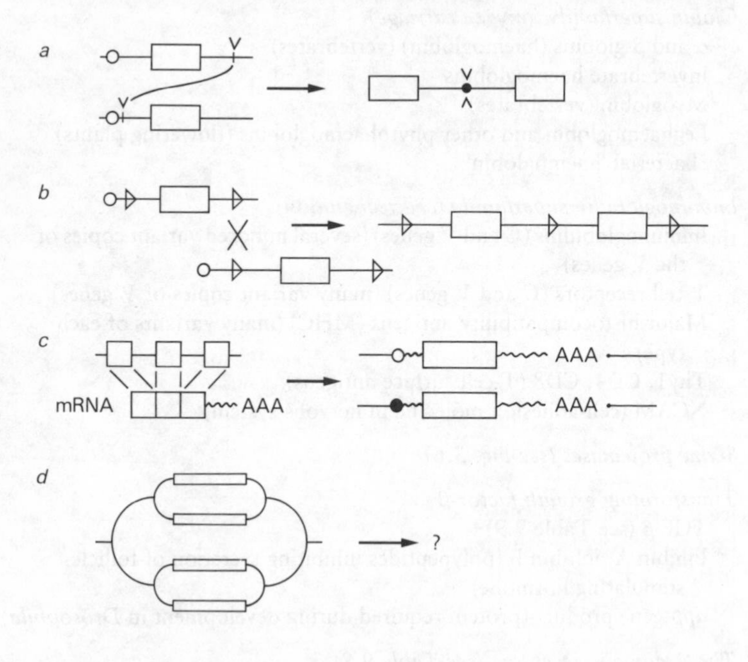

Fig. 3.4 Various ways in which gene rearrangement and duplication can occur. *a*, Random breakage and reunion; *b*, recombination between homologous repetitive sequences in the genome; *c*, retrotransposition – reverse transcription of mRNA and integration of copies at random; *d*, over-replication. (From N. Maeda and O. Smithies, *Annual Review of Genetics*, **20** (1986) 81–108.)

When this is combined with extensive alterations in their coding sequences, a protein with quite new properties and physiological role can be created.

Ever since it became possible to determine the amino acid sequence and three-dimensional structure of proteins, the apparent relatedness of some proteins has been recognized. For example, genes encoding the globins of vertebrate and invertebrate haemoglobins, the muscle oxygen-carrier myoglobin, and the legume nitrogen-fixing nodule protein leghaemoglobin and other phytohaemoglobins have been grouped in the same multigene family and are believed to be related by direct descent and divergence from a single ancestral globin gene.

In determining family relationships between proteins, no less than between whole organisms, the complicating issue of convergent evolution has to be faced. The similarities between some proteins in structure and function may be the result of convergence as a result of selection for the same function, arriving at a similar solution to a problem from different points of departure in two independent lines of descent. DNA sequences are more likely than protein sequences to be able to distinguish between convergence and direct descent and the position of introns is a particularly useful feature. An intron in the same position in two corresponding sequences from different species almost invariably indicates common ancestry, as the likelihood of this occurring by chance in two independent lineages is infinitesimal. DNA sequences that show a measure of similarity due to descent from a common ancestral sequence are said to show **homology**, the degree of homology usually being expressed as a percentage of nucleotide positions that are identical. (Molecular biologists often also use the term homology in a looser sense, simply to indicate a degree of similarity between two DNA sequences.)

The eukaryotic globin multigene family is of undisputed common ancestry by all criteria and illuminates many aspects of gene evolution. The relationship of the plant haemoglobins (e.g. legume leghaemoglobin) to the rest is now widely accepted. It is not yet clear, however, whether they represent animal genes that were transferred to higher plants (specifically legumes) at a recent stage of evolution, or whether plant and animal haemoglobins derive separately from a gene already present in a common ancestor. The recent discovery of haemoglobins in plants other than legumes may throw light on this issue. A bacterial 'haemoglobin' has also been discovered. Its gene resembles the eukaryotic globin genes in some respects, but, like most bacterial genes, does not possess introns. The nucleotide sequence outside the important active site region is so dissimilar from the eukaryotic globins that it is probably impossible to decide whether it represents an entirely independent evolution of a globin-like protein or whether it is related and pushes back the evolutionary origins of the eukaryotic globins to before the prokaryote/eukaryote divide. This would place the globins amongst the oldest gene families. Another possibility is that the gene was picked up by a bacterial ancestor from a eukaryote.

Within the globin family, comparisons between the globin genes of mammals, birds and amphibians reveal changes in base sequence, gene duplications and dispersion to different chromosomes all occurring during vertebrate evolution. In mammals for example, the genes encoding the globin subunits of the various haemoglobins that are made at different times in embryonic and adult life are organized in two families – the α-globin gene cluster and the ß-globin gene cluster, which reside on different chromosomes (Fig. 3.5). Each cluster is composed of several closely related genes that encode either α-type globin subunits or ß-type globin subunits (adult human haemoglobin is made up of 2 α and 2 ß subunits, embryonic haemoglobins of two other identical α-type and ß-type subunits (Fig. 3.5)). Separated α and ß gene clusters are present in mammals and birds, but in amphibians α and ß genes occur together in the same cluster. Only a single type of globin is made by lampreys and hagfishes, whose ancestor branched off the vertebrate line early in vertebrate evolution. This indicates that gene duplication and divergence into α and ß genes occurred after this time. All α-type genes in a particular species are more similar to each other than they are to the ß

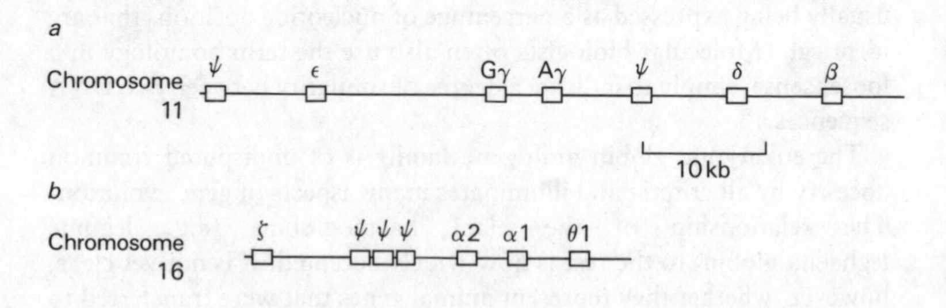

Fig. 3.5 Human globin genes. *a*, β-globin cluster; *b*, α-globin cluster; ψ, inactive pseudogene. The ß cluster comprises ϵ (embryonic), γ (foetal, two types), δ, (foetal and adult) and ß (adult). The α cluster comprises ζ (embryonic), α1 and α2 (foetal and adult) and θ1. Pseudogenes are inactive copies of genes that have become mutated in such a way that they are not expressed. θ1 globin is an α-like globin discovered in primates only in 1986. It is produced early in human embryonic development but its role is not known.

genes and vice versa, showing the more recent diversification within the two families.

The various members of the globin multigene family show their common origin fairly clearly. They all still specify rather similar proteins with the same basic physiological function – oxygen carriage. But it is now possible by computer analysis of DNA sequences to trace relationships that are not immediately evident from the proteins themselves. **Gene superfamilies** are now recognized that contain members with very different functions, but whose genes betray a common evolutionary origin (Table 3.2).

Sophisticated computer programs can scan nucleotide sequences and pick out those that could have evolved from each other by extensive additions, deletions, duplications or rearrangements, as well as the more easily recognized simple base changes that give rise to the replacement of one amino acid by another. Families of receptors, ion channels, structural proteins and enzymes are all being revealed and it may eventually be possible to trace the origin of the many thousands of genes in a typical vertebrate genome back to a few thousand ancestral genes.

Gene superfamilies can be very complex in their relationships. The immunoglobulin gene superfamily in vertebrates comprises the V genes, C genes, etc. that make up the antibody and T-cell receptor gene families in mammals (see Chapter 12) and, on the wider scale, takes in several other cell-surface proteins found on immune system cells, and the histocompatibility antigens. A basic 'immunoglobulin domain' is repeated in modified form in all these proteins. The way in which they have evolved – by exon shuffling, intron insertion, repeated duplication, point mutation, exchange of material between repeated genes and distribution to different chromosomes – is only now beginning to be resolved. Like the globin genes, the immunoglobulin genes can be tentatively traced back to a small number of genes in an ancestral vertebrate by looking at their present arrangement in extant vertebrates from sharks to mammals.

The serine proteases are another well-studied superfamily whose relationships are beginning to be disentangled (Fig. 3.6 and Table 3.3). In their 'basic' form serine proteases are exemplified by some proteolytic enzymes (e.g. trypsin and chymotrypsin). The enzymatic function resides in a particular region (domain) of the protein. (They are called serine proteases because of the crucial participation of a serine residue at the active site.) The serine protease domain is also a component of many other proteins, including members of the mammalian blood clotting and clot-dissolving systems. These include plasminogen (the

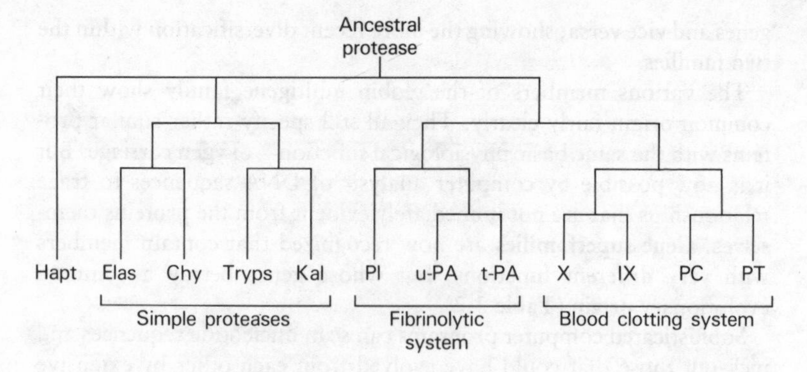

Fig. 3.6 Relationship of some of the mammalian serine protease superfamily as estimated from their protein sequences. Hapt, haptoglobin (a haemoglobin-binding protein from blood with no proteolytic action); Elas, elastase; Chy, chymotrypsinogen; Tryps, trypsinogen; Kal, kallikrein; Pl, plasminogen; u-PA, urokinase; t-PA, tissue plasminogen activator; X and IX, blood clotting factors X and IX; PC, proconvertin; PT, prothrombin. (Adapted from L. Patthy, *Cell*, 41 (1985) 657–653.)

precursor of the clot-dissolving enzyme plasmin), tissue plasminogen activator (TPA), which splits plasminogen, and the blood-clotting factors XII, VII, IX and X. These are all proteases. The most recent member of this family to be discovered, apolipoprotein Lp(a), forms part of the protein component of cholesterol-transporting lipoprotein particles in the blood, and high levels of Lp(a) are linked to a greater risk of developing atherosclerosis (by some as yet unknown mechanism).

The serine protease domain itself is of ancient evolutionary origin. Exon shuffling, gene duplication, and divergence as a result of mutation could have generated the members of the superfamily during mammalian evolution (Table 3.3). A very recent analysis of the nucleotide sequence encoding the active site amino acids reveals yet further complexities of descent. It suggests two quite independently derived lines of descent for the eukaryotic serine proteases which have subsequently become mingled by exon shuffling, etc. Apolipoprotein Lp(a) is probably the last member of the family to have evolved. It has only been found in primates so far and could derive from a copy of the plasminogen gene, to which it has many resemblances. Although the protein contains a serine protease domain it does not have proteolytic activity, since, unlike plasminogen, it cannot be cut to form an active enzyme by tissue plasminogen activator because of an altered amino acid at the cleavage site.

Lp(a) poses an interesting evolutionary question. It appears to be an example of an evolutionary development that does not confer any particular advantage on the species carrying it; for example, it does not for example seem to extend the efficiency or regulatory flexibility of the lipoprotein system. It may simply be an example of a random duplication and divergence that happens to produce a protein that can be fitted into an existing system without disturbing it, and is therefore not immediately selected against.

Table 3.3 Animal proteins with sequence segments in common

EGF (epidermal growth factor)-type segment
- EGF
- TGF-α
- LDL lipoprotein receptor
- Clotting factor IX
- Clotting factor X
- Protein C (component of alternative complement pathway)
- Tissue plasminogen activator (TPA)
- Urokinase
- Complement component C9
- *Notch* gene product (fruitfly) } proteins required during
- *lin-12* gene product (nematode) } development

C9-type segment
- Complement component C9
- LDL lipoprotein receptor
- (Fibronectin?)
- *Notch* gene product
- *lin-12* gene product

Fibronectin 'finger'
- Fibronectin
- TPA

Proprotease 'kringle'
- Plasminogen
- TPA
- Urokinase
- Prothrombin

'Finger' and 'kringle' refer to particular types of modular units of protein construction. In addition, the serine proteases (see text) contain a common protease domain.

Transposons

In any species, whether of bacteria or higher organism, the genes on a given chromosome are arranged in a fixed linear order. Genes do not normally move around the chromosomes and the chromosomal rearrangements that occur from time to time often have deleterious effects and are associated with pathological states. During evolution, however, DNA rearrangements on the grand scale have occurred and even present-day genomes are not so static as was once assumed. Blocks of DNA sequence can be repositioned within a chromosome by a variety of mechanisms.

Both eukaryotes and prokaryotes contain DNA sequences that can not only move themselves from one site to another, but can also carry adjacent sequences with them. **Transposable elements** or **transposons**, as this rather diverse class of DNA elements is called, were first discovered in the 1940s in maize (*Zea mays*), where they are responsible for the spotted and streaked colouring of the seed in some strains. In maize, they are known as **controlling elements**, from the genetic consequences of their movement. The controlling elements of maize remained a genetic curiosity for thirty years until similar mobile sequences were found in the bacterium *E. coli* in the 1970s and transposons became respectable.

The movement of a DNA sequence from one site to another is known as **transposition**. Unlike some types of chromosomal rearrangement (such as translocation and transversion) in which a chunk of chromosome is physically repositioned elsewhere, transposition usually results from the insertion of a duplicate copy of the transposon, which remains at its original site (Fig. 3.7). This type of transposition generally only occurs to another site within the same DNA molecule (or chromosome) but subsequent recombinational events can often transfer transposons to other DNAs.

Unlike more conventional DNA recombination, transposition does not require any similarity between the transposon and its target sequence, and most transposons can insert anywhere on a chromosome. Transposable elements carry the genes required for their own transposition although, like viruses, they may use their 'host's' enzymes (such as DNA polymerase for replication) as well. They have no independent existence outside the chromosome.

In bacteria the basic transposable modules are known as *insertion elements (IS)*. They are characteristically between 800 and 5000 nucleotides long. When they insert into a new site they generate short

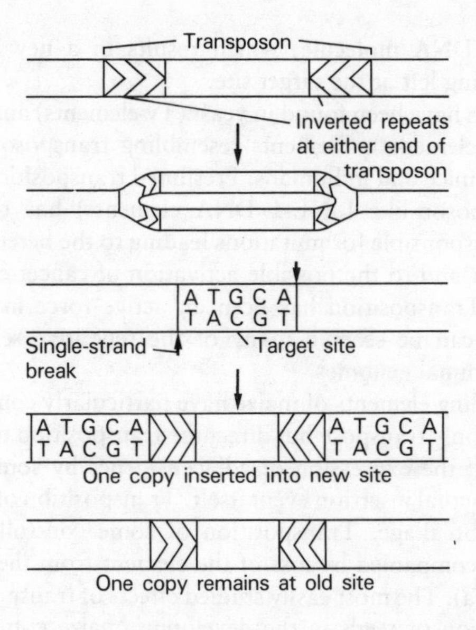

Fig. 3.7 Duplicative transposition. The transposon is replicated and one copy is inserted into a new site. The process is mediated by a transposase encoded by the transposon. Insertion generates a direct repeat (from 4–11 base pairs, depending on the transposon) of the original site sequence on each side of the inserted transposon. No free transposon DNA is produced.

direct repeated DNA sequences at either end and mutational insertions caused by transposition are identified by these characteristic footprints. Bacterial transposons are not only of academic interest. Larger bacterial transposons carry drug-resistance genes for a variety of antibiotics. Many bacteria have become multiply resistant to a wide range of commonly used antibiotics. The drug-resistance genes are typically carried together on large plasmids (independently replicating circles of DNA outside the host chromosome). These can be transmitted from one bacterium to another and thus spread multiple resistance rapidly throughout the population. Some of the most common resistance genes turn out to be parts of transposons, and this may have a bearing on the ease with which multiple drug-resistance plasmids are generated. Some of these larger transposons are a composite of two IS elements, or very similar elements, flanking a stretch of DNA encoding the additional genes, and transpose by duplicative transposition. Other types of drug-resistance transposon colonize new sites by non-specific recombination with another DNA molecule, which results in a new copy of the transposon being left at the target site.

Transposons have been found in yeast (Ty elements) and *Drosophila* (copia and P elements). Elements resembling transposons have also been found in mice and in humans. Presumed transposition in humans (of the transposon-like LINE-1 DNA elements) has recently been shown to be responsible for mutations leading to the hereditary disease haemophilia A and to the possible activation of cancer-causing genes (oncogenes). Transposition has been an active force in the past, as its footprints can be seen in many of the repetitive sequences that characterize animal genomes.

The controlling elements of maize have particularly complex effects. Some can not only transpose but direct the transposition of others, and also switch off the expression of adjacent genes by some means unrelated to the actual insertion event itself. Transposition often results in chromosome breakage. Transposition of some controlling elements (e.g. Ds) is accompanied by loss of the element from the original site (unlike bacteria). The most easily studied effects of transposition are on the pigmentation of seeds in the developing maize cob. Episodes of transposition near pigmentation genes in the cells of the developing endosperm result in pigmented spots and streaks, the progeny of cells in which transposition has occurred. In this case, transposition on one chromosome of a homologous pair has happened to inactivate a dominant allele controlling pigment production, allowing the recessive (in this case pigmented) phenotype to emerge. Strains of maize showing spotting and streaking were selected and maintained by the indigenous peoples of America.

The retroviruses, an unusual class of RNA viruses, behave very similarly to transposons, generating direct repeats when they integrate a DNA copy of their genome into the host cell chromosome and having internal similarities of structure. The close similarity between yeast Ty elements and retroviruses is now well established. One view of the evolution of these viruses is that they may be 'transposons' which by picking up cellular genes have also acquired an independent existence outside the chromosome. Conversely, some present-day transposon-like sequences in animal genomes could represent modified retroviruses – these are called **retrotransposons**.

Basic transposable elements such as IS elements are prime examples of 'selfish DNA' that appear to have no other function than to propagate themselves. Nevertheless, their random activities, and those of other DNA replicating and recombinational mechanisms, may have shaped genomes to an extent that we are only now beginning to realize.

CONTROL OF GENE EXPRESSION

With very few exceptions, each of the cells of a multicellular organism contains a full complement of genes. These specify a repertoire of thousands, and in the case of vertebrates probably tens of thousands, of different proteins. To build and maintain a living cell, let alone a multicellular animal or plant, each protein must only be made when and where it is needed. **Gene expression** – the execution of genetic instructions – is therefore under strict control.

In an individual cell, gene expression is regulated to meet the varying demand for enzymes, structural proteins and RNAs for metabolism and growth throughout its life. The second-to-second control of intermediary metabolism (e.g. the routine biochemical reactions of cellular respiration, and the synthesis and interconversion of small molecules such as nucleotides, amino acids, sugars and fatty acids) and the rapid responses that cells make to many external stimuli do not require direct genetic intervention, but are regulated by complex networks of interactions between metabolites, enzymes and other regulatory proteins (see, for example, Chapter 9).

Some external stimuli do, however, trigger changes in gene expression that can be readily studied in cultured cells. These have provided a starting point for studies on eukaryotic gene expression. The B cells of the immune system, for example, only start to make and release large quantities of immunoglobulin antibody at a certain point in the immune response, in reply to a battery of signals provided by incoming antigen and other immune system cells. The metallothionein gene (which specifies a protein involved in the transport of essential metal trace elements such as Cu and Ni) is switched on in response to heavy metals or glucocorticoid hormones. There are genes in all organisms, from bacteria to humans, that are temporarily expressed in response to a sharp rise in temperature – the 'heat-shock' genes whose function remained until recently a complete mystery (see Table 3.4). Light, steroid hormones, cyclic AMP and interferon are just a few of the stimuli that are known to switch on specific genes in appropriate cells.

Differentiated cells

Short-term regulation of gene activity is common to both unicellular and multicellular organisms. But embryonic development and the cooperative division of labour between specialized cells make additional demands of control mechanisms in multicellular organisms. During embryonic development cells **differentiate** from unspecialized precursors to become nerve, muscle, bone, xylem or phloem cells, and so on. A fully differentiated cell is able to express only a small subset of all its genes and much of the genome appears to be permanently switched off. For example, the reticulocyte precursors of red blood cells are the only cells out of the more than 200 different types in the mammalian body that make haemoglobin, although all the others also contain a full set of globin genes. In contrast most cells make a common set of basic metabolic enzymes and structural proteins.

Why some genes are active in one cell type but not others is one of the questions that can now be tackled at the biochemical level and some answers are beginning to emerge (see later sections). As a general rule no irreversible structural change in DNA or chromatin has occurred; the silenced genes have not been lost or irreversibly inactivated. A nucleus from a fully differentiated intestinal cell of a tadpole is 'reprogrammed' and supports the development of a complete animal when transplanted into the oocyte in exchange for the oocyte's own nucleus (see Chapter 10).

Its repertoire of 'expressable' genes is programmed into a differentiated cell before it becomes visibly specialized. How and even when this occurs is in most cases completely unknown, but with current advances in molecular biological technique, appears a far less intractable problem to solve than it did some years ago.

Orderly embryonic development and multicellular life depend on the phenomenon of **cell memory**. Once a cell has become committed to a particular pathway of embryonic development it never under normal circumstances goes back to a previous state, and it also passes on its committed state to all its descendants. (This is dealt with in more detail in Chapter 10.) Fully differentiated cells also (when they retain the ability to divide) pass on their particular specialized state to their progeny. Liver cells make more liver cells and fat cells more fat cells. The ability of plant tissue to produce undifferentiated cells and even a complete new plant from a single differentiated cell is a major exception to this rule. In most animals, the only cell that can give rise to a new animal is the female gamete – the ovum. We have as yet little hard evidence on how cell memory is established, how it is presumably erased during gamete formation, and how plant cells circumvent it.

One pertinent question in multicellular organisms is whether each gene is regulated entirely individually at the level of transcription throughout development, or whether there are means of shutting down or activating whole sets of genes by a generalized mechanism, perhaps depending on the way DNA is packaged into chromatin. Given the large number of genes that have to be regulated together, this is an attractive option. Present evidence does indeed suggest that genes that are characteristically expressed in a particular cell type are in a different packing state that renders them available for transcription (even when they are not being actively transcribed), compared with those that are never expressed.

The ultimate aim of much research on the regulation of gene expression is a solution to the problem of cell memory. A working assumption is that it is essentially the result of a particular pattern of active (or potentially active) genes being established in a cell lineage during development and then being maintained and passed on. How this may be achieved through the upheavals of repeated rounds of DNA replication and cell division is still unknown. (For more on this, see below, Chromatin structure and gene expression, and DNA methylation.)

The long-term changes in gene activity that accompany cell differentiation do not usually involve any permanent restructuring of the DNA itself. There are some notable exceptions however, one of which is the extensive rearrangement of DNA in the B cells and T cells of the immune system that precedes the expression of antibody and T-cell receptor genes (see Chapter 12). Other well-studied examples in eukaryotes are the DNA rearrangements at the mating type locus in yeast, which result in switches in mating type, and those that underlie the ability of parasitic trypanosomes to evade the immune system by expressing a sequence of different surface antigens (antigenic variation).

Control of transcription

To make a protein, a gene has to be transcribed, the RNA processed (in eukaryotes) to form messenger RNA and the mRNA translated. The whole procedure is encompassed in the term **gene expression**. There are several control points along this pathway, the first of which is the initiation of transcription. Because, whatever else happens to it, a gene must be transcribed to be expressed, most work at the molecular level has focused on how **initiation of transcription** is controlled.

Gene expression is, however, subject to other controls. A gene may be transcribed, but is the RNA processed? Is an mRNA exported to the cytoplasm and once there is it translated? Regulation at **post-transcriptional** stages, especially those of RNA processing and export, may well be as important as initial transcriptional control in eukaryotes. Vertebrate cells at least seem to expend considerable effort in transcribing RNA they never use, as only a very small quantity of nuclear RNA ever reaches the cytoplasm as mRNA. Intron removal from long pre-mRNA transcripts accounts for some but not all of this discrepancy. In the frog

Xenopus, RNA characteristic of an early embryonic stage (the blastula) is still being transcribed much later in differentiated cells of the tadpole intestine, but the mRNA complement of these two tissues, which represents the RNA actually leaving the nucleus, is very different. It is, however, technically difficult to find out what is actually going on inside the nucleus, even when the molecular biology of individual steps such as transcription and RNA processing can be studied in cell-free extracts.

Until very recently most of what we know about the control of transcription came from work in bacteria. Whether a gene is transcribed or not depends on the actions of **regulatory proteins** (repressors, activators and transcription factors) that bind to control points in DNA outside coding sequences and in doing so either activate RNA polymerase to start transcription, or prevent it from doing so. Once launched on its way an RNA polymerase molecule proceeds on automatic pilot until it reaches DNA signals that mark the end of the **transcription unit** – the stretch of DNA that is transcribed into a continuous RNA. A fundamentally similar method of regulating transcription is gradually being confirmed for eukaryotic genes but in a more elaborate form.

A full description of the control of transcription in eukaryotic cells also has to take account of several features lacking in bacteria. Eukaryote DNA packaged into chromatin is not always available for transcription even if all the necessary biochemical machinery is present. The higher order structure of chromatin, as well as the direct interaction of gene regulatory proteins with DNA sequences, is therefore likely to be involved in controlling gene expression.

The bacterial operon: the paradigm of transcriptional control

Bacteria lead a precarious if successful existence that depends on fast growth and multiplication, and on responding extremely rapidly to changes in available nutrients in their vicinity and not wasting valuable metabolic energy in transcribing genes and synthesizing proteins that are not immediately required. In response to these pressures, they have evolved a streamlined gene organization and an elegant and efficient mechanism of coordinating the expression of functionally related sets of genes.

In bacteria and their phages, the coding DNA sequences (the genes) specifying the components of a particular functional pathway (e.g. the enzymes needed for the breakdown or synthesis of some compound) are often clustered together and are *co-transcribed* as a single unit when

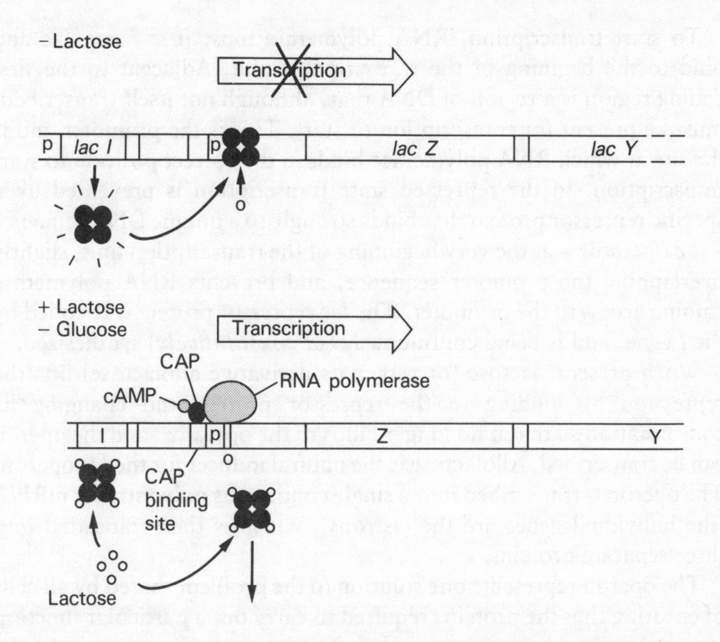

Fig. 3.8 Regulation of expression of the *lac* operon. See text for the regulatory action of the repressor encoded by the *I* gene. Some bacterial promoters also require the additional action of positively-acting regulatory proteins to launch the RNA polymerase on its way. The *lac* promoter is one of these, and even if metabolic repression is lifted it is only activated if a regulatory protein – CAP – is bound to a specific site adjacent to the promoter. CAP is indirectly responsive to the presence of glucose; if glucose is present, CAP does not bind to its control site and the operon is not transcribed. This prevents the operon being unnecessarily activated in the presence of lactose if plenty of glucose, which the bacteria can use directly as an energy source, is also present.

their products are needed. Their transcription is typically controlled by a regulatory protein encoded by a separate adjacent gene. This protein acts at control sites at or preceding the beginning of the transcription unit. The protein-coding sequences, control sites and the regulatory gene together comprise an **operon**.

The *lac* operon (Fig. 3.8) illustrates the general principle. (The relationship of control to coding regions varies in detail between different operons.) It specifies three proteins required for the uptake and utilization of the sugar lactose. When no lactose is available, the operon is *repressed*, in which state it is not transcribed. When lactose is present, the repression is lifted – the operon is said to be *induced* – and transcription begins.

To start transcription, RNA polymerase must first recognize and bind to the begining of the transcription unit. Adjacent to the first coding region is a region of DNA that, although not itself transcribed, must be present for transcription to start. This is the **promoter** and is the site at which RNA polymerase binds in the correct position to start transcription. In the repressed state transcription is prevented by a specific **repressor** protein that binds strongly to a unique DNA sequence – the **operator** – at the very beginning of the transcription unit, slightly overlapping the promoter sequence, and prevents RNA polymerase gaining access to the promoter. The *lac* repressor protein is encoded by the *I* gene, and is being continuously (or *constitutively*) synthesized.

When present, lactose (or rather its derivative allolactose) lifts the repression by binding to the repressor protein and changing its conformation so that it no longer binds to the operator, and the operon can be transcribed. Allolactose is the natural **inducer** for the *lac* operon. The operon is transcribed into a single continuous **polycistronic mRNA** (the individual genes are the **cistrons**), which is then translated into three separate proteins.

The operon represents one solution to the problem shared by all cells of ensuring that the proteins required to carry out a particular function are produced together when needed. The *lac* operon is an example of a simple negatively regulated control loop – it will be expressed unless switched off by repressor protein and remains in the repressed state unless its specific inducer is present. Other types of control are found in other operons. For example, those specifying the enzymes of biosynthetic pathways (e.g. the synthesis of amino acids) are normally switched on, but if the cells are supplied with the required amino acid, transcription is shut down by a feedback mechanism in which the amino acid – the endproduct of the pathway – combines with a specific repressor, this time changing its conformation so that it binds to the operator and blocks transcription.

The operon type of gene organization is not found in eukaryotes, where genes for enzymes of the same metabolic pathway are often dispersed widely throughout the chromosomes. The general principle exemplified by sequence-specific protein/DNA and protein/protein interactions as a way of selectively switching genes on and off is, however, shared by eukaryotes.

Bacterial Promoters

To initiate transcription RNA polymerase must bind strongly to DNA and 'open' the DNA helix. The promoter region is where this occurs. Promoter regions in bacteria can be defined quite precisely by *in vitro* tests that delimit the minimum length of DNA preceding or around the startpoint of transcription that is required for a gene to be transcribed.

Promoters for different operons differ in their overall nucleotide sequence but many share several 'conserved' features which must be present for the promoter to function. These conserved sites appear to mark out binding sites for regulatory proteins and RNA polymerase.

Two short DNA sequences recur in the same or very similar form and location in many promoters and are essential for transcription to be initiated. One is located about 35 base pairs upstream of the start of transcription, and is often called the **-35 site**. The **Pribnow box** (typified by TATAAT), is centred around 10 base pairs preceding the startpoint of transcription. It represents the part of the promoter over which the polymerase binds tightly and opens the DNA helix to start transcription. Both sites are believed to be recognized and bound by the sigma subunit of bacterial RNA polymerase. The sequence of the intervening DNA seems unimportant, but the distance between the two sites is always more or less the same. It seems that the topography of the promoter, rather than its overall sequence, is its important feature, bringing the two sites into a particular spatial relationship when the DNA is in its natural double-helical state.

Promoter function is in general quite independent of the coding sequences it controls. In genetic engineering, for example, bacterial promoters are hitched up to the coding sequences of animal genes so that the animal gene products can be made in bacteria, as eukaryotic promoters are not recognized by the bacterial RNA polymerase and associated transcription factors.

Control of gene expression in eukaryotes

Eukaryotic genes also possess regulatory sites flanking (and in some cases even within) the gene itself, without which they either cannot be transcribed at all or become unresponsive to signals that normally regulate their expression.

In protein-coding genes these control sites are generally located 'upstream' of the coding sequences, to the 5′ side of the startpoint of transcription (see Figs 3.1 and 3.9). Specific control sites are recognized by virtue of their nucleotide sequence by a variety of sequence-specific DNA-binding proteins known as **transcription factors** or **transcriptional regulators**, whose binding stimulates (or in some cases represses) transcription. A site analogous to the bacterial Pribnow box, the **TATA box** or **Hogness box**, has been found in many eukaryotic promoters,

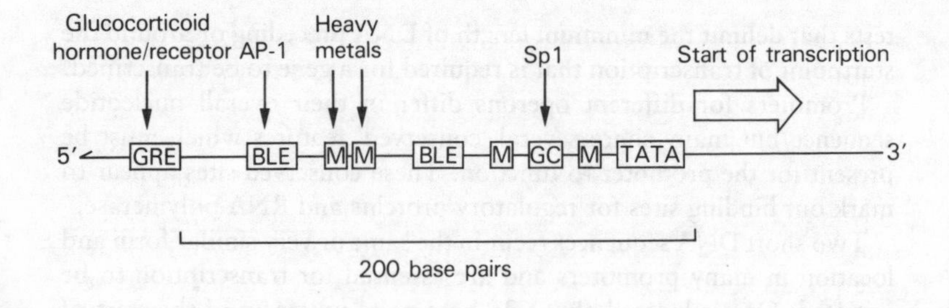

Fig. 3.9 Multiple control elements in the 5′ upstream region of the mouse metallothionein A gene (adapted from W. Lee *et al.*, *Nature*, 325 (1987) 368). TATA, TATA box; M, heavy-metal response element; GC, GC-rich element; BLE, basal level element; GRE, glucocorticoid hormone response element. AP-1 and Sp1 are transcription factors. Constitutive levels of transcription require the TATA box, the GC box and the BLE to be activated. Higher levels of transcription are induced by additional activation of the M or GRE elements. The sequence recognized by AP-1 is also sometimes called the 'phorbol response element'.

this time centred some 25 base pairs preceding the transcription startpoint. A second conserved site has also been found in certain eukaryotic promoters at around 70–80 base pairs preceding the start of transcription, and is known as the **CAAT box**. A third ubiquitous control element – the **GC** box – a GC-rich decanucleotide motif, which is sometimes repeated several times in promoter regions has also recently been identified.

Control Sites in Eukaryotic DNA and Gene-regulatory Proteins

Eukaryotic RNA polymerases are unable to position themselves correctly at the beginning of a gene and start transcription without the cooperation of proteins bound elsewhere in the control region, sometimes thousands of nucleotides away. It is therefore difficult to define a strict 'promoter' site analogous to those in bacteria. On average a protein-coding mammalian gene requires a few hundred nucleotides of flanking sequence in order to be transcribed and correctly regulated. The TATA box is believed to be the site at which RNA polymerase II (see below) binds tightly to initiate transcription. In general, however, the whole of the regulatory region in eukaryotes is often called the 'promoter region'.

In eukaryotic cells, unlike bacteria, different classes of gene are transcribed by different RNA polymerases (I, II and III). Protein-coding genes are transcribed by RNA polymerase II and their promoter regions are typically found at the beginning of the gene in the 5′ non-transcribed flanking sequences. Genes for rRNA are transcribed by polymerase I in the nucleolus, and genes for tRNA, 5S RNA (a ribosomal RNA) and many small nuclear RNAs by RNA polymerase III. The promoter for 5S RNA is located *within* the transcription unit, downstream of the startpoint of transcription.

To inititate transcription, each of these polymerases requires other proteins, which variously bind to DNA, to each other and to the polymerase to form a **transcriptional complex** that starts the polymerase on its way (see, for example, the 5S RNA gene transcriptional complex in Table 3.4). Some of these proteins are general transcription factors required for the expression of many genes, others are more selective. Eukaryotic transcription seems to be regulated not by single 'gene-specific' regulatory proteins, but by permutations of a much larger number of control proteins (Fig. 3.9 and Table 3.4). The fully regulated control of many eukaryotic genes may require the binding of half-a-dozen or more different proteins.

The control sites in DNA are often known as ***cis*-acting control elements** – that is, sites that are required to be on the same DNA molecule and in reasonable proximity to the coding sequences whose transcription they control; the proteins that interact with them are *trans*-acting – that is they are produced by genes that may be located some distance away (in eukaryotes even on another chromosome).

Enhancers

Cis-acting control sites called **enhancers** have been found in many eukaryotic genes. These are the crucial genetic switches responsible for activating their associated genes only in particular cell types or in response to a specific stimulus. When an enhancer is activated by the binding of a cognate transcription factor there is a dramatic stimulation of transcription (often several hundredfold) of any adjacent gene. In the control region diagrammed in Fig. 3.9, the BLE and GRE sites are enhancers. Enhancers can also be repressed, effectively shutting down transcription.

They were among the first of the eukaryotic gene control sites other than the TATA and CAAT boxes to be discovered and their unusual properties stimulated much new thinking about the mechanism of eukaryotic gene regulation. Unlike other control sites, which have to be in a fixed relationship to, say, the start of transcription, enhancers are

Table 3.4 Some control sites in eukaryotic genes and the regulatory proteins that interact with them

Control site(s)	*Regulatory protein (s)*
General transcription	
The 'GC' box in many genes	Sp1 (produced by many cell types)
The 'CAAT' box in many genes	CTF-1 (NF-1) (produced by several cell types)
Specific regulatory elements	
Sequences within the promoter for the 5S RNA gene	A gene-specific transcription factor, TFIIIA. This binds to DNA and is then bound by two more proteins TFIIIB and C, forming a transcriptional complex to which RNA polymerase III binds. TFIIIA is a zinc metalloprotein, the Zn atoms folding the protein chain into several 'fingers' that contact the DNA. This structural motif is shared by several other gene-regulatory proteins.
Immunoglobulin heavy-chain gene enhancer (located in an intron)	NF-muE1, NF-muE2, only in B cells (the only cells that express immunoglobulin genes)
Immunoglobulin light-chain (kappa) enhancer	NF-kappaB (in B cells)
Immunoglobulin gene-promoter 'octamer' sequence (ATTTGCAT). The octamer element is also present in promoters for some other genes, such as histone H2B (histones are produced in all cell types)	NF-A2 (in lymphoid cells only) NF-A1 (in many cell types)
The 'GAL' promoter in yeast	GAL4 protein: recognizes a 'gene-specific' element in the promoter
Several other gene-specific transcriptional activators are known from yeast	

Table 3.4 contd.

Control site(s)	*Regulatory protein (s)*
'Response elements' shared by different genes	
'Phorbol-ester response element'. Found in several genes (e.g. within the SV40 enhancer region and the mouse metallothionein IIA gene promoter region). Genes containing this element can be identified by their responsiveness to phorbol esters which activate the cognate transcription factor. The physiological stimuli activating the element have not yet been identified. As yet undiscovered genes containing this control element may be involved in control of cell division as the product of an oncogene *(jun)* mimics the normal activator AP-1	AP-1, AP-3 (depending on cell type)
Glucocorticoid, progesterone, oestrogen, etc. response elements	The cognate hormone/receptor complex
Thyroid hormone response element	Thyroid hormone/receptor complex
The heat-shock response element in the heat-shock genes. The heat-shock genes are expressed in many animal cells and bacteria in response to a sudden rise in temperature or other stress. The role of heat-shock proteins remained a mystery until recently. There is now some evidence that they may be involved in helping proteins to fold into their final conformation or unfolding proteins that have to be translocated across membranes	HSTF (heat-shock transcription factor)

still effective if experimentally repositioned in reverse orientation and even when relocated in different positions around, or even within, the genes they affect. Some are even effective when replaced several thousand base pairs away from either end of the gene.

Even in their natural state, enhancers occur in several different positions. They are often located a hundred or so base pairs upstream of the start of a transcription unit, but have also been found in the middle of transcription units, as in the immunoglobulin genes, where their activation is responsible for the massive increase in antibody production that occurs in plasma cells. They have also been found immediately before the startpoint of transcription and even, as in the chicken ß-globin gene, in the 3′ flanking regions at the end of the gene. On further analysis, some enhancers have been found to consist of several separate control sequences, containing binding sites for a variety of enhancer-specific transcriptional activator proteins.

Enhancers are variously responsible for stimulating gene expression in response to steroid hormones, interferon, and numerous other stimuli. In some green plants several genes specifying photosynthetic enzymes are known to be only switched on in response to light and this light-sensitivity is caused by enhancer elements in their control regions.

Cell-specific control of gene expression can also be mediated through enhancers. Transcriptional regulators that recognize the enhancers of highly cell-specific genes, and are only produced by the appropriate cell type have already been identified in a few cases (see Table 3.4). By working back from these transcriptional activators to their genes and possibly to further hierarchies of regulatory genes that control their expression, molecular biologists may be able to pinpoint genetic events that mark the important choices cells have to make between one differentiation pathway or another during development.

When enhancers were first found in eukaryotic genes, no bacterial control region with similar properties was known, but enhancer-like control elements have since turned up in bacteria as well.

Coordination of Eukaryotic Gene Expression

Bacteria express a set of genes together as required by transcribing them into one mRNA. There is no evidence of this type of gene organization in eukaryotes.

Because each eukaryotic gene represents a separate transcription unit, one way of coordinating the expression of several genes in response to a given signal could be through common 'response elements' for the signal. DNA sequences that confer specific responsiveness to steroid hormones, interferon, heavy metals, light and cyclic AMP (amongst others) have been found. These 'response elements' are activated by specific transcription factors that are themselves activated by the stimulus.

One set of transcriptional activators of particular biological interest are the receptors for steroid hormones, the morphogenetic hormones thyroxine and insect ecdysterone, and retinoic acid, also recently identified as a developmental morphogen (see Chapter 10). These compounds are believed to exert their effects by switching on sets of genes. The hormone forms an intracellular complex with a protein receptor and this complex then binds to response elements in DNA (see Chapter 9). Which genes are activated and how this leads to the long-term changes stimulated by these hormones is not yet known, but several genes containing response elements at which the hormone/receptor complex binds strongly and switches on transcription have been found.

Some genes are known to be developmentally regulated – that is, they are expressed only at a particular stage (or stages) during development. Many are likely to be genes intimately involved in the developmental process itself; little is yet known about them in most organisms (but see Chapter 10). A few well-known proteins and enzymes are, however, also subject to developmental regulation – for example the globins in mammals. Embryonic, foetal and adult red blood cells contain different haemoglobins as the result of expressing different pairs of globin genes (see above, Gene families and superfamilies). Determining how these genes are switched on and off during development may uncover control machinery of more general application to gene regulation during embryogenesis.

Many layers of gene control are being revealed in the cells of multicellular organisms. Questions still to be answered are, for example, how is the production of transcription factors themselves controlled? Are the appropriate cells already primed or are the regulatory proteins synthesized only in response to the stimulus? The heat shock transcriptional activator, for example, appears to be present in unstressed cells, but only binds to the specific control site after the cell is stressed. Steroid hormone receptors are also already present in the cytoplasm of cells competent to respond to steroid hormones. The potential ability to respond to these stimuli must therefore be programmed into the cell at an earlier stage in differentiation.

Mechanisms for starting transcription

The mechanism by which transcriptional regulators control the initiation of transcription is not yet entirely clear. Some eukaryotic transcription factors can act at considerable distances from the start of transcription – more than 1000 nucleotides in a few instances. There are several competing theories at present, all of which have some evidence to support them. Their adherents have been labelled by one leading scientist in the field as respectively 'twisters, sliders, oozers and loopers'. The twisters propose that when regulatory proteins bind to DNA this induces some conformational change in the DNA, possibly a local unwinding, that is transmitted to the polymerase binding site making it easier for RNA polymerase to bind. Sliders propose that the regulatory proteins induce a local binding of RNA polymerase. This then tracks along the DNA until it encounters a 'promoter' proper. Oozers suggest that binding of one regulatory protein aids the binding of another at an immediately adjacent site, and so on until the start of transcription is reached, and, presumably, RNA polymerase is finally bound and activated.

Loopers suggest that regulatory proteins bound at one site interact directly with other regulatory proteins which bind at their sequence-specific sites and eventually with RNA polymerase, to form a multiprotein transcriptional activating complex. This complex then induces RNA polymerase to bind tightly to the promoter and start transcription. If the regulatory site(s) and the promoter are far apart, the intervening DNA simply loops out to bring the proteins together (Fig. 3.10)

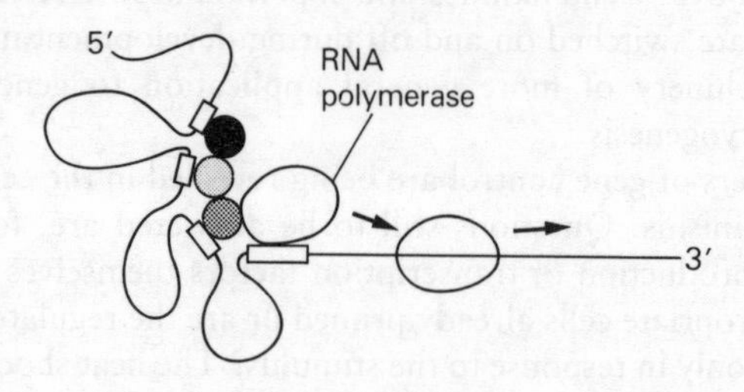

Fig. 3.10 Formation of a transcriptional complex by DNA looping.

DNA looping is an attractive hypothesis, for which there is now some good evidence. Many gene-regulatory proteins are known to consist of two distinct 'domains', one of which recognizes and binds to the DNA sequence, while the other interacts with another protein. Artificially created control regions in which DNA looping can be visualized in the electron microscope show that it is a physical possibility. It provides an explanation for the properties of some control elements (see above, Enhancers) that are still effective in stimulating transcription even if switched around and repositioned at the other end of a gene.

Chromatin structure and gene expression

In eukaryotic cells, changes in chromatin structure are also involved in gene expression. To be transcribed, the DNA comprising a gene has to be locally unpacked from its chromosomal state, and probably also unwound from at least some of the nucleosomes into which it is generally packaged (see Chapter 2: Chromosome structure). Little is yet known about how easy it is for transcriptional activating complexes to gain access to the DNA in different states of chromatin packing. Highly condensed chromatin (as in mitotic chromosomes and the inactivated X-chromosome in mammals) is not transcribed and regions of localized condensation may be a way in which the specialized cells of higher organisms permanently shut down those parts of their genome that they do not express.

One way in which genes may be kept 'open' and available for transcription against this sort of background, is through a long-term association with transcriptional complexes that prevents the gene being packed away, and which only needs the addition of some stimulus-specific regulatory protein for transcription to begin.

An association of this sort seems to lie behind the differential expression of the two types of 5S RNA genes that are present in the frog *Xenopus*. Of these two almost identical genes, one is expressed in both the oocytes and somatic cells throughout the animal's lifetime, and the other is only expressed in oocytes. Both are present in multiple copies (400 and 20 000, respectively). Both require binding of the gene-specific transcription factor TFIIIA (plus two more proteins that bind to TFIIIA) before RNA polymerase can start transcription. Chromatin containing the somatic-type genes isolated from somatic cells contains the transcriptional factors bound to it and only requires RNA polymerase to start transcribing. Chromatin containing the oocyte-type genes on the other hand is not complexed with the transcription factors, is unable to recruit them and cannot be transcribed. This seems to be due to its association with histone H1, which packs up the nucleosome subunits of chromatin tightly, presumably rendering the DNA inaccessible to the transcription factors.

The pattern of chromatin packing is believed to be established in the very early embryo, as a result of the weaker affinity of the oocyte-type promoter for TFIIIA. When amounts of TFIIIA are in limited supply, as they are at this stage, there is strong competition for it and the somatic genes win. Oocyte-type genes that have failed to recruit TFIIIA are then kept repressed by being tightly packed up in nucleosomes.

Detailed proposals of this kind are at present only available for a few model systems such as the 5S genes. But in differentiated cells generally there are differences in the packing state of chromatin containing, on the one hand, genes that are typically transcribed in that cell and on the other, genes that are permanently switched off. 'Inactive' chromatin is much less susceptible to digestion with certain DNases than is transcribable ('activated') chromatin, suggesting that in activated chromatin the structure has opened out in some way, allowing the nucleases access to the DNA. Chromatin containing globin genes from chick red blood cells is much more sensitive to digestion than is, for example, the chromatin containing the gene for ovalbumin (an egg white protein), which these cells do not express.

A self-perpetuating state of chromatin packing established by conditions early in development and then transmitted from generation to generation of the cell's progeny, long after the original stimulus has disappeared, is an attractive model for explaining the phenomenon of cell memory. How the change from inactive to potentially active chromatin (and vice versa) is achieved may differ from case to case. Factors influencing the packing state of chromatin include chemical modification of DNA itself and the modification of histones as well as the binding of gene regulatory proteins. The overriding question, to which there are few answers as yet, is how a particular pattern of active and inactive chromatin can be sustained through repeated cycles of DNA replication and cell division. Chromatin is a complex association of DNA and many different types of proteins. We know how DNA copies itself, but we do not yet know how, or even if, newly replicated DNA strands reconstitute exact duplicates of previous associations with proteins such as the transcriptional complexes discussed above.

DNA methylation

Chemical modifications to DNA may be one such way of labelling active and inactive genes, through the phenomenon of **DNA methylation**. DNA from many species is methylated, chiefly at C nucleotides occurring in 5′ CpG3′ pairs. These occur relatively rarely overall, and an intriguing feature of vertebrate genomes is the clustering of CpGs into **CpG-rich islands** several hundred base pairs long and associated with certain types of gene. Most CpG in the genome is methylated – a methyl group is added to the cytosine residue by so-called maintenance methylases after the DNA has been synthesized. In marked contrast, however, the CpG islands are unmethylated, thus forming a distinct and recognizable feature in the genome.

They have been found so far associated chiefly with 'housekeeping' genes (e.g. for several metabolic enzymes, actin, tubulin, histocompatibility genes, and some ribosomal proteins) where they occur around the startpoint of transcription, but not with highly tissue-specific or stimulus-specific genes such as those for myoglobin (a muscle protein), nerve growth factor, the ß-globin family (but they do exist in the α-globin genes) and growth hormone. The biological significance of CpG islands is not yet clear. One proposal is that they identify genes that have to be constantly available for transcription as opposed to genes that are under much tighter control. They have escaped methylation possibly by having the appropriate transcriptional regulator proteins permanently bound at these sites. There is some evidence that methylated cytosines are associated with genes that are permanently or temporarily switched off.

How methylation can influence the transcribability of a gene is not yet known. Its attraction lies in the fact that the methylation state of DNA can be transferred to newly synthesized DNA on replication, thus providing a way in which a cell could retain and pass on a particular pattern of activated and inactive genes. In newly duplicated DNA, methylation occurs for preference at unmethylated Cs which form part of the sequence

. . . CmG . . . old strand
. . . G C . . . new strand

thus providing a way of passing on the pattern of methylation.

The idea that DNA methylation is involved in long-term gene control finds some support from the fact that chemical treatments that block methylation also seem to allow cells to switch genes on that normally would not be expressed, and from experimental evidence that genes whose CpG islands have been artificially methylated are no longer transcribed *in vitro*.

Methylation cannot be a universal answer to gene control, however, as insects manage very well without it – their DNA is not methylated. For vertebrates, and especially mammals, it may well be one way in which cells remember which genes are inactive.

FURTHER READING

See general reading list (Strickberger; Watson et al.; Russell; McLean et al.).

P.A. SHARP. Splicing of mRNA precursors, *Science*, 235 (1987) 766.

T. MANIATIS and R. REED, The role of small nuclear ribonucleoprotein particles in pre-mRNA splicing, *Nature*, **325** (1987) 673–8. A review of the role of small RNPs in spliceosomes.

D. SOLNICK, Trans-splicing of mRNA precursors, Cell, **42** (1985) 157–64.

N. MAIZELS and A. WEINER, In search of a template, *Nature*, **334** (1988) 469–470. A brief review of the newly-discovered phenomenon of RNA editing in trypanosomes. This amazing phenomenon, where nucleotides are added to mRNA in a consistent fashion *after* transcription, to produce a final mRNA from which a protein is synthesized, was reported too recently to be included in this book.

D.J. FINNEGAN, Transposable elements in eukaryotes, *International Review of Cytology*, **93** (1985) 281–326.

C.-S. LIU, D.A. GOLDTHWAIT and D. SAMOLS, Identification of Alu transposition in human lung carcinoma cells, *Cell*, **54** (1988) 153–9. The first direct demonstration of transposition in human cells.

D. BALTIMORE, Retroviruses and retrotransposons: the role of reverse transcription in shaping the eukaryotic genome, *Cell*, **40** (1985) 481.

B. FURIE and B.C. FURIE, The molecular basis of blood coagulation, *Cell*, **53** (1988) 505–18. The biological roles of the blood-clotting serine proteases.

S. BRENNER, The molecular evolution of genes and proteins: a tale of two serines, *Nature*, **334** (1988) 528–30. A further twist in the evolutionary history of the serine proteases.

N. MAEDA and O. SMITHIES, The evolution of multigene families: human haptoglobin genes, *Annual Review of Genetics*, **20** (1986) 81–108. A detailed example of the evolution of a gene cluster and the interchange of genetic information between its members.

L. HOOD, M. KRONENBERG and T. HUNKAPILLAR, T cell antigen receptors and the immunoglobulin supergene family, *Cell*, **40** (1985) 225–29.

K.R. HINDS and G.W. LITMAN, Major reorganization of immunoglobulin V H segmental elements during vertebrate evolution, *Nature*, **320** (1986) 546–9, and commentary by W. Gilbert (p. 485, same issue).

M.S. BROWN and J.L. GOLDSTEIN, Teaching old dogmas new tricks, *Nature*, **330** (1987) 113–14 and references therein. A commentary on some suprising new discoveries amongst the plasma lipoproteins.

M. PTASHNE, *A genetic switch: gene control and phage lambda*, Blackwell, Oxford, 1986. An account of the intricacies of prokaryotic gene regulation in one of the best studied examples.

T. MANIATIS, S. GOODBOURN and J.A. FISCHER, Regulation of inducible and tissue-specific gene expression, *Science*, **236** (1987) 1237.

D.D. BROWN, The role of stable complexes that repress and activate eukaryotic genes, *Cell*, 37 (1984) 359–65.

M. PTASHNE, Gene regulation by proteins acting nearby and at a distance, *Nature*, **322** (1986) 697–701.

N.J. SHORT, Flexible interpretation, *Nature*, **334** (1988) 192–3, and references therein. A commentary on some recent work on transcription factors.

R. CHIU et al., The c-fos protein interacts with c-jun/AP-1 to stimulate transcription of AP-1 responsive genes, *Cell*, **54** (1988) 541–52, and references therein. An entrance to the literature on the recent identification of the *jun* oncogene product as the transcription factor AP-1 and its role in the cell. See also references in Chapter 11 for the role of *fos*.

A. RAZIN and A.D. RIGGS, DNA methylation and gene function, *Science*, **210** (1980) 604–10.

CHAPTER 4

RECOMBINANT DNA TECHNOLOGY

Recombinant DNA technology embraces a variety of techniques for isolating, analysing and manipulating individual genes, which together have revolutionized the study of genetics during the past decade by providing at last a direct view of the gene.

The great advances in genetics in the 1950s and 1960s that followed the identification of DNA as the genetic material and the determination of its structure were all made in the absence of any way of directly confirming the central tenet of the new genetics. There was ample indirect evidence to prove that the informational content of DNA lies in the order of its nucleotide subunits, and that this is reflected in the amino acid sequence of the proteins it specifies. But, until the early 1970s, no way had been found of isolating and identifying a piece of DNA physically corresponding to a particular gene. The nature of the genetic code, the existence and role of messenger RNA, the mechanism of translation and the basic principles of gene regulation in bacteria were all established indirectly, without being able to determine the nucleotide sequence of the DNA involved.

The problem lies in the great size of DNA molecules, and the chemical similarity of one DNA molecule to another. The DNA molecules from the chromosomes of eukaryotes are millions of base pairs long and each contains thousands of different genes. Although each DNA is quite unique in its informational content, from a chemist's point of view one DNA molecule is much like another, a featureless string of only four chemically distinguishable subunits repeated at random many thousands of times. It presents no unique landmarks that can be used to pull out particular pieces of DNA. Another aspect of the problem is that the DNA representing a typical protein-coding gene is present in only a few copies per cell and represents an infinitesimally small fraction of a sample of DNA from plant or animal tissue, or even from a bacterial culture. Even if it had been possible to isolate a particular piece of DNA, to obtain any plant or animal gene in sufficient amounts for analysis seemed an impossibility.

Ways were needed to isolate identifiable DNA fragments of manageable length in sufficient quantity for chemical analysis. By the beginning of the 1970s the tools were at last to hand in the form of an exceptional group of DNA-cutting enzymes from bacteria – the restriction enzymes – which snip out particular pieces of DNA from the heterogeneous mixtures of DNA molecules in natural DNA samples. Restriction enzymes were also used to construct recombinant DNAs from these fragments, in which form they could be multiplied to give sufficient DNA for analysis.

These techniques have transformed basic research, as well as forming the basis for a multimillion pound biotechnology industry.

RECOMBINANT DNA

Artificial **recombinant DNA** is a DNA that has been constructed outside a living cell by joining two or more pieces of DNA from different sources to produce a novel combination of genes. It usually, but not always, involves a combination of DNA from widely different species, such as the splicing of human or animal genes into a bacterial or yeast DNA.

These techniques enable small fragments of DNA to be introduced into bacterial or yeast cells as part of a bacterial or yeast DNA, where it will then be replicated under the control of the host cell to provide a virtually unlimited source of cloned copies of that DNA (see below, DNA cloning). This is an essential step in isolating individual genes from the large genomes of plants and animals. With further manipulation, microorganisms can be persuaded to express the foreign genes they carry, to mass-produce medically and commercially useful proteins that are difficult or impossible to obtain by any other means (see below, The production of useful proteins). Pure cloned DNA can be introduced into plant and animal cells, including the fertilized eggs and early embryos of mammals, and, from these, **transgenic** organisms can be raised that carry potentially useful genes from a wide variety of species (see below, The genetic modification of animals and plants).

Behind these well-publicized feats of **genetic engineering** lies a wide range of techniques for isolating, identifying and altering individual pieces of DNA that have been developed during the past 15 years and now go under the collective title of recombinant DNA techniques or recombinant DNA technology (see following sections).

Cloned DNA can be altered to order *in vitro* by novel mutational techniques and then reintroduced into cells to be translated into new proteins with slightly altered properties. This is the basis of the emerging art of **protein engineering**, the second generation development of recombinant DNA technology. This technique is predicted eventually to produce novel enzymes and other proteins of industrial or medical importance to order, and to create 'hybrid' proteins with quite new properties – such as 'enzymatic' antibodies that combine in one molecule an enzyme activity and an antibody's exquisite specificity for a particular target. Mutation to order has also proved a powerful technique for studying the ways in which gene expression is regulated in the cells of multicellular organisms.

Recombinant DNA technology has made possible some completely new approaches to old problems – such as the mapping of gene locations on the chromosomes and the study and detection of inherited genetic disease. As well as the topics in this section, some uses of recombinant DNA techniques in cancer research, immunology and in developmental biology are also mentioned elsewhere. One of the most useful aspects of recombinant DNA to biologists in all fields is the ability it gives to produce large amounts of rare or hard-to-purify proteins by translation of cloned genes, and it is also now easier in most cases to determine the amino acid sequence of a protein by working backwards from the nucleotide sequence of the DNA that specifies it.

Microorganisms into which novel DNA has been introduced are often referred to as **recombinant microorganisms** and the proteins produced from such recombinant organisms as **recombinant insulin, recombinant interferon,** etc. to distinguish them from the same proteins isolated directly from animal tissue. Recombinant DNA is often abbreviated as rDNA. (As this is also a long-standing abbreviation for the DNA specifying ribosomal RNAs, the abbreviation rtDNA has also been suggested to avoid confusion.)

In a wider sense, the terms recombinant DNA and recombinant organism are now also applied to any DNA that has been altered by direct manipulation outside the cell, and to any organism into which that DNA is introduced, even if the same effects might have been achieved (given infinite patience and good luck) by the natural processes of mutation and genetic recombination.

This distinction is not simply of academic importance. In most countries, there are regulations governing recombinant DNA work over and above those that relate to the usual handling of microorganisms or release of 'natural' organisms to the environment for biological control purposes (see below, Regulation of recombinant DNA work). Whether an organism is classed as a 'recombinant' genetically engineered organism, or one bred by 'natural' means can make a great difference to both bureaucracy's and the public's perception of the comparative 'hazards' of such organisms even when the final result is biologically virtually identical.

ISOLATING AND IDENTIFYING GENES

Restriction enzymes

Restriction enzymes, more correctly called **restriction endonucleases**, are the molecular scalpels without which none of today's powerful techniques for analysing and manipulating DNA would be possible. They are a group of exceptional DNA-cutting enzymes from bacteria, which differ from other nucleases in only cutting a DNA chain at a precise point within short, precisely defined nucleotide sequences (**restriction sites**) (Table 4.1).

Table 4.1 Examples of restriction enzyme sites

Enzyme	Site
HpaI *(Haemophilus parainfluenzae)*	5′ G T T ↓ A A C C A A ↑ T T G
EcoRI *(Escherichia coli)*	G ↓ A A T T C C T T A A ↑ G
HindIII *(Haemophilus influenzae)*	A ↓ A G C T T T T C G A ↑ A
BamI *(Bacillus amyloliquefaciens)*	G ↓ G A T C C C C T A G ↑ G

Over a hundred different restriction enzymes have been isolated and each cuts at a different sequence. Restriction enzymes are named after the bacterium they come from. In their native bacteria restriction enzymes degrade incoming "foreign" DNA such as that of bacteriophages. Their own DNA is protected against attack by chemical modification, usually by methylation of A or C residues within the restriction sites. The sites at which restriction enzymes cut have one feature in common. They possess a particular type of symmetry. If the sequence is cut at the centre of symmetry (dotted line in the example of HpaI above, and one half (hypothetically) rotated so that the two DNA strands change places, the sequence on each strand forms a perfect *palindrome* that is, it reads the same from left to right.

DNA molecules are extremely long and chemically very featureless, being made up of only four different subunits repeated many thousands of times in a chemically random sequence. This is a formidable obstacle to isolating a particular fragment of DNA or determining its sequence. Any attempt to cut a DNA preparation from natural sources into more manageable pieces by most chemical and enzymatic treatments leads to a mixture of anonymous random fragments, which it is impossible to analyse further. The restriction sites that occur by chance in all DNAs provide natural chemical guideposts along the DNA.

Because most restriction sites are only a few base pairs long they will occur in any DNA of sufficient length, whatever its origin. In the DNA of the small animal virus SV40, for example, the restriction site for the endonuclease EcoRI (a restriction enzyme from *Escherichia coli*) occurs once, that for the enzyme HpaI (from *Haemophilus parainfluenzae*) four times and that for the enzyme HindIII (from *Haemophilus influenzae*) eleven times. A particular restriction enzyme will therefore always cut a given DNA at the same point (or points), generating a set of specific DNA fragments (**restriction fragments**) that can be sized and separated by **gel electrophoresis** (Fig. 4.1).

Electrophoresis is a general technique for separating DNA fragments or proteins on the basis of their movement through a sheet of polyacrylamide or agarose gel in an electric field. The distance they move in a

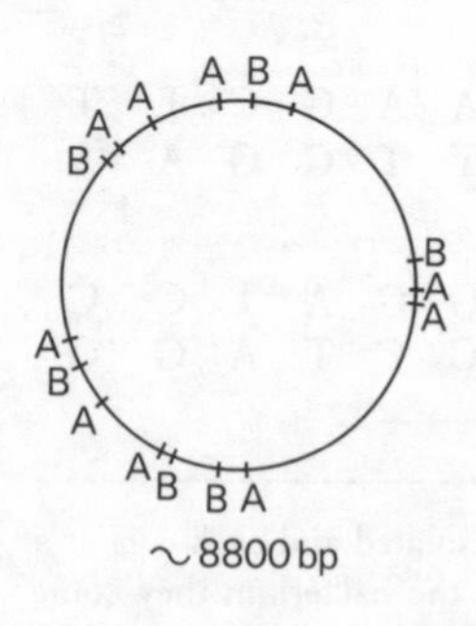

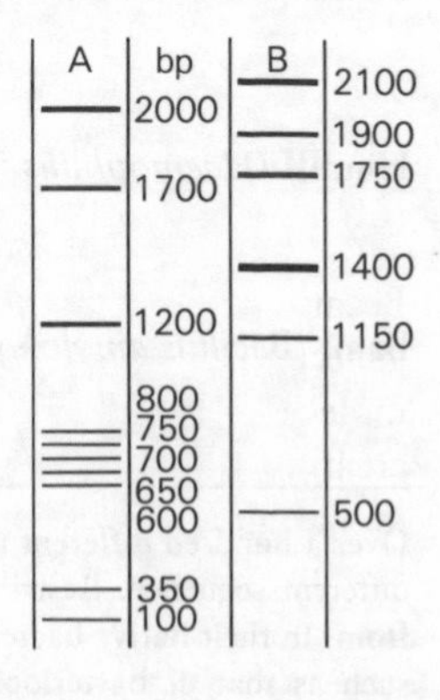

Fig. 4.1 Arrangement of restriction sites on a hypothetical circular DNA and the pattern of restriction fragments obtained on gel electrophoresis after complete digestion with restriction enzyme A or B. To show up the DNA bands the gel is stained with the dye ethidium bromide which fluoresces orange on binding to DNA. The lengths of the fragments in base pairs (bp) are determined from standards of known size.

given time depends on their size and electric charge. Smaller DNA fragments move quicker than larger ones and so a mixture of fragments of different lengths separates out into discrete bands in the gel, each band consisting of fragments of the same length. Each band therefore consists of a purified source of identical DNA fragments. These can be accurately sized by comparison with the movement of fragments of known length (standards), and on suitable gels fragments differing in length by only one nucleotide can be separated (see below, DNA sequencing).

Conventional gel electrophoresis is used to separate fragments of DNA tens to thousands of base pairs long. New developments (pulsed-field gel electrophoresis) now also allow much larger DNAs of chromosome length – millions of base pairs long – to be separated.

Restriction fragments from any source can be inserted into a suitable carrier or **vector** DNA to produce an artificial **recombinant DNA** that is multiplied in rapidly dividing bacterial cells to produce a virtually unlimited supply of a given piece of DNA.

Some restriction enzymes simply cut the two strands of the DNA straight across at the centre of symmetry (see Table 4.1, cleavage sites are arrowed). Others, however, make a staggered cut in the DNA, cutting each strand symmetrically with respect to the other and generating in the process two DNA fragments which have complementary single-stranded ends. BamHI, whose restriction site is illustrated here, is an enzyme of this type, cutting

    ↓
5′ . . . G G A T C C . . . 3′

3′ . . . C C T A G G . . . 5′
                          ↑

Because the nucleotide sequence on either strand is symmetrical this generates two pieces of DNA with 'sticky ends'. If two *different* DNA molecules are cut in this way, the pieces will rejoin to each other, creating a 'hybrid' or 'chimaeric' DNA. The source of the DNA is immaterial, DNA from species as far apart as bacteria and humans can be joined up in this way. This feature of restriction enzyme cutting was used to generate the first recombinant DNAs.

DNA cloning

DNA cloning is a means of isolating individual DNA fragments from a mixture, and multiplying each fragment to produce enough material for analysis and further use.

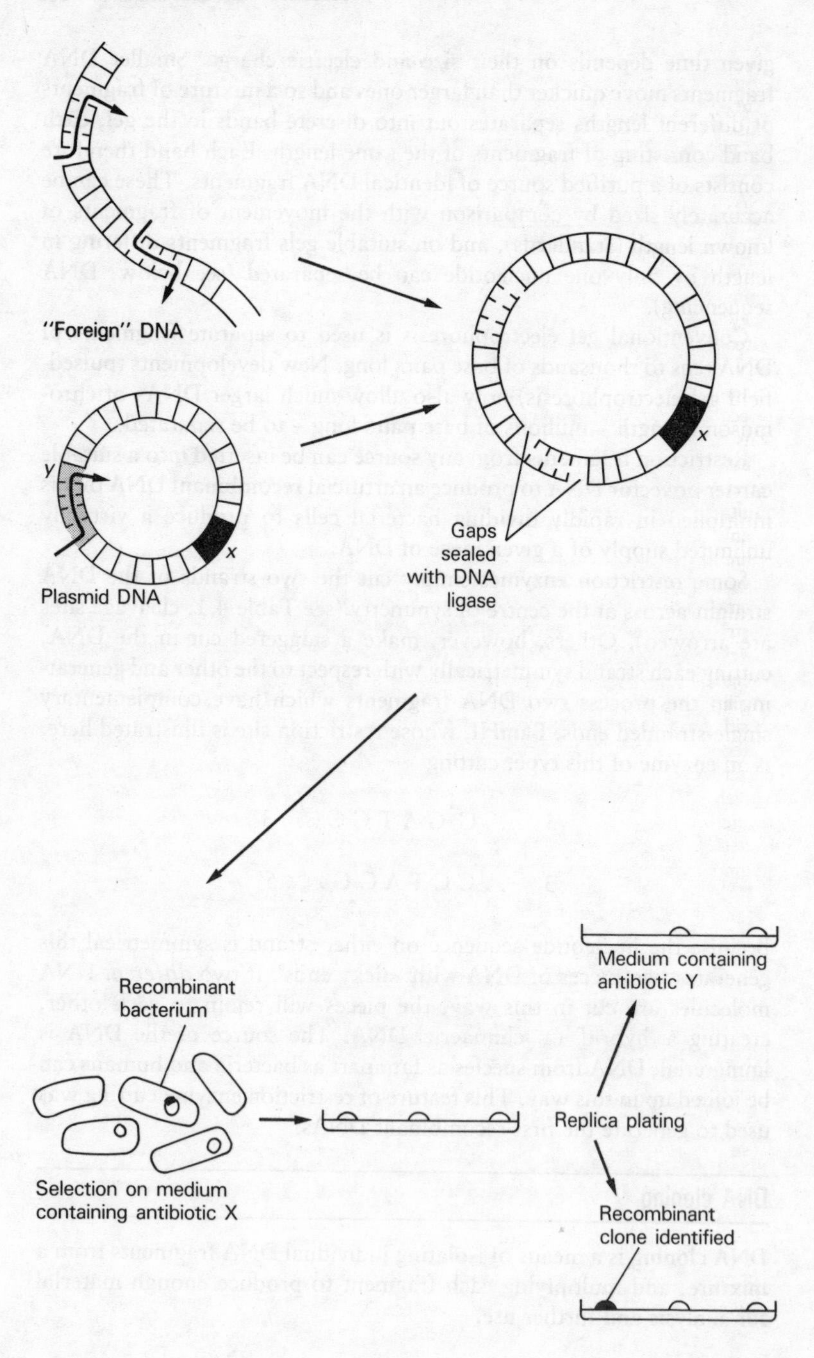
"Foreign" DNA
y
x
Plasmid DNA
x
Gaps
sealed
with DNA
ligase
Recombinant
bacterium
Selection on medium
containing antibiotic X
Medium containing
antibiotic Y
Replica plating
Recombinant
clone identified

Fig. 4.2 Simplified DNA cloning strategy. This diagram illustrates cloning with a plasmid vector. The vector is cut with a restriction enzyme especially chosen so that it cuts the DNA circle only once, producing either two complementary 'sticky ends' (illustrated) or a 'blunt end' to which any blunt-ended DNA fragment can be spliced. The DNA to be cloned is treated in the same way, either generating fragments with the same 'sticky ends' that will pair with those on the plasmid to produce a hybrid DNA, or a straight end that can be joined up by the enzyme DNA ligase, which will join any two pieces of DNA that have been cut straight across. This technique is useful when two pieces of DNA must be joined precisely at a predetermined place. The two DNAs are also sometimes joined by 'linkers', short synthetic oligonucleotide chains that are added to both opened plasmid and DNA fragments to form artificial sticky ends. They are usually constructed so that they form restriction sites when paired up, allowing easy retrieval of the cloned DNA. Gaps in the DNA backbone are joined up by DNA ligase to produce a complete recombinant DNA molecule. Bacterial cells are treated with this mixture which will contain recombinant DNA, unrecombined plasmids and unrecombined restriction fragments. In suitable conditions bacterial cells take up added DNA (see text). The DNA concentration is adjusted so that cells are unlikely to take up more than one DNA molecule. The bacteria are then cultured on solid media at a low enough concentration so that each resulting bacterial colony represents the multiplication of a single bacterium. All colonies containing plasmid DNA are identified by selecting for marker X, which is commonly an antibiotic resistance gene. Only cells containing plasmids will therefore grow in the presence of the antibiotic. The tiny subset of colonies that contain recombinant plasmids can then be identified by a variety of techniques. In the case illustrated the vector also carries resistance to a different antibiotic – Y. Gene *Y*, however contains the restriction site at which the vector is cut. So insertion of a foreign DNA at this restriction site inactivates gene *Y* and bacteria containing recombinant plasmids can be distinguished from those carrying intact plasmids by their sensitivity to antibiotic Y.

Recombinant colonies are then grown on and stored for further use. Recombinant bacteria containing particular genes can be identified by hybridization with radioactively labelled complementary nucleic acid probes (see below, Identification of restriction fragments). To recover cloned DNA, vector DNA extracted from the bacterial culture is treated with a suitable restriction enzyme to release the required fragment which can then be separated from the remainder of the DNA.

A simple DNA cloning strategy is illustrated in Fig. 4.2. DNA fragments are enzymatically spliced *in vitro* into small carrier DNA molecules to produce a recombinant DNA. The carrier DNAs or **cloning vectors** are derived from small self-replicating DNA molecules such as the plasmids and phages that naturally infest bacterial cells (see below, Plasmid and phage vectors and their bacterial hosts). Once inserted into such a carrier, the 'foreign' DNA fragment can be introduced into a suitable host cell, where it will be maintained and replicated along with the carrier DNA. For routine DNA cloning to initially isolate and amplify a gene, specially bred strains of *Escherichia coli* are used as hosts.

Transformation

Recombinant constructs containing a foreign gene are introduced into bacterial cells by **transformation**. Bacterial cells naturally take up DNA added to their culture medium and, if the DNA is compatible, they will express the genes it contains, acquiring new properties. This is the phenomenon of transformation noticed first in the 1920s and which eventually led to the identification of the 'transforming principle' as DNA. (Bacterial transformation should not be confused with malignant or neoplastic transformation, which is the change seen in cultured mammalian cells when treated with carcinogens, tumour viruses or oncogenes – see Chapter 11.) On its own, however, a fragment of mammalian or non-bacterial DNA will not replicate or be maintained in the bacterial cell as it does not contain replication and other regulatory signals recognizable by the bacterium, and is soon broken down by bacterial nucleases. Only when incorporated into a bacterial carrier containing the appropriate signals can the DNA be maintained and replicated.

The mechanism of transformation is still something of a mystery as a large molecule such as DNA should not theoretically be able to penetrate the cell membrane. Those bacteria that have taken up the added DNA are selected by the new properties they express; in recombinant DNA work an antibiotic resistance marker gene is commonly incorporated into the vector, so that only those bacteria that have taken up the recombinant plasmid can grow on medium containing the appropriate antibiotic.

DNA libraries

DNA libraries are collections of cloned DNAs. There are two types of library – **genomic libraries** representing the entire genetic complement (genome) of individual animals, plants, viruses or bacteria, and **cDNA libraries** representing the complete mRNA complement of a single type of cell (see below). Genomic libraries are usually constructed by randomly shearing chromosomal DNA into suitably sized fragments (of around 20 000–40 000 base pairs) either by restriction enzymes or mechanical means, and then cloning the individual fragments into a suitable cloning vector, usually a specially modified phage lambda as the DNA is more easily stored in this form. The random cloning of an entire genome is often called **shotgunning**. To ensure that the whole genome has been incorporated into recombinant DNAs, a typical library of a human genome (3.5×10^9 base pairs) will contain between a quarter and half a million overlapping separate clones.

Once a DNA library has been established, clones containing a gene of interest can be identified by screening the collection by hybridization with specific nucleic acid probes (see below, Identification). Modern equipment now makes it possible to screen large numbers of clones relatively rapidly, an important practical consideration as hundreds of thousands of clones may have to be screened to pick out ones containing a single-copy gene from a human genomic library.

The overlap between individual clones is crucial to techniques such as **chromosome walking** which allows one to proceed from a known site on the chromosome to an unidentified gene nearby for which no specific probe is yet available. This is the situation for many genes implicated in inherited genetic defects whose gene product and function are totally unknown, but whose approximate position on the genetic map has been established by classical genetic methods (see below, Gene mapping).

cDNA and cDNA Libraries

For many genetic engineering applications DNA isolated directly from the cell (genomic DNA) is not suitable. This applies in particular to cases where bacteria are to be modified to produce eukaryotic gene products. Many eukaryotic genes contain introns, and although bacteria can produce a primary RNA transcript from such genes they do not possess the machinery for splicing out the introns to produce a functional protein-coding mRNA. If the products of these genes are to be synthesized by bacteria, a continuous ready spliced coding sequence is needed as starting material. This is provided by a complementary DNA (**cDNA**) copy of the appropriate messenger RNA made using the enzyme **reverse transcriptase**. As its name suggests this enzyme catalyses a process like transcription in reverse, with RNA providing the template for the synthesis of a complementary DNA strand. This single-stranded cDNA can then be converted to double-stranded DNA by DNA replication *in vitro*.

Double-stranded cDNA can be manipulated like any 'natural' DNA. It can be cloned, mapped with restriction enzymes and its DNA sequence determined in the routine way – procedures that cannot be carried out on RNA. In this way, the detailed organization and sequence of messenger RNAs can be deduced. It can also be used as a probe for identifying corresponding RNA or genomic DNA sequences or as the starting material for a tailor-made 'gene' for construction of genetically engineered bacteria or other cells.

The total mRNA complement of a particular cell type can be captured and preserved by using it as a template to synthesize cDNAs, which are then cloned. The collection of clones is known as a **cDNA library**. The mRNA of a cell represents those protein-coding genes that are actively being expressed. A cDNA library is therefore a convenient way of isolating genes that are characteristically expressed only in a particular cell type. cDNA libraries also only capture protein-coding genes, which is a considerable advantage over genomic DNA libraries containing many clones of repetitive non-coding DNA. They do not, however, capture intron sequences or the non-transcribed gene regulatory sequences such as promoters and enhancers, which then have to be fished for in genomic libraries. cDNA libraries are now used widely as the source of protein-coding DNA, which is sequenced to provide a short-cut to the amino acid sequences of hard-to-purify proteins.

Synthetic DNA

Recombinant DNA work often makes use of **synthetic DNA**, short polynucleotide chains built up chemically nucleotide by nucleotide in any sequence required and then converted to a double-stranded DNA by enymatic replication *in vitro*. The chemical procedure has now been automated for DNA chains up to 100 nucleotides or so. Synthetic DNA is sometimes used in place of naturally derived DNA to direct the synthesis of short peptides in genetically-engineered bacteria. A synthetic gene for the small peptide hormone somatostatin was used in the first genetically engineered *E. coli* to produce a mammalian protein. Short synthetic oligonucleotides are also used to introduce restriction sites or mutations at predetermined positions in a DNA.

Identification of restriction fragments

Whether isolated as bands on a gel or as cloned recombinant DNAs, restriction fragments are still anonymous pieces of DNA. The most common way of identifying them is to screen the gel or the collection of clones with specific nucleic acid **probes** that recognize and pick out only a particular DNA sequence. Probes are short lengths of DNA or RNA of known sequence or origin, labelled with a radioactive or chemical tag for easy detection of the DNAs to which they attach. The probes recognize their target DNAs by the exquisitely specific phenomenon of **nucleic acid hybridization**: this is where a single strand of DNA or RNA pairs up with an exactly complementary strand to form a correctly base-paired double-stranded molecule.

The two strands of the DNA double helix are held together by relatively weak hydrogen bonds between the bases of opposite strands. These weak bonds allow the strands to separate easily so that DNA can replicate and be transcribed. In the cell, strand separation relies on enzymes and other special proteins, but if isolated DNA is gently heated or subjected to certain chemical treatments this is sufficient to disrupt the hydrogen bonding and the two strands fall apart. Single-stranded DNA will reform a double helix only by base-pairing with any DNA or RNA of exactly complementary sequence.

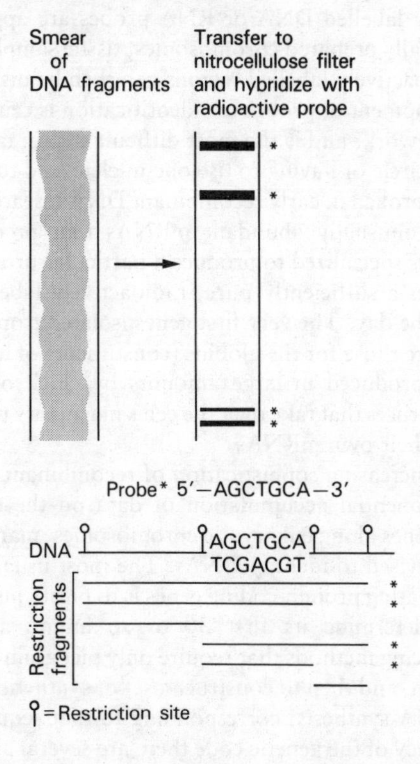

Fig. 4.3 Identification of restriction fragments from a digest of complete genomic DNA. DNA is extracted from cells and treated with an appropriate restriction enzyme to generate a continuum of random fragments which form a continuous smear on gel electrophoresis. The DNA from the gel is imprinted onto a sheet of nitrocellulose by a 'blotting' technique (known as Southern blotting after its inventor). Once transferred to nitrocellulose the DNA can be selectively hybridized with a radioactive nucleic acid probe (RNA or cDNA) corresponding to the required gene. This reveals the position of any fragments containing part or all of the gene (lower panel), and which can then be cut out of a duplicate gel for cloning and further analysis.

Nucleic acid hybridization is central to recombinant DNA work by providing the means of recognizing and isolating a given piece of DNA from a mixture of fragments. A routine method of identifying restriction fragments is illustrated in Fig. 4.3. DNA hybridization is so precise that clones containing a required gene can be picked out from 'libraries' of hundreds of thousands of DNA clones.

In situ hybridization techniques are used to identify the position of a particular DNA sequence on a chromosome, or the spatial and temporal pattern of specific mRNA synthesis in tissues (see Chapter 10). Radioactively labelled DNA or RNA probes are applied to preparations of specially prepared chromosomes, tissue samples, embryos, etc. and the radioactively labelled regions can then be distinguished.

The development of probes for identification is central to all recombinant DNA work, and is the most difficult step in most applications. The vicious circle of having to use one nucleic acid to identify another was initially broken in early recombinant DNA research by using as the first probes unusually abundant mRNAs that predominate in, for example, cells specialized to produce a particular protein. These could be isolated in a sufficiently pure, radioactively labelled form by the methods of the day. The very first genes isolated from eukaryotic cells were therefore those for the globins (constituents of haemoglobin) and ovalbumin (produced in large amounts by chick oviduct cells) and genes from viruses that take over the cell's machinery to synthesize large amounts of their own mRNAs.

With the increasing sophistication of recombinant DNA techniques and the exponential accumulation of data on the organization and location of genes along eukaryotic chromosomes, many technical tricks have been devised to identify DNAs. The most usual procedure nowadays for isolating protein-coding genes is to purify just sufficient of the protein to determine its first 30 or so amino acids by modern microsequencing methods that require only picogram (10^{-12} g) quantities of protein, and then to construct a set of synthetic DNA probes (by chemical DNA synthesis) corresponding to that sequence. Because of the redundancy of the genetic code there are several alternative nucleotide sequences that could correspond to the amino acid sequence. One of these will correspond to the genomic DNA sequence and will pick out the relevant clone from a library of cloned DNAs. Even a partial sequence is usually selective enough to pull out only the required clone. For very large genes, where the coding sequence may be spread across several clones, the overlap between clones can be used to pull out the next one by using each clone as a probe for the next.

When a gene of interest is isolated and cloned from one species it can then often be identified in other species using the cloned DNA as the probe. Even though the DNA sequences of corresponding genes from even quite closely related species are unlikely to be identical, the conditions in which the probe is hybridized can be 'relaxed' so that it picks out closely related but not identical sequences. Relaxed hybridization is similarly used to screen a library for related but not identical sequences within the same genome, often identifying hitherto undiscovered members of a gene family (see Chapter 3).

Restriction analysis

Restriction sites provide the landmarks for finding one's way around a DNA molecule. They allow a newly isolated DNA to be characterized and located in its wider DNA context, and they are also invaluable for distinguishing one DNA from another.

The first step in characterizing an isolated DNA fragment is to construct a physical map (a **restriction map**) of the different restriction sites along its length (Fig. 4.4).

A restriction map is pieced together by digesting portions of the DNA sample with different restriction enzymes. Each digestion produces a typical set of restriction fragments of different lengths. Those covering the area of interest are identified by an appropriate nucleic acid probe (see above). Fragments obtained by digestion with one enzyme are then digested with one or more of the other restriction enzymes (and vice versa) to obtain the location of different restriction sites relative to each other.

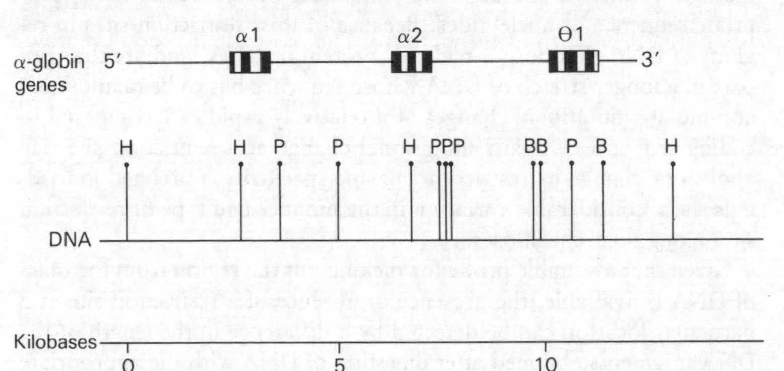

Fig. 4.4 Restriction map of the α-globin region from orang-utan. H, HindIII; P, PvuII; B, BamHI. (After J. Marks, J-P. Shaw and C-K. J. Chen, *Nature*, 321 (1986) 785–788.)

Restriction maps themselves tell us nothing about the biological function of a stretch of DNA. They simply provide a map of the exact location of restriction sites relative to each other and the distance in base pairs between each one. They are, for example, an essential preliminary to DNA sequencing. Once a restriction map has been established, conveniently sized overlapping subfragments can be isolated for sequencing, allowing a DNA sequence of quite extensive regions to be built up.

Restriction enzymes always cut at specific sites in DNA, producing a characteristic and invariant pattern of DNA fragments even from a sample of uncloned DNA. Two DNAs differing in, say, the position of one of these sites can therefore readily be distinguished by the different patterns of fragments they give when treated with the same enzyme. This provides the basis for very powerful techniques that have been developed to distinguish two DNAs from each other without the need for cloning individual fragments or for detailed sequencing. This type of restriction analysis is now widely used, for example, in the clinical diagnosis of genetic disease, as it allows the abnormal form of a gene to be detected directly from a sample of the patient's DNA (see below, Genetic disease). Some other applications of restriction analysis are given below.

Restriction Fragment Length Polymorphisms (RFLPs)

In eukaryotic DNA the sites at which restriction enzymes cut have no biological function and are simply a chance occurrence of the appropriate sequence of nucleotides. Because of this, restriction sites in regions of DNA that do not code for protein or RNA, and are thus not part of a longer stretch of DNA whose sequence has to be maintained, accumulate mutational changes at a relatively rapid rate compared to coding sequences. As any mutational change at a restriction site will abolish or change its restriction enzyme specificity, unrelated individuals show considerable variation in the number and type of restriction sites along their chromosomes.

Given that a suitable probe for picking out the region from the mass of DNA is available, the presence or absence of a restriction site at a particular location can be detected by a difference in the length of the DNA fragments obtained after digestion of DNA with the appropriate enzyme. The variability in restriction sites between individuals has therefore come to be known by the unwieldy name of **restriction fragment length polymorphism** or **RFLP** for short. It is often also called **restriction site polymorphism**.

RFLPs have many practical uses. The presence or absence of a restriction site at a particular chromosomal location is inherited according to mendelian rules and provides a marker that can be used in gene mapping. RFLPs are often used to distinguish between different alleles (gene variants) carried on homologous chromosomes in a diploid cell. If, for example, one of these chromosomes carries a defective gene in close association with a restriction site (as determined by their linked inheritance in a family) and the other (normal) chromosome lacks the restriction site, the presence of the restriction site can often be used as a preliminary diagnostic test for the presence of the defective gene in a foetus at risk.

DNA fingerprinting

The general technique of restriction analysis has been exploited to develop a method of **DNA fingerprinting** by which any individual can be unambiguously identified from his or her DNA. It can be used in forensic work to identify a suspect from bloodstains or semen, and can also be used to establish family relationships – such as disputed parentage – with a certainty not possible with conventional blood and tissue-typing tests. It has already been used in the UK to convict a rapist and murderer and in successful appeals against immigration decisions.

The first DNA fingerprinting technique to be developed is the one now commercially available in the UK. It relies on identifying a particular class of DNA sequences present in all individuals but which are particularly variable in number and size from one person to another. These are the **minisatellites** – short sequences (typically around a dozen base pairs) each serially repeated in near identical copies many thousands of times in the human genome. The repeats are arranged in blocks distributed throughout the chromosomes, each block containing a highly variable number of repeats. The total number of blocks and their sizes are virtually unique to each individual and constitute their DNA fingerprint.

To reveal an individual's DNA fingerprint, DNA isolated from as little as a single drop of blood is digested with a restriction enzyme chosen so that it cuts fairly frequently in human DNA except within the minisatellite sequence. The resulting DNA fragments are separated and sized by gel electrophoresis and screened for the minisatellite sequences with a complementary nucleic acid probe.

The probe reveals a different pattern of DNA bands for each person. Using a probe for only one type of minisatellite there is a small possibility that two individuals will have exactly the same pattern. (Identical twins will, of course, have identical patterns.) If two or more different probes (detecting different families of minisatellite sequences) are used, this possibility becomes vanishingly small. The technique is an important advance as it provides an unambiguous *positive* identification, rather than the probabilities that the other tests offer. Blood and tissue typing can definitely eliminate a suspect, but can never provide 100 per cent positive identification.

To ascertain family relationships, the DNA fingerprint of the child is compared with those of its putative parents, or with other family members from whom the parental fingerprint can be deduced. Minisatellite blocks are inherited in the same way as ordinary genes. This means that a child's total minisatellite DNA will contain some blocks from its mother and some from its father, but no others. Brothers and sisters will have some blocks in common, and all the blocks found in children of the same parents will also be found in the parents' DNA collectively.

Although first developed for use in humans, DNA fingerpinting has also been taken up by zoologists as a means of determining actual as opposed to apparent paternity in wild populations. Extensive illegitimacy, if found, could have important implications for sociobiological theory and for population and ecological genetics.

Mitochondrial 'Eve'

Restriction analysis has also been used to trace the ancestry of present-day human populations. The tiny circular DNA molecules contained inside the mitochondria of human cells provide a novel means of tracing the descent and relationships of human populations. Unlike the chromosomes of the cell nucleus, which are inherited from both parents and are also the product of extensive reshuffling by recombination in preceding generations, mitochondrial DNA is inherited only from the mother, through the mitochondria present in the cytoplasm of the egg, which replicate themselves and their DNA as the cells of the developing embryo grow and divide. (The sperm contributes virtually no cytoplasm to the fertilized egg.)

Mitochondrial DNA cannot undergo genetic recombination and is therefore inherited unchanged through the female line for many generations. Throughout time, however, mutations inevitably accumulate so that the mitochondrial DNAs of unrelated individuals now differ in nucleotide sequence (particularly in non-coding regions) and these differences can be detected by restriction analysis.

Comparisons of restriction patterns of mitochondrial DNAs from large numbers of people from different ethnic groups is therefore yet another way of gaining information on human ancestry, origins and dispersal. Interpretation of the results follows the general principle that

DNAs with very similar restriction patterns (i.e. more similar in sequence) have diverged more recently than those which are less similar, and that population groups with the greatest amount of variation within them are probably of more ancient origin (they have been around longer to accumulate more mutations).

A pioneering study of human mitochondrial DNA carried out in recent years focused on people representing five broad geographic regions: sub-Saharan Africans (represented in the study chiefly by black Americans), Asians (China and the Far East), Caucasians (Europe, North Africa and the Middle East), aboriginal Australians and aboriginal New Guineans. The first results provide another strand of evidence to add to palaeontological evidence for an exclusively African origin for present-day humans. When the mitochondrial DNAs are grouped by degree of similarity they form two main lines of descent, both stemming from Africa, one containing exclusively Africans and the other some Africans and all other groups. More speculatively, the researchers reconstructed a hypothetical 'ancestral' human mitochondrial DNA, and suggested the existence of a mitochondrial 'Eve', who, they calculate, could have lived in Africa some 200 000 years ago and from whom we are all descended.

DNA sequences

The final step in analysing any DNA is to determine the linear order of its component nucleotides – variously termed the **DNA sequence, base sequence** or **nucleotide sequence.**

The DNA sequence of a gene provides a wealth of information that can be used to probe deeper into the function of its product and the way in which gene expression is regulated. Even for genes whose protein product has not yet been identified and whose function is unknown, a comparison of the DNA sequence with those of known genes and proteins can sometimes provide valuable clues to its identity and the function of its product.

In every branch of biology, cloned and sequenced DNAs are being used as a short-cut to the amino acid sequences of proteins and to produce large amounts of a protein in order to study its biological function more easily. Structural motifs within a sequence can be identified representing binding sites within proteins for regulatory molecules such as cyclic AMP, sites for phosphorylation by protein kinases, sites at which carbohydrate side groups may be added, etc. DNA sequences of the genes for difficult-to-isolate membrane receptor proteins have provided much of our present understanding of their detailed structure and disposition in the membrane (see e.g. Chapter 6: Ion channels). A sequence is also essential before any sophisticated chemical reconstruction can be carried out to produce a DNA tailored for a particular purpose.

Worldwide, thousands of bases of DNA sequence data are being accumulated daily, to add to the millions of base sequence already banked on computer in several centres. Organized data banks, which can be accessed by researchers, have held amino acid sequences and protein structures for many years, and are proving even more necessary for DNA.

Two sequencing methods for DNA are used, developed independently in the mid-1970s by Fred Sanger and his colleagues at Cambridge (the chain termination method) and by Allan Maxam and Walter Gilbert at Harvard (the chemical cleavage method). They both depend on generating from the DNA to be sequenced, a nested set of DNA fragments each terminating before a known nucleotide and which form a series differing by a single nucleotide in length. These can then be separated according to size by gel electrophoresis, and the sequence read off the gel (Fig. 4.5). Fully automated DNA sequencing is now being developed.

The tiny single-stranded bacteriophage ΦX174 made scientific history when its 5375 bases were the first complete genome to be sequenced at Cambridge in 1977. It has been followed by the DNA sequences of many other phages, viruses (e.g. SV40, adenovirus and hepatitis B) and the complete genomes of yeast and mammalian mitochondria (around 85 000 and 16 500 base pairs respectively) and plant chloroplasts (12 000 base pairs from a liverwort and 156 000 base pairs from tobacco).

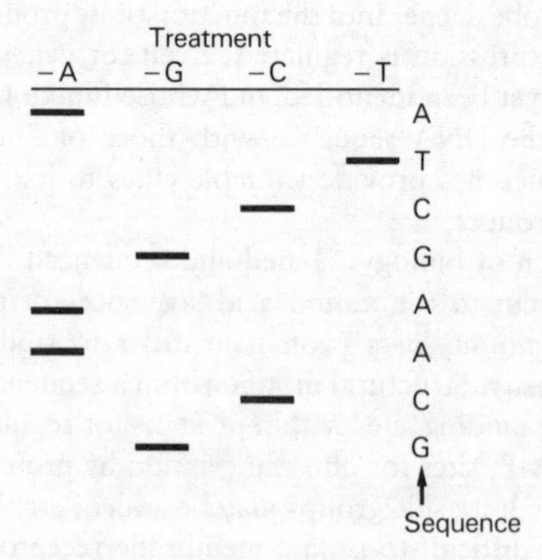

Fig. 4.5 Schematic diagram of pattern of fragments on a sequencing gel.

RNA genomes can also be sequenced after preliminary transcription into DNA by the enzyme reverse transcriptase. In this way, RNA tumour viruses have been sequenced as well as other important RNA viruses such as polio, influenza and the AIDS virus – HIV.

DNA sequences have provided many surprises. Notable are the overlapping genes (see Chapter 3) of ΦX174 and departures from the 'universal' genetic code in mitochondria and protozoa (see Chapter 2: The genetic code).

THE NUTS AND BOLTS OF GENETIC ENGINEERING

Plasmids and phages

The carrier DNAs into which 'foreign' DNAs are spliced to make a recombinant DNA for introducing into a new host cell are often known as **vectors**. The first vectors used were small self-replicating circles of plasmid DNA from the bacterium *Escherichia coli*. Plasmids also occur in many other bacterial species. Naturally occurring plasmids typically carry a handful of genes that, for example, specify resistance to antibiotics, the production of proteins (colicins) lethal to other bacterial species, or useful adaptive enzymes.

Plasmids are typically carried in one or more copies per bacterial cell. Plasmid DNA is easily isolated and separated from the bacterial chromosomal DNA by virtue of its small size. A piece of DNA (up to a certain size limit) spliced into plasmid DNA *in vitro* and reintroduced into a bacterial cell is then multiplied along with the plasmid DNA as it replicates. Some plasmids reach very large numbers of copies per cell, which together with the rapid division of bacterial cells – once every 30 minutes or so – soon produces a large number of copies of the required DNA. The plasmid DNA is then reisolated from the bacterial culture and the 'foreign' DNA recovered by various means. Bacterial viruses – in particular phage lambda, a virus that infects *Escherichia coli* – have also been developed as vectors. Each recombinant phage infecting a single bacterial cell produces many thousands of progeny, each bearing the 'foreign' DNA.

The only parts of a plasmid or phage that are required in its role as a **cloning vector** are its **origin of replication**, a specific signal encoded in the DNA that is recognized by the host cell's replication machinery, and for plasmid-derived vectors one or two 'marker' genes (these are usually antibiotic resistance genes) by which cells that contain the recombinant DNA can be easily selected from those that do not. Parts of plasmids and phages that allow them to spread from one bacterium to another are removed in the interests of safety – for example, to prevent any escaped plasmid from infecting the normal bacterial inhabitants of the human body. From the original plasmid and bacteriophage vectors large numbers of purpose-built carrier DNAs have been derived for a wide variety of applications (see below).

Vectors derived from plant bacterial plasmids (the Ti and Ri plasmids of *Agrobacterium* species) have extended recombinant DNA techniques to plants (see below, and Genetic modification of animals and plants), to the benefit of plant molecular biology which is now rapidly catching up with its more advanced animal and bacterial counterparts.

Bacterial hosts

The most commonly used bacterial host for gene transfer, both for commercial application and for research, is *Escherichia coli*. It is a natural inhabitant of the human gut and has been for years the workhorse of molecular genetics. The strains used in recombinant DNA work are modified so that they cannot survive outside the laboratory. Other bacteria have also been developed as recombinant DNA hosts but not to the same extent. Species of *Bacillus* are one potential target as in their unengineered form they are already widely used in the large-scale production of enzymes and fine chemicals (such as the proteolytic enzymes in biological washing powders). Genes can also now be introduced into other bacteria of agricultural interest such as the nitrogen-fixing rhizobia and species of soil *Pseudomonas*, and into *Agrobacterium* for subsequent transfer to plant cells (see below). Antibiotic-producing streptomycetes, which despite their superficial resemblance to fungi are in fact prokaryotes, are another potential target for genetic engineering.

Expression Vectors

No bacterial cell will however express an 'unprocessed' plant or animal gene as it does not recognize many of the DNA signals by which eukaryotic gene expression is activated. It is now usual to hitch a cDNA copy of the coresponding mRNA, trimmed to contain only the protein-coding sequence, to an efficient host cell promoter. Vectors that have been constructed to contain suitable promoters and other regulatory sequences in the correct relationship to inserted 'foreign' DNA are generally known as **expression vectors**. Expression vectors have also been developed for use with yeast and mammalian host cells.

Other vector/host combinations

Genetically engineered bacteria cannot of course be used to study the control of eukaryotic genes and neither are they ideal for producing some mammalian proteins (see below). The unicellular eukaryotic microorganism *Saccharomyces cerevisiae* (baker's yeast) is now increasingly used in studies of eukaryotic gene expression, and commercially to produce recombinant proteins. Vectors for use in yeast are derived from yeast plasmids. **Shuttle vectors** are vectors specially constructed from bacterial and yeast vector components, that enable the genes they contain to be initially cloned in *E. coli* and then directly transferred to yeast cells for expression studies.

Long DNA fragments of 500 000 base pairs or more, much larger than those that can be accepted by plasmid or phage vectors, can also now be cloned in yeast cells by incorporation into an 'artificial chromosome' (YAC or **yeast artificial chromosome**) containing a yeast replication origin and yeast telomeres (DNA that seals the ends of the chromosomes and which is needed for successful replication).

Vectors derived from animal viruses, chiefly the DNA virus **SV40** (simian virus 40), **retroviruses** (see below) and, for some vaccine applications, **vaccinia virus**, provide replication origins and regulatory signals that allow cloning and expression in animal cells (see below and Chapter 12: Vaccine development).

Gene transfer into animal and plant cells: basic techniques

The genetic modification of animal and plant cells lagged behind that of bacteria, the chief obstacle being that eukaryotic cells do not take up added DNA as readily as do bacteria.

Animal Cells

Transfection and microinjection

Various ways of introducing DNA into animal cells and making it more likely that it will be incorporated permanently into the chromosomes and correctly expressed have been developed over the past decade.

In suitable conditions, cultured animal cells can now be induced to take up added DNA, albeit at a fairly low frequency. Although the process is analogous to that of transformation in bacteria, where animal cells are involved it is usually called **transfection**, to avoid confusion with the existing use of 'transformation' to denote the changes seen in cultured mammalian cells when they convert to a cancerous growth pattern. Reliable techniques for transfecting animal cells have only been devised within the past 10 years. The entry of DNA fragments into the cell is aided by placing the cells in a strong electric field (**electroporation**) or by using a calcium precipitate of DNA.

DNA can also be introduced by direct **microinjection** into the nucleus with a fine glass micropipette. In order to be stably maintained and be expressed in a eukaryotic cell, introduced DNA must not only cross the cell membrane but also enter the nucleus. Injection into the nucleus therefore increases the chances that the introduced gene will become permanently established in the chromosomes.

Whole chromosomes or parts of chromosomes can also be introduced into animal cells by fusing two cells to make a **somatic cell hybrid**.

Retrovirus vectors

A way of increasing the chances that the introduced genes will become part of the chromosomes is to splice them into vectors that insert into the chromosomes. These vectors are derived from retroviruses and some DNA tumour viruses such as SV40. Retroviruses are unusual RNA viruses that naturally integrate DNA copies of themselves into the chromosomes where they are passed on from cell to cell and inherited just like cellular genes (see Chapter 11: RNA tumour viruses). They contain sequences (the so-called long terminal repeats (LTRs)) that promote their integration at random into chromosomal DNA.

Parts of the viral DNA responsible for integration can be spliced to a 'foreign' gene to make a recombinant DNA that is more likely to become permanently established in the chromosome than the gene on its own. Unwanted bits of the viral genome – for example those that direct the production of infectious virus particles – are deleted. The modified virus DNA is introduced into cells by transfection or microinjection. Ingenious strategies based on retroviral vectors now hold out the best hope of making gene introduction into human somatic cells efficient and reliable enough to be able to correct some genetic defects (see below, Human gene therapy).

Plant Cells

Protoplasts and plant tissue culture

Plant cells are normally encased in a thick wall which prevents many of the procedures for introducing new genes, such as transfection or cell fusion, that are possible with animal cells. But if the wall is removed by gentle treatment, the naked **protoplast** can be kept alive in culture for a short time. Protoplasts can take up DNA, can fuse with each other to form hybrid cells, and will eventually regrow the cell wall and divide to form an undifferentiated mass of **callus** tissue. In some plant species at least, a complete new plant can be regenerated from this callus tissue in culture after treatment with plant growth hormones.

Protoplasts from different species, or even different genera, can be fused, and this is a promising approach to introducing genes (for disease resistance, drought tolerance, improved nutritional value of plant proteins, etc.) from one species into another.

Genes can be introduced into intact plant tissue, usually leaf disks in tissue culture, through the medium of genetically modified forms of the plant pathogenic bacterium *Agrobacterium*. New plants containing the introduced gene are regenerated from the infected tissue.

Agrobacterium and its plasmids

The bacterium *Agrobacterium tumefaciens* infects many 'broad-leaved' or dicotyledonous plants causing tumorous galls ('crown gall') on leaves and stems. This bacterium contains a plasmid – the **Ti plasmid** – that carries genes required for the transformation and continued proliferation of infected cells. During infection, *A. tumefaciens* transfers a portion of this plasmid – the **T-DNA** – into the cell's chromosomes, where it becomes stably integrated.

This transfer of DNA from bacterium to plant makes *Agrobacterium* a natural carrier for introducing foreign genes. Any DNA spliced into the T region of the Ti plasmid is automatically integrated into the host plant genome. A decade of genetic engineering of the Ti plasmid has produced tailor-made vectors that still transfer their DNA but do not, for example, cause tumours. Many genes have been introduced into whole plants and plant cells in culture either by infection with the bacterium itself or by introduction of the plasmid DNA alone.

However, agrobacteria naturally infect only dicotyledonous plants (e.g. tobacco and tomatoes and other vegetables), and the important monocotyledonous crop plants (e.g. cereals and maize) have not been particularly susceptible up to now to genetic engineering with Ti plasmids. Ways of introducing T-DNA into monocots are now being sought, and some success is being achieved. The host specificity of agrobacterial infection seems to be due in part to chemicals produced by the plant itself when wounded – agrobacteria generally enter through wounds. The wound exudate apparently activates a bacterial gene necessary for infection and agrobacteria have, for example, been induced to infect a monocot (e.g. the yam, *Dioscorea bulbifera*) and produce a gall by treating the bacteria beforehand with wound exudates.

Plasmids similar to Ti plasmids occur in another *Agrobacterium* species, *A. rhizogenes*, which causes a proliferation of fine rootlets ('hairy root disease') on the plants it infects. Like the gall tissue produced by *A. tumefaciens*, 'hairy root' tissue also contains DNA derived from a bacterial plasmid – the **Ri plasmid** – that has become incorporated into the plant's chromosomes. Ri plasmid DNA is also used as a carrier for introducing foreign genes into dicotyledonous plants.

Plant virus vectors

The use of viruses as vectors for introducing genes into plant cells (cf. the SV40 vectors developed for mammalian cells) is restricted by the fact that there are few DNA viruses that infect plants: most plant viruses are RNA viruses. One DNA virus, maize streak virus, is a potential vector of especial interest as it infects such an important crop.

GENETIC ENGINEERING

Producing proteins from genetically engineered cells

Many proteins of medical and commercial importance are now produced from genetically engineered bacterial, yeast and cultured mammalian cells into which the required gene has been introduced, rather than by expensive and laborious purification from animal or human cadaver tissues in short supply (Table 4.2).

For proteins used therapeutically – such as insulin, growth hormone and the blood-clotting factors – this enables the human protein rather than an animal substitute to be used, and also eliminates the risk of virus transmission from the original tissue along with the extracted protein. In the past, this has led to contamination of growth hormone with a slow-acting virus that eventually causes neurological damage, and to the anti-haemophilia factor VIII with the human immunodeficiency virus (HIV) that causes AIDS. It is also virtually the only way of producing large quantities of proteins that occur normally only in minute amounts in tissues (such as the lymphokines of the immune system, Chapter 12).

The first 'recombinant' mammalian proteins were produced in bacteria. But although bacteria grow rapidly, and potentially can produce enormous quantities of protein, they have several drawbacks when it comes to producing mammalian and other eukaryotic proteins.

One is that all secreted mammalian proteins, amongst which are most of medical interest, are glycoproteins – that is, they carry carbohydrate side chains attached at specific sites on the protein. The sugar

Table 4.2 Some proteins and peptides of proven or prospective medical use which are now produced from recombinant cells

Hormones
Insulin
Somatostatin
Growth hormone

Growth factors (see Chapter 9)
EGF
PDGF
Transforming growth factors
Lymphokines (e.g. interleukin-2) *(see Chapter 12)*
Tumour necrosis factor (TNF)

Interferons
alpha
beta
gamma

Others
Atrial natriuretic peptide (atriopeptin, ANF, ANP): a possible antihypertensive agent
Tissue plasminogen activator (t-PA, TPA): a thrombolytic agent, recently passed for clinical use
Blood-clotting factors (anti-haemophilia factor VIII and others)

Viral and other parasite antigens for use in vaccines (see Chapter 12)
Viruses (e.g. hepatitis B, AIDS, rabies)
Leprosy bacillus antigen for prospective use in vaccines
Parasite antigens for prospective use as vaccines (e.g. malaria and schistosome antigens)

---

side chains are often essential for full biological activity of the protein. In the eukaryotic cell, the addition of these sugar side chains (glycosylation) occurs in the endoplasmic reticulum (see Chapter 7). Bacteria do not possess an endoplasmic reticulum and although they have a mechanism for glycosylating their own proteins often do not glycosylate the mammalian proteins that they are synthesizing from transplanted genes. Neither will bacteria secrete mammalian proteins without the addition of a bacterial 'signal' sequence, which has later to be removed. This has led to the search for alternatives to bacteria for recombinant protein production.

Much effort is now going into developing genetically engineered yeast, cultured mammalian cells and even simple organisms such as silkworms, to mass-produce useful mammalian proteins. Several recombinant proteins including insulin and interferon are already produced commercially from mammalian cells, and an anti-hepatitis vaccine is now produced from hepatitis antigens produced in yeast cells.

For commercial production, the transplanted gene is never in its original state. Various modifications are made to allow it to be transcribed and translated as efficiently as possible in its prospective host cell. All that is required from the original source is the protein-coding sequence; all the other DNA sequences needed to control its expression are supplied on the carrier DNA into which the gene is spliced. Genes intended for bacterial cells are hitched up to an efficient bacterial promoter, those intended for mammalian cell cultures to appropriate cellular (or animal virus) promoters and enhancers.

The final recombinant 'gene' is usually a mosaic of elements from different sources. This mosaic is constructed by successive modifications *in vitro* to the vector DNAs. Many 'ready-made' vectors containing the requisite control elements, into which a cDNA or synthetic gene can be slotted as required, are now in use.

The genetic modification of animals and plants

The deliberate improvement of the genetic makeup of crop plants and animals has been going on for thousands of years through selection by farmers and, during the past 100 years, by increasingly sophisticated breeding programmes. The development of recombinant DNA technology, however, with its potential for directly altering the genetic makeup of plants and animals in ways not previously possible, looks set in the future to revolutionize plant and animal breeding.

The techniques of gene transfer pioneered in bacteria have over the past decade been extended to plants and some animals – mammals and a few other species. Many strains of **transgenic** plants and animals bearing genes from other species now exist. The particular developmental programme of mammals lends itself to *in vitro* fertilization and embryo replacement in ways not immmediately possible with other animals. This section deals entirely with transgenic mammals. Genetic modification in other types of animal has hardly been explored except in the case of the fruitfly *Drosophila* where rather different techniques are used to create particular genotypes for genetic and developmental studies.

Transgenic plants that produce their own insecticides from transplanted bacterial genes (see below) are only one example of the new generation of crop plants that are being developed by a combination of recombinant DNA techniques and the longer-established methods of plant cell culture. An important limitation in conventional plant breeding programmes is that plants of different species do not naturally interbreed. Many modern crop varieties are of a different species from their wild relatives, or have become so divorced from them over thousands of years of selection that normal interbreeding is not possible. The wild relatives of crop plants, however, constitute a vast reservoir of potentially useful genes for disease resistance, lerance to cold or drought, nutritional quality, etc. Although plant breeders have devised several ingenious ways of crossing the species barrier, a breeding programme using standard techniques can take years to introduce a desirable trait from a wild plant into a cultivated relative.

The problems usually do not lie in any incompatibility in the genes of the two species at the level of their DNA. DNA is a most accommodating molecule and, as recombinant bacteria show, living cells can accept genes directly transplanted from any source and, with a little extra genetic engineering, be persuaded to produce the proteins they specify. The new genetic techniques can now be used to introduce useful genes from plants and bacteria directly into plant cells, from which, by the well-tried methods of plant cell culture, new transgenic plants can be regenerated.

The current acquisition of seed companies and plant breeding companies by the giants of the agrichemical industry reflects the potential ability of genetic engineers in the long-term to produce seeds with their own inbuilt resistance to fungal and insect pests, and to make more economic use of fertilizers (or even, a more remote possibility at present, fix their own nitrogen). Such plants would not need the current heavy applications of pesticides and fertilizers.

Of more immediate value, DNA technology has also provided genetic probes for valuable genes that can be used to screen large numbers of seedlings or cultured cells in more conventional breeding programmes, so that it is not necessary to wait months before finding out, through laborious testing of mature plants, whether the desired genes have been passed on in the genetic lottery.

In animals, the techniques of gene transfer are most advanced in the laboratory mouse. Work with larger, less prolific domestic animals has progressed more slowly but strains of transgenic pigs, sheep, goats and cows have been established. When advances in reproductive technology that include *in vitro* fertilization, storage of frozen embryos and surrogate motherhood are combined with gene transfer directly into fertilized eggs and very early embryos, a powerful new genetic technology has been created whose potential is only just beginning to be explored.

Genetic manipulation in animals aims to incorporate the new gene into the germline (the reproductive cells) so that it can be passed on from generation to generation just like any other gene, and a strain of stable transgenic animals can be bred. Such manipulation of the human germline – for example, to correct a genetic defect – may be possible in the long-term, but is not in prospect, for both ethical and technical reasons (see below). However, there are some inherited genetic diseases affecting humans whose effects may be alleviated by transfer of an unaffected gene into somatic cells after birth (see below, Human gene therapy). Although there are considerable technical problems to be solved, the first trials of somatic gene therapy in humans are undoubtedly not far away.

Transgenic animals (mammals)

Transgenic animals are reared from embryos into which new DNA has been introduced by genetic manipulation of a fertilized egg or very early embryo outside the mother. If the new DNA becomes stably incorporated into the chromosomes at this early stage, it will be present in a large number, if not all, the cells of the developing embryo, including the reproductive cells (germline). After birth, those animals that carry the new gene (**the transgene**) can easily be distinguished by analysing a sample of their DNA. A strain of transgenic animals carrying the new gene in their germline, and therefore able to pass it on to their progeny who will then carry it in all their cells, can be bred from these original carriers (Fig. 4.6).

There are various routes for introducing DNA into a fertilized mammalian egg or very early embryo *in vitro*. (In this connection, the mammalian **early embryo**, sometimes called the **pre-embryo**, refers to the as yet totally undifferentiated ball of cells that develops within a few days of fertilization.) Up to the **blastocyst** stage (at which the mass of dividing cells has formed a hollow ball ready to implant into the wall of the uterus) the embryo can be reintroduced into the womb.

DNA can be injected straight into the nucleus of the fertilized egg (microinjection) or alternatively into the centre of the blastocyst where it is taken up by at least some of the cells and becomes incorporated into their chromosomes. Another technique is to modify compatible, undifferentiated embryonic stem cells by transfection or retrovirus-carrier infection, and then replace them in the embryo (which can be frozen to arrest its development and stored until needed).

Mammalian pre-embryos at this stage are remarkably robust to interference. The cells that will give rise to the embryo itself have not yet been set aside (see Chapter 11). Unlike the embryos of other species

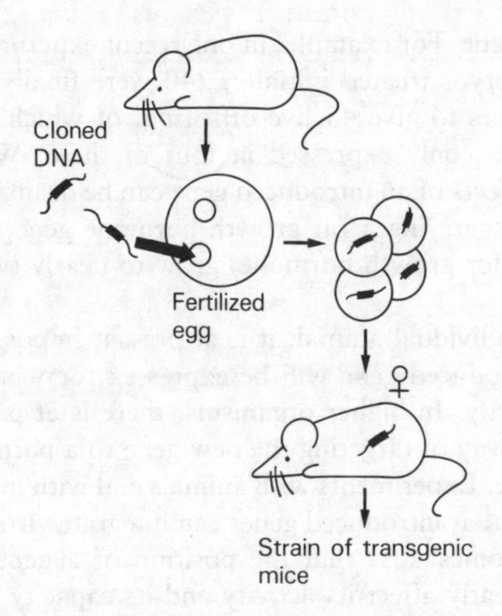

Fig. 4.6 Procedure for obtaining transgenic mice.

they still retain the potential to replace removed cells or incorporate additional ones without affecting subsequent development.

So far, these techniques have been applied to laboratory and some domestic animals, chiefly the laboratory mouse. Work on farm animals has produced transgenic pigs, sheep, goats and cows, but as yet only on an experimental basis. One goal for animal breeders is to be able to produce an animal with required characteristics – such as better reproductive performance or disease resistance – without the present lengthy breeding process. There are still considerable problems to be overcome before this goal might be achieved. These are largely being investigated in the transgenic mouse, which as well as being a testing ground for techniques destined for use in farm animals and human gene therapy is also of interest in its own right in fundamental studies on the control of gene expression.

The most immediate problem is that it is not yet possible to predict where in a genome the introduced DNA will settle. If it disrupts resident gene function, the modified embryo may abort or be born with congenital malformations or other defects. Given that DNA can be introduced without interfering with the expression of the remainder of the genome, there is still the question of whether the introduced genes function effectively, that is, whether they are transcribed into RNA and translated only at the proper times during embryonic and adult life. At present, the absolute success rate per treated egg or embryo is low.

Only a very small proportion of treated embryos end up expressing an introduced gene. For example, in one recent experiment, of around 390 mouse embryos treated initially, 240 were finally transferred to surrogate mothers to give 45 live offspring, of which 17 carried the transgene. It was only expressed in four of these. When it works, however, the effects of an introduced gene can be dramatic. A strain of transgenic mice carrying a rat growth hormone gene (in addition to their own gene for growth hormone) grow to nearly twice the size of normal mice.

But, for an individual animal, it is at present impossible to predict whether an introduced gene will be expressed, or whether it will be regulated correctly. In higher organisms, there is at present no completely reliable way of targeting the new gene to a particular place on the chromosome. Experiments with animals and with mammalian cells in culture show that introduced genes can integrate virtually anywhere in the chromosomes, and that the position of a gene on the chromosome can greatly affect its activity and its capacity to be properly regulated. If it comes within the orbit of another gene regulatory system it may be permanently inactivated or become active in the wrong tissues. However there are numerous cases where genes appear to have become permanently incorporated and are being correctly expressed and regulated.

There are ways of making it more likely that the gene will be expressed in the required tissues and at the appropriate time. This involves the use of a 'fusion gene' as the introduced material. These are synthesized *in vitro* by hitching up the required coding sequences to the control regions (promoter, etc.) from a host gene that is expressed only in the desired tissue. The host promoter should respond to the biochemical cues it normally responds to and switch on its associated protein-coding sequences, which may be derived from any source, bacterial or eukaryotic, in the required fashion. This type of construct has been used recently to obtain a transgenic mouse strain that secretes sheep milk proteins in its milk, as a preliminary test of the feasibility of modifying the protein content of cow's milk (see below).

A more general problem in the application of direct gene transfer to farm animals is that most of the characteristics one might wish to modify, such as size, reproductive characteristics, growth rate, etc. are the product of many genes acting independently. The individual genes contributing to such polygenic traits are in most cases not known, and it is extremely difficult, if not impossible, to disentangle their individual effects. They may well exist as blocks of **co-adapted genes**, where a radical alteration in one gene often upsets the function of the block of genes as a whole. Such traits are therefore not easily amenable to improvement by artificial genetic manipulation. A start has been made however, by introducing extra copies of their own growth hormone genes into pigs in the hope of producing a strain of pigs that will grow faster without the use of the steroid hormones and antibiotics that are used commercially at present to promote growth.

Another idea is that of engineering the genome so that a cow, for example, will produce desirable 'foreign' proteins in its milk. This approach could be used either to improve the protein quality of milk for consumption, or as another way of producing large amounts of commercially valuable proteins that could be easily isolated from milk.

One problem in transgenic work is that the introduced DNA may actually cause mutations by inserting into host genes and disrupting their action. Although this is a grave disadvantage in work aimed at gene therapy it has been exploited by geneticists as a rapid way of producing new mutants that might be useful in mapping large genomes and in developmental research. Mammalian developmental research generally is severely hampered by a dearth of informative mutations. An advantage of mutations caused by transgenes over those identified purely genetically is that the gene affected by the mutation can be easily recovered using the transgene as a probe, and then used to isolate the normal gene from a DNA library for sequencing, gene mapping, etc.

Human gene therapy

The ability to introduce new genes into mammalian cells raises the possibility of being able to correct genetic defects in humans by introducing a copy of the normal gene into the appropriate cells. The techniques are at present not sufficiently reliable to be applied to *in vitro* fertilized early human embryos to correct genetic defects, even if such procedures were considered ethically acceptable or even necessary, given the prospective ability to detect genetically defective eggs and implant only those not carrying the defective gene. Proposed legislation would ban any experimental genetic manipulation of fertilized eggs in the UK, and similar legislation has been proposed in most other countries. The first targets for gene therapy in humans are tissues such as the easily extracted, self-renewing and transplantable cells of the bone marrow that produce all the cells of the blood and the lymphocytes of the immune system. Defects in the bone marrow precursors of these cells are specifically responsible for the inheritable anaemias – sickle-cell anaemia and the thalassaemias – and for heritable immune deficiencies. Genetically corrected bone marrow might also be able to alleviate the symptoms of other conditions that arise from a general lack of a particular enzyme. Some inheritable defects in bone marrow cells are in principle already treatable by a transplant of healthy marrow from a matched donor, but the problems of tissue matching mean that it cannot be offered to everyone who might benefit from it. A transplant of a patient's own corrected cells would overcome some of the problems presently associated with bone marrow transplantation (Fig. 4.7).

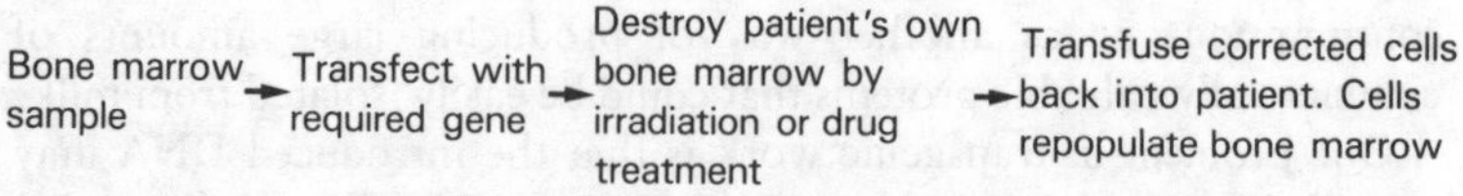

Bone marrow sample → Transfect with required gene → Destroy patient's own bone marrow by irradiation or drug treatment → Transfuse corrected cells back into patient. Cells repopulate bone marrow

Fig. 4.7 Scheme for gene therapy with corrected bone marrow cells.

A year or so ago it was widely predicted that bone marrow replacement would soon be ready for clinical trial in humans. However, there are technical problems of getting a sufficient number of the treated cells to take up the gene and express it correctly, so that at the time of writing, human gene therapy is still untried.

The goal in gene therapy is ideally a clean swap of the good gene for the defective resident genes or at the least to target the incoming gene to a predetermined site on the chromosome where it will be correctly regulated and not interfere with the expression of other genes. Many of the genetic diseases that might be alleviated by corrected bone marrow are caused by defective genes that either make no product at all or an ineffective product. In most of these cases only one copy of a good gene need be introduced and the defective gene need not be removed.

Straightforward replacement of a resident gene by its introduced counterpart occurs routinely in bacteria and yeast. Unfortunately, although a similar process can occur in mammalian cells, it is for practical purposes overshadowed by the far greater likelihood that the introduced gene will become incorporated at random in the chromosomes. Although genes integrated at random often appear to function and be regulated normally, in many instances they do not, and the unpredictability of the results presents problems for gene replacement in humans. Nothing is known about the long-term effects over a human lifetime of such an anomalously positioned gene.

However, in the case of genetic therapy involving somatic cells these problems could in theory be overcome by screening for acceptable insertion before replacing cells in the body. At present, the chief technical problem is in devising acceptable methods of getting genes into a sufficient number of cells in a bone marrow sample. The self-renewing stem cells that have to be corrected if the marrow is to continue to generate corrected cells throughout the individual's lifetime have not yet been isolated with any certainty, and are calculated to make up only a tiny proportion (0.01%) of the cells in a sample of bone marrow. If it becomes possible to obtain enriched cultures of stem cells the prospects for reliable gene replacement in bone marrow could look brighter. Vectors likely to provide the required high rates of gene introduction are those derived from retroviruses. Human bone marrow cells can now be successfully infected with specially engineered retroviruses carrying an appropriate gene, which, when once inside the bone marrow cells, insert a DNA copy of themselves and their added gene into the chromosomes but are unable to produce infectious virus.

The first attempts at gene therapy via corrected bone marrow are predicted to be aimed at adenosine deaminase (ADA) deficiency, which produces severe combined immunodeficiency usually leading to early death in childhood, and Lesch–Nyhan disease, a rare X-linked condition caused by a deficiency of the enzyme HGPRT (hypoxanthine:guanosine phosphoribosyltransferase) and leads to a bizarre syndrome of uncoordinated involuntary movements and aggressive self-mutilation. Although the effects of HGPRT deficiency are most marked in the brain, it is possible, but by no means certain that a circulating source of HGPRT might alleviate symptoms. Both these enzymes are produced constitutively in most cells and do not present any special problems of gene regulation. Both these conditions are mercifully rare. Much more common are the hereditary anaemias caused by defective globin genes. Until recently however, these most obvious candidates for genetic correction have proved somewhat intractable. The globin genes represent a very tightly-regulated gene system: two separate globin genes have to be expressed to make haemoglobin, they are only expressed in the precursors of red blood cells, and it is not yet known exactly how globin genes are regulated.

Genetic modification of plants and their associated bacteria

Unlike cells and tissues from an adult animal, many plant tissues retain their capacity to regenerate a complete new plant throughout the life of the plant. Cell and tissue culture as a means of plant propagation is widely used commercially, and it is now possible to introduce novel genes into cultured plant cells and tissues (Fig. 4.8).

Plants bearing new genes introduced from other plant species and from bacteria have been raised from individual plant cells and plant tissue genetically modified in culture. The practical application of these techniques to important crop plants still awaits solutions to several

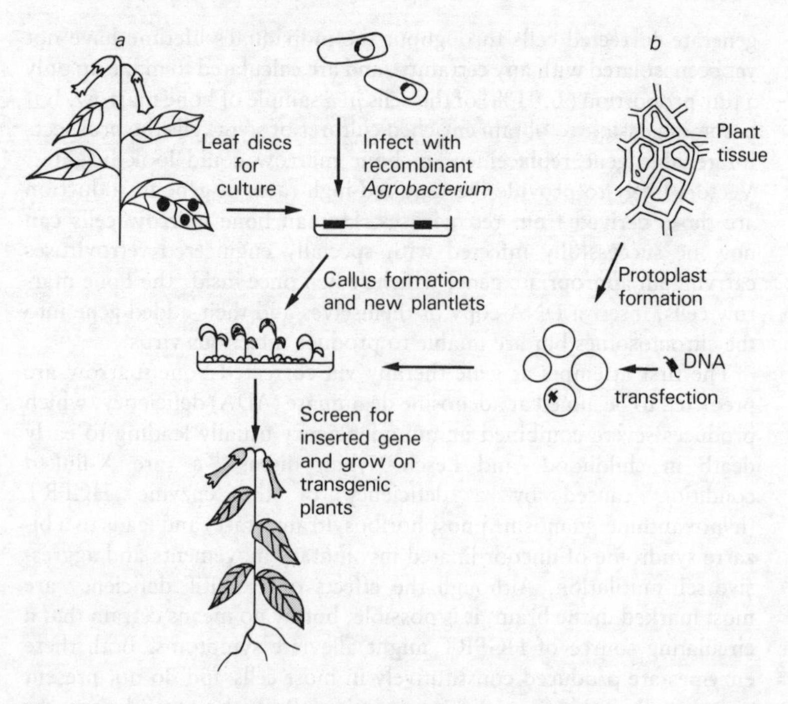

Fig. 4.8 Generation of a transgenic plant from (*a*), leaf discs infected *in vitro* with recombinant *Agrobacterium*, or (*b*), protoplasts transfected with DNA.

problems. Finding suitable conditions for regenerating entire plants from cultured cells and tissues is still largely a matter of trial and error for different species. The staple cereal crops of wheat, rice and maize are monocotyledons and have proved much more difficult than dicotyledonous plants (such as potatoes and tobacco) to grow and modify in culture. Tobacco, tomatoes, petunias and potatoes at present figure largely in accounts of genetically engineered plants as they are easy to manipulate with *Agrobacterium* vectors and to regenerate from tissue culture. The first successful transfers of foreign DNA into cultured cereal cells, by transfection of protoplasts, were only made very recently and at the time of writing it is not yet possible reliably to regenerate transgenic cereal plants from a single genetically modified cell, although it is undoubtedly only a matter of time before this will be achieved.

One reason that monocots have lagged behind dicots in the genetic-engineering business is that they are not naturally susceptible to infection with *Agrobacterium*, the best-developed carrier of foreign DNA for plant genetic manipulation. Ways of introducing DNA into cereal cells using novel routes and other carriers are now being developed.

As well as introducing new genes, it is now also possible to engineer the genome so that some of the plant's own genes are permanently switched off, or their expression changed in some way.

Pest Resistance

The increasing cost and environmental problems caused by heavy applications of fungicides and pesticides makes built-in resistance against insect pests and the fungal and viral pathogens that afflict crop plants a major aim of any plant-breeding programme. Plants have many ways of resisting attack. Many possess physical defences against insect damage in the form of leaf hairs and spikes, and others produce noxious chemicals that deter herbivorous insects. They can produce general chemical responses that limit fungal infection, and in some cases genes responsible for resistance against specific fungal pathogens have been identified. In most cases, natural pest and disease resistance is the product of many genes working together and so are not immediately amenable to genetic engineering techniques. Even where single-gene resistances have been identified, fungal pathogens, with their capacity for rapid variation, soon evolve to overcome them (and also to develop resistance to many of the fungicides used against them) which is why the lifetime of most modern disease-resistant varieties is usually short.

For these reasons the conventional methods of plant breeding, backed up by protoplast fusion and tissue culture techniques, are unlikely to be superseded immediately. Recombinant DNA, however, offers the possibility of attempting some novel strategies and a few examples are outlined below. One important consideration in introducing totally new genes into crop plants or in persuading them to make much larger amounts of some proteins than normal, is that this does not make the plant unpalatable or harmful to eat.

As well as altering the plants themselves, much effort is focused on altering the genetic makeup of the bacterial inhabitants of the plant and its immediate environment, and in improving the effectiveness of natural pathogens of the pests.

Insect resistance

Strategies being developed by genetic engineers to protect crops against attack by insect pests range from incorporating genes for natural 'insecticides' into the plant to producing more effective forms of the viruses and other microorganisms that attack the pests.

Several types of pest-resistance genes have already been transferred experimentally. One is a bacterial gene encoding an insecticidal toxin produced by the common soil bacterium *Bacillus thuringiensis*. The toxins produced by *B. thuringiensis*, which are lethal to the caterpillar stage of many insect pests, are already in use as natural insecticides. They are proteins and are processed to toxic form only in the insect's gut. From the point of view of safety to humans, animals and beneficial insects they are the exemplar of a safe insecticide. The bacteria produce large crystals of toxin, which have been produced commercially and used as insecticides for many years, especially in the production of organically grown crops.

Plants have been raised recently that incorporate the bacterial gene specifying the toxin. They produce the protein in their tissues with no apparent effects on their own growth, and in experimental conditions show considerable pest resistance. However, the toxin is sensitive to ultraviolet light and is soon inactivated. In an alternative approach to the problem, the genes for *B. thuringiensis* toxin have also been transferred to bacteria of the genus *Pseudomonas*, common inhabitants on the leaves of many plants, in the hope that live *Pseudomonas* producing the toxin may also act as a deterrent.

Another approach to insect resistance is to incorporate genes for **proteinase inhibitors**. These are small proteins occurring naturally in some plants which, when taken in by caterpillars, interfere with the activity of digestive proteinase enzymes such as trypsin. The caterpillar is unable to digest and make use of the plant protein and starves to death or fails to develop. A gene for the trypsin inhibitor of cowpea (*Vigna unguiculata*) has recently been introduced experimentally into tobacco (*Nicotiana*), where it significantly deters tobacco's own herbivorous pests, such as the tobacco budworm. (This particular trypsin inhibitor appears to have no effect on human trypsin.)

Another avenue to protecting plants against insect attack is to exploit the viruses and bacteria that naturally infect the insects. This approach is being investigated in the UK at present to control the caterpillars of the moth *Panolis flammea*, a pest of the introduced lodgepole pine (*Pinus contorta*), now planted in large numbers in Scotland. In some parts of Britain the caterpillars are subject to natural control by a baculovirus parasite. The virus, however, is absent from the Scottish plantations. Introduction of the natural baculovirus as a biological control agent has been under trial for the past 2 years. It might be made more effective, however, if engineered to contain genes specifying, for example, a caterpillar-killing toxin. A controlled release of virus carrying a harmless genetic marker introduced by recombinant DNA techniques was made in 1986 to monitor the virus's natural spread in the insect population and rate of disappearance. This was the first release of a genetically engineered organism into the environment in the UK (see below, Regulation of recombinant DNA work). It has been followed up in 1987 by a further controlled release of virus containing 'suicide' genes, which limit its life in the environment. The aim eventually is to produce a virus that is more effective against its insect pest, but which retains its natural strict host-specificity and inability to persist in the absence of the host.

Virus Resistance

Virus diseases pose a considerable problem to agriculture and horticulture as they cannot be controlled by chemicals. Viruses spread by sap-sucking or leaf-eating insects can to some extent be prevented by controlling the insect population or by growing stocks for propagation in areas free of the insect carrier. Virus-free stocks of many ornamentals, soft fruit, etc. are now also produced by propagation from tissue cultures of virus-free shoot tips. Natural resistance to viruses also occurs and can sometimes be incorporated by cross-breeding.

Several new molecular genetic approaches to virus resistance involve a detailed knowledge of virus molecular biology and the use of recombinant DNA techniques. One approach depends on incorporating the gene for the virus's own coat protein into the plant's DNA. (A viral particle consists of a core of RNA or DNA surrounded by a protective protein coat, often composed of only one type of protein – most common plant viruses are RNA viruses.) Viral RNA entering the cell on infection is then immediately mopped up by the coat proteins produced by the cell, preventing its multiplication. Coat protein genes for tobacco mosaic virus (in this case, DNA copies of the virus RNA) have proved a partial protection at least to tobacco mosaic infection.

Another strategy is directed against those viruses that are sometimes accompanied by 'satellites' – small circular RNAs that are not part of the virus and which appear to lessen the severity of infection. The satellites are of considerable biochemical interest in their own right as they belong to the diverse group of 'ribozymes' – RNAs with enzymatic activity (see Chapter 5). DNA copies of the satellite RNA incorporated into the plant genome are activated to produce large amounts of satellite RNA on infection by their corresponding virus and have been shown to provide protection against, for example, the tobacco ringspot virus.

Herbicide Resistance

Genes specifying resistance to some herbicides have recently been introduced successfully into plants, with the long-term aim of possibly developing crop plants resistant to broad-spectrum herbicides to simplify routine weed-spraying programmes. The first target was resistance to glyphosate, a comprehensive weedkiller, and used resistance genes from other plants and bacteria.

Nitrogen Fixation

Engineering crops like cereals to fix their own nitrogen at present remains a long-term aim. The natural nitrogen-fixing symbiosis between rhizobia (the nitrogen-fixing bacteria) and the roots of leguminous plants depends on the action of many plant as well as rhizobial genes to establish the symbiosis and accommodate the complex biochemistry involved in nitrogen fixation by the bacterial enzymes. Conventional genetics has, however, been deployed for some time in improving the efficiency of nitrogen fixation within the important legume crops (such as soybean) by, for example, extending the host range of particularly efficient rhizobial species, and can now be supplemented by recombinant techniques. The genes involved in symbiosis can now be analysed more readily thanks to the ability to isolate and clone plant and bacterial genes and to study their products and the factors that regulate their expression. Many of the plant and bacterial genes that are needed to establish symbiosis have now been cloned and their gene products identified.

Free-living, soil nitrogen-fixing bacteria, such as *Klebsiella pneumoniae* and *Azotobacter vinelandii*, contribute greatly to the fertility of soils and are also now amenable to genetic modification.

Pseudomonas

Species of the bacterium *Pseudomonas* are also targets for agriculturally directed genetic engineering. Varieties of *Pseudomonas syringae* and other pseudomonads are chiefly harmless inhabitants on the leaves of many plants and in the soil (although some forms of *Ps. syringae* and other species are plant pathogens). They are readily amenable to genetic manipulation.

The best known example of a genetically engineered *Pseudomonas* is the so-called 'ice-minus pseudomonas'. Some forms of *Ps. syringae* have a gene specifying an ice-nucleation protein, which is a component of the bacterial cell membrane and induces ice crystal to form around the bacteria at around −4 °C. This causes frost damage to plants at warmer temperatures than if the protein is absent. In attempts to decrease late-frost damage to susceptible plants such as strawberries and potatoes, genetically engineered *Pseudomonas* lacking the ice-nucleation gene were developed in the United States and put forward as long ago as 1982 for preliminary trials of their ability to survive and be effective in the field. (A form of *Ps. syringae* lacking this gene also occurs naturally and makes up a small proportion of the normal population.) The ice-minus pseudomonas is intended to compete with and prevent the ice-forming *Pseudomonas* strains colonizing leaves and blossoms.

In 1982 ice-minus pseudomonas was set to be the first genetically engineered microorganism to be tested in the field. However, it then enjoyed some notoriety as legal action by anti-genetic engineering groups in the USA delayed field trials for several years. Other species of non-pathogenic *Pseudomonas* (e.g. *Ps. fluorescens* and *Ps. aureofaciens*) are also being developed as carriers of potentially useful genes.

THE STUDY AND DIAGNOSIS OF GENETIC DISEASE

In the past most of our knowledge of human genetics has come from the study of heritable genetic defects and diseases. Several hundred heritable conditions involving defects in single genes are now recognized, some merely inconvenient to those affected, such as red–green colour blindness, whereas others are fatal in childhood if not treated. Most of the latter are very rare, but a few, such as sickle-cell anaemia, cystic fibrosis and Duchenne muscular dystrophy (Table 4.3), are relatively common in certain populations.

The single-gene defects are recognized by their strict mendelian inheritance in affected families. Most heritable diseases only show the full-blown symptoms in homozygous individuals – that is, someone who has inherited a defective gene from both parents. Heterozygotes, who have inherited a defective gene from one parent and a normal gene from the other, often show few if any symptoms, although their carrier state can sometimes be detected by biochemical tests, and now by direct analysis of their DNA.

This type of mutant gene (more precisely a mutant allele – see p. 19) is in genetic terminology said to be recessive to the normal allele. Potentially lethal recessive mutations are able to persist in the population because, even though homozygotes usually die before they reach reproductive age, as long as the defective gene remains rare heterozygotes far outnumber homozygotes.

There are far fewer dominant inherited conditions, where the disease shows up in individuals carrying only one defective allele. Dominant

Table 4.3 Some heritable genetic conditions

Duchenne muscular dystrophy

Duchenne muscular dystrophy (DMD) is an X-linked recessive muscle-wasting disease. It affects about 1 in 3500 newborn boys, becomes apparent in childhood and usually proves fatal by the early twenties and thirties. There is a less severe Becker muscular dystrophy (BMD) which is apparently caused by a defect in the same gene.

The chromosomal region involved in DMD has now been revealed at the molecular level and is the longest gene so far recorded at nearly 2 million nucleotides long. Much of the 2 million nucleotides is probably taken up in introns – one of over 100 000 base pairs has already been found. Individual cases of DMD show a range of different mutations in this region, including many large deletions.

The complete gene has not yet been sequenced at the time of writing, but from translation of cloned cDNA produced from the 14-kb long mRNA, and from the corresponding gene in mice, the normal DMD gene is believed to encode a large protein of molecular mass approx. 400 000, which has been named dystrophin. This is present in normal muscle but missing from muscle of DMD patients. Antibodies against the DMD protein have identified a possible candidate in a large protein that is located at the muscle cell membrane. Whether, and how, the absence of this protein is related to the muscular degeneration characteristic of the disease is the next, and very difficult problem to be resolved.

Cystic fibrosis

Cystic fibrosis (CF) is caused by a recessive defect in a gene recently located to chromosome 7. The main symptoms in homozygotes are excessive production of mucus in the lungs, leading to congestion and susceptibility to infection, and abnormal secretion from the pancreas and other glands. If untreated, affected children die in childhood from infection or lung failure. The chance of survival to adulthood has been much improved in recent years by physiotherapy and antibiotic treatment.

The CF gene has only very recently been located and at the time of writing has not yet been isolated. Preliminary DNA-based diagnostic tests are based on a closely linked 'marker'. Affected homozygous fetuses and heterozygous carriers can also be detected by the presence of abnormally high levels of the product of another unrelated gene – the CF antigen gene – in amniotic fluid and blood.

There is biochemical and physiological evidence that the basic defect in CF lies in biochemical pathways regulating chloride transport across the plasma membrane of epithelial cells (i.e. those lining the gut, glands, lungs etc.). The cells accumulate excessive amounts of chloride, which leads to symptoms such as abnormal fluid production in the lungs, increased salt

Table 4.3 contd.

content of sweat and disturbances in pancreatic secretion. Biochemical research has already come very close to identifying the cause of this defect and the isolation of the CF gene and identification of its product will be of great value in helping to close the gap, as well as opening the way for DNA-based diagnostic tests using the CF gene itself.

Huntington's chorea

This is one of the few dominant inherited diseases, i.e. it shows symptoms in individuals who have inherited one copy only of the defective gene. Heterozygotes for Huntington's remain symptomless until middle age and then suffer a tragic rapid mental deterioration. A DNA-based diagnostic test using markers closely linked to the HC gene on chromosome 4 and which is able to identify heterozygotes correctly in about 90% of cases has been developed, but the exact location and identity of the HC gene itself is not yet known. It is unusual amongst dominant heritable conditions in that homozygotes show no more severe symptoms than heterozygotes.

Huntington's chorea illustrates in a particularly tragic way the psychological and ethical dilemmas that these powerful diagnostic tests pose. The availability of a diagnostic test for Huntington's confronts those at risk, but still showing no symptoms, with a difficult choice. If the test shows that they carry a defective gene, they are then faced with a certain foreknowledge of an inevitable mental and physical deterioration, which will begin at some unpredictable time in the future. In a small percentage of cases the test may give a false result or not be able to resolve the issue unambiguously.

Manic-depressive illness

Manic-depressive psychosis (bipolar affective disorder) is one of several mental conditions which appear to have a strong underlying predisposing genetic component. Its genetic basis is difficult to study in the general population, however, as it can apparently be caused independently by defects in several different genes, and not all those carrying a presumed defective gene develop the disease. A study of an affected family in the inbred community of the Old Amish in Pennsylvania, however, has traced a predisposition to manic-depressive illness to a defect in a single gene located on the short arm of chromosome 11. An inherited tendency to manic-depressive illness determined by single genes elsewhere on the chromosomes – in some cases possibly on the X chromosome – has also been reported from other families.

The predisposition is inherited as a dominant trait, but unlike the classic heritable genetic diseases, not all those carrying the apparently defective allele become ill – in genetic terminology it shows incomplete penetrance (the penetrance of an allele is its measurable effect in those who carry it).

Table 4.3 contd.

Familial Alzheimer's disease

Familial Alzheimer's disease (FAD) is a rare inherited form of the much more common Alzheimer's disease (formerly called senile dementia), a degenerative disease of the brain affecting around 5% of people over 65. FAD sufferers develop the condition much earlier – in their 40s or 50s. Early onset of Alzheimer's is inherited as a dominant trait determined by a single gene (the FAD gene) located on chromosome 21. The FAD gene has not yet been isolated or its product identified. It is now the subject of much research in the hope that identifying the product of the FAD gene may throw light on the underlying causes and possibly suggest ways of treating the sporadic form of the disease.

Alzheimer's disease is characterized by a profound decline in memory and other cognitive functions such as the ability to perform a complex task, often accompanied by changes in behaviour and personality. The symptoms are thought to be at least partly due to deficits of some neurotransmitters, in particular to a decline in levels of acetylcholine resulting from the selective degeneration of cholinergic neurones in areas of the forebrain. Attempts to alleviate Alzheimer's symptoms by trying to increase acetylcholine levels by drug treatment have not been successful so far, however.

Neuronal degeneration results in typical structures known as senile plaques, areas where neurones have been destroyed and which also contain at their centre deposits of amyloid, a complex proteinaceous material. One of the amyloid constituents, a small polypeptide of 4000 M_r, has been the subject of molecular genetic analysis. It seems to be an abnormal breakdown product of a much larger amyloid-precursor protein (AP) which is probably, according to the most recent evidence, an abundant membrane protein in normal neurones. Another abnormal constituent of senile plaques is inorganic aluminosilicate. Aluminium does not normally cross the blood–brain barrier; how it does so in Alzheimer's brains, and what part, if any, it may play in the disease process is still unknown.

For some time it seemed likely that the FAD gene might be the amyloid-precursor gene, to which it shows close linkage, but this has now been ruled out.

---

inherited diseases can persist in the population only if they allow those affected to reach reproductive age and pass the gene on to their children. Huntington's chorea and polycystic kidney disease are two of the best known. The predisposition to some cancers is also inherited as a dominant trait (see Chapter 10: Familial cancers). Huntington's chorea, for which the underlying defect is not yet known, characteristically leads to rapid general mental deterioration from middle age onwards.

Recessive conditions for which the underlying defect is already known include many 'inborn errors of metabolism' caused by a lack of a particular enzyme as a direct result of a mutation in the gene that specifies it. These include phenylketonuria (a deficiency of phenylalanine hydroxylase, which causes permanent mental retardation due to phenylalanine accumulation if not controlled by diet), alcaptonuria (a harmless deficiency of homogentisate oxidase in which the only symptom is that a person's urine turns black on exposure to air), Tay–Sachs disease (a deficiency in a lysosomal enzyme hexosaminidase A, which is lethal early in life if not treated) and adenosine deaminase deficiency (which leads to severe combined immunodeficiency because of the effects of this particular enzyme deficiency on the precursor cells of the immune system). The hereditary anaemias such as sickle-cell anaemia and the thalassaemias are also recessive defects, in this case all caused by structural or regulatory defects in the globin genes that specify the protein components of the oxygen-carrying haemoglobin molecule.

One of the commonest genetic diseases in the caucasian population – cystic fibrosis – is also recessive. At the time of writing the basic defect in cystic fibrosis has not yet been finally identified, although there is strong biochemical evidence implicating a defect in the regulation of chloride transport across the membranes of epithelial cells (see Table 4.3). The defective CF allele has recently been located and is now in the process of being isolated. If its gene product can be identified, this should provide a clue to the exact nature of the regulatory defect.

The conditions mentioned above are due to genes located on the autosomes (i.e. any chromosome other than the X and Y sex chromosomes). Diseases caused by genes carried on the X chromosome show a different and characteristic pattern of inheritance. Recessive X-linked genetic diseases such as haemophilia A (caused by non-production of blood clotting factor VIII), Duchenne and Becker muscular dystrophies and Lesch–Nyhan syndrome (caused by a deficiency of hypoxanthine:guanine ribosylphosphotransferase or HGPRT) are transmitted through symptomless heterozygous female carriers (females carry two X chromosomes) and show up only in males who inherit the defective X chromosome (males carry only one X chromosome, inherited from the mother, and a Y chromosome inherited from the father). See Fig. 2.6 for a typical haemophilia pedigree.

In all recessive genetic diseases where the underlying cause is known, the defect results in non-production of the gene product or production of a non-functional product. The heterozygous carriers show few if any symptoms as in these cases one normal gene is sufficient to produce enough of the required protein.

Most heritable conditions are rare – phenylketonuria, which is treatable by special diet if detected early enough and for which all babies in Britain are screened at birth, has an incidence of 1 in 10 000 newborns. Cystic fibrosis, the most common serious genetic disease in caucasians, although rare in other groups, has an incidence of around 1 in 2500 births and a carrier incidence of about 1 in 20. Duchenne muscular dystrophy affects 1 in every 4000 newborn boys.

Some heritable conditions are so common in certain populations, although rare in others, that it is likely that the heterozygous state confers some positive benefit. The classic case is sickle-cell anaemia, in which the defective gene is carried by 1 in 6 of the population in West Africa and in 1 in 10 of American blacks, with the homozygous condition occurring at around 1 in 100 and 1 in 400 births, respectively. In other populations sickle-cell anaemia is very rare. The thalassaemias, anaemias caused by aberrant haemoglobin production, are rare outside the Mediterranean region but carrier rates can reach as high as 1 in 8 in these populations and immigrant communities from the region who have settled elsewhere. The persistence and high frequency of the defective sickle-cell anaemia allele in West African populations are linked to the protection the heterozygous state confers against falciparum malaria. The same explanation has been proposed for the relatively high frequency of the thalassaemia alleles in the previously malarial Mediterranean region. A similar adaptive explanation has been put forward for the relatively high frequency of the Tay–Sach's gene (up to 11 per cent heterozygote frequency) in Ashkenazi Jewish populations – in this case, a (highly speculative) link between the heterozygote state and resistance to tuberculosis has been proposed.

The genetic disease now present in the human population is the result of mutation in the more-or-less distant past. Recessive mutations can persist for generations, and even become relatively widespread if the heterozygote has an advantage. Mutation is still occurring, as the appearance of hereditary defects in hitherto unaffected families testifies. In some types of genetic disease, new mutations are believed to account for a large proportion of cases (e.g. up to a third in cystic fibrosis and haemophilia). At the DNA level the mutational changes found range from alterations in a single nucleotide, as in sickle-cell anaemia where the change of a single nucleotide in the gene for ß-globin results in the replacement of valine by glycine at a crucial position in the protein, to extensive deletions characteristic of Duchenne muscular dystrophy.

Prenatal diagnosis

There is no effective long-term treatment for many of the most serious genetic diseases, and this has placed great emphasis on the search for ways of reliably detecting carriers, and on the prenatal detection of foetuses at risk early enough to be able to offer the option of abortion.

Prenatal diagnosis of affected foetuses is now possible for many genetic diseases. In some cases the characteristic lack of a particular enzyme, or the presence of abnormal metabolites or marker proteins can be determined from a sample of amniotic fluid taken by **amniocentesis**, which first becomes possible around 15 weeks of gestation. There are other hereditary defects, however, that are not diagnosable this way, and in which the presence of defective genes can only be detected by direct examination of foetal DNA. This can be obtained either from foetal cells from the amniotic fluid or by sampling of the placenta (**chorionic villus sampling**). Where genetic probes are available it is now possible to detect whether a foetus is homozygous or heterozygous for a defective gene as early as 10 weeks by chorionic villus sampling. Amongst the conditions that can now be diagnosed in this way are the thalassaemias, Duchenne muscular dystrophy, haemophilia and, by the time this book is published, almost certainly cystic fibrosis. (Haemophilia and sickle-cell anaemia have been detectable for some time by foetal blood sampling later in pregnancy; cystic fibrosis has also become detectable by amniocentesis in recent years.)

A common method for detecting a defective allele at the DNA level is by a comparison of its **restriction pattern** with that of the normal allele. The mutational alteration in the nucleotide sequence of the defective gene usually abolishes one or more restriction enzyme sites (see Restriction enzymes; Restriction analysis). By digesting the DNA with various restriction enzymes, differences between the normal and defective form of the gene in a particular family are often immediately apparent, and homozygotes and heterozygotes for the defective gene can be distinguished. Where simple restriction analysis is not informative, very sensitive methods are now being developed that can detect even a single nucleotide change in a gene.

When the causal gene itself has not yet been identified, a reasonably reliable diagnostic test can sometimes be made by testing for the presence of an identifiable marker (usually a known restriction enzyme site) that is found only on the defective chromosome and is so near the gene itself that it is invariably inherited with it. This linkage is determined by testing as many family members in as many generations as possible, and as well as being of diagnostic use, is often the first step to isolating the gene itself – by chromosome walking and allied techniques (see below).

The development of generally informative DNA-based diagnostic tests that could eventually be used not only to help families known to be at risk but for more general screening, can be difficult. In many cases, a genetic disease can be caused by several different molecular defects in the same gene. If a test is not to be restricted to one particular family or group of families with the same underlying molecular defect, it has to be able to detect all possible defective variants. Where the normal and defective forms of the gene are well characterized and there is only a limited number of defective variants in the population, the detection rate can be very high, nearing 100 per cent. In other cases, however, such as Duchenne muscular dystrophy, where DNA-based diagnosis is still in the early stages, only 50 per cent of known defective genes can at present be detected by these tests.

However, it will eventually become possible in principle to screen the general population (or at least all prospective parents) for carriage of some of the more-common defective alleles. For conditions such as cystic fibrosis and haemophilia A, where up to a third of all new cases are believed to represent new mutations this would be the only way of detecting all those at risk of producing an affected child. Whether such large-scale screening ever becomes a practical possibility, will however depend on both technical advances and economic considerations. The availability of prenatal screening for congenital disabilities can also place undue psychological and social pressures on parents to abort a 'defective' foetus, even where the degree of handicap cannot be estimated, or where the condition can be treated, albeit expensively.

The genetic basis of non-heritable disease

Identifying the gene concerned in some rare heritable conditions may throw light on the causes of much commoner, non-heritable diseases and congenital defects that show very similar symptoms. For non-heritable Alzheimer's disease, atherosclerosis, manic-depressive illness, cleft-palate and some forms of cancer for example, there are rare 'familial' heritable forms which have been traced to a defect in a single gene (see Table 4.3). (See also Chapter 6: Receptor-mediated endocytosis, for familial hypercholesterolaemia and atherosclerosis and Chapter 11: Familial cancers; see also Chapter 12 for the link between a predisposition to certain diseases and an individual's histocompatibility antigens.)

GENE MAPPING

Genetic maps

Gene mapping is one of the oldest exercises in genetics. Long before geneticists had the least idea of what genes consisted of or what they did, they found that certain characters tended to be inherited together, in apparent violation of the mendelian law of independent segregation. Genes were found to fall into distinct **linkage groups**, each group representing a set of loci that do not behave entirely independently of each other in experimental crosses.

A linkage group represents genes carried on the same chromosome. The linear order of the genes within each linkage group can be determined by the degree of recombination that occurs between them when homologous chromosomes are paired at meiosis (see Fig. 2.3). The further apart they are, the greater the likelihood of recombination. By counting the proportion of recombinants in the progeny, the distance between the loci in arbitrary 'map units' can be calculated. The order of genes on a chromosome can be determined by making pairwise comparisons between the inheritance of different alleles at each of three linked loci. Linkage groups determined genetically from experimental crosses can be assigned to their corresponding chromosomes by **cytogenetic** studies. This is easiest in organisms such as *Drosophila*, where the location of a mutant allele can often be determined by the consequent disruption in the fine banding pattern of the polytene chromosomes (see Chapter 2: Chromosome structure). The maps obtained by recombinational analysis are called **genetic maps**, and represent the order of genes, but not necessarily their precise physical distance apart.

For prolific organisms such as the fruitfly *Drosophila*, where many gene loci have been identified by mutation, recombination frequencies and thus distance apart in map units can be calculated with some precision. Recombinational analysis in humans is less easy. Only genes that are naturally present in the population in at least two different variants (alleles) with easily distinguishable phenotypes can be mapped in this way. Even when a suitable family is found, the true frequency of recombination between pairs of loci is difficult to determine accurately because of the small number of offspring in each generation. Nevertheless, around a hundred loci have been roughly mapped by these methods over the years, mostly for genes encoding enzymes. In some cases data from many different families can be pooled to give statistically reliable results.

With the advent, first, of **somatic cell genetics**, and more recently, recombinant DNA techniques, human gene mapping has been transformed, and detailed maps of the human genome are for the first time becoming a reality. Of the 900 or so genes now assigned a place on the human genetic map, two-thirds have only been mapped in the past 5 years.

Gene maps have many uses. In general, a detailed map makes it easier to clone newly identified genes by working along from the nearest known landmark. Much of the impetus for developing a map of the human genome comes from the need to explore links between genetic constitution and disease. The maps of the human genome now in preparation will make it easier to locate and isolate genetic loci involved in disease, not only for the inherited single-gene defects, but for far-commoner conditions, such as diabetes, heart disease and cancer. These and many other common diseases are believed to have predisposing genetic components, and cancer is now known to involve alterations in specific genes in the cancer cells themselves (see Chapter 10: Oncogenes). Once 'disease genes' have been cloned they can be used as the basis for DNA-based diagnostic tests to detect those at particular risk, and to study the underlying genetic contribution to many common diseases.

Mapping genes to chromosomes

Recombinational analysis cannot by itself determine which linkage group corresponds to a particular chromosome. The X and Y chromosomes are the exceptions as some mutations in genes on these chromosomes have an easily recognizable pattern of inheritance (e.g. the transmission of the X-linked defect haemophilia through female carriers and its appearance as a disease solely in males).

Gene loci can be mapped to a chromosome through mutations which, as well as causing an obvious phenotypic effect cause visible cytogenetic defects such as a change in length and/or banding pattern of the chromosomes that carry them. Once several loci on each chromosome have been established independently in this way, other loci can be assigned by their linkage to these chromosome markers. In *Drosophila*, the existence of giant polytene chromosomes (see Chapter 2: Chromosome structure) with their distinctive banding patterns makes cytogenetic analysis a powerful mapping method. In humans (and other mammals) only relatively large chromosomal alterations can be detected and additional ways of assigning genes to chromosomes have had to be devised. These make use of **somatic cell hybrids** between human and other mammalian cells (usually those of mouse or hamster).

Somatic Cell Hybrids

Mammalian somatic cells from two different species can be induced to fuse by a variety of treatments to give a cell with a hybrid nucleus containing both sets of chromosomes. The chromosomes of the two different species can be distinguished from each other by their staining patterns. During culture, these hybrid cells tend to lose chromosomes of one or the other species. The loss of individual human chromosomes can be correlated with the loss of the cell's capacity to synthesize a particular human protein, thus assigning the gene for that protein to the lost chromosome. Originally this technique could only be used for genes whose gene product is known and can be easily assayed and distinguished from the corresponding animal protein. With the advent of recombinant DNA techniques, however, it is now possible to locate not only known protein-coding genes, but any length of DNA for which a specific probe is available.

The positions of genes relative to each other can sometimes be ascertained in somatic cell hybrids by **fragmentation mapping**. When the hybrid cells are irradiated, breaks in the chromosomes are produced. The loss of an identifiable fragment, correlated with a lost ability to produce say, a particular enzyme, locates the gene for that enzyme on the lost fragment.

In the past few years, as well as phenotypically detectable gene variants, additional markers in the form of restriction length polymorphisms (RFLPs) (see above) have been used to provide extra landmarks on genetic maps. More than 800 sites detected by RFLPs and specific nucleic acid probes have now been mapped, and the numbers are increasing rapidly. As the human gene map becomes fuller the easier it is to fit new genes into place.

In situ Hybridization

Any DNA sequence for which a specific nucleic acid probe is available can be mapped by direct *in situ* hybridization of the radioactively or chemically labelled probe to intact chromosomes in specially prepared cells. The label marks out the location at which the probe has bound to its corresponding chromosomal DNA sequence.

New techniques for gene mapping

Much of the progress since 1980 has been the result of techniques based on recombinant DNA. As more genes are cloned and sequenced, genetic maps are being gradually correlated with **physical maps** built up from ordered sequences of the DNA fragments cloned in DNA libraries.

DNA libraries consist of collections of DNA fragments representing all the DNA in a genome. There is considerable redundancy and the fragments overlap. They can be arranged in an ordered series by using one fragment as a probe to recognize the overlap and pull out the next, and so on. Each series represents the sequence of DNA along an individual chromosome, with fragments containing known genes providing the necessary genetic guideposts along the way. Such maps are at present being prepared for several species including humans. When complete they will be of great value in linking genetic loci whose mutant character and position on the chromosome, but not much else, is known, to their corresponding cloned DNAs. The actual length of a gene in base pairs and in some cases the distance from one gene to the next can then be determined from the physical map, grounding the genetic map firmly in reality. So far, around 400 mapped human genes have been cloned and their sequences determined.

Chromosome Walking

In the absence of any clues as to what a gene may specify, virtually the only way to isolate it is to first map it genetically and then work along the chromosome from the nearest identifiable site using ordered cloned fragments of DNA and sampling the DNA sequence along the way – this is called **chromosome walking.** The initial mapping often depends on chance – for example, the discovery of a patient whose disease is linked to an obvious cytogenetic defect. The recent isolation of the human DMD locus (at which defects cause Duchenne muscular dystrophy) depended on such a lucky chance.

The largest piece of DNA that can be cloned in a conventional DNA library using, for example phage lambda as vector, is around 40 000 kilobases. Chromosome walks covering hundreds of thousands of bases therefore assume heroic proportions. As such distances commonly have to be travelled over to get from a known marker to the site of the 'unknown' gene, new techniques that cover the ground quicker are being devised. **Chromosome jumping** consists of using specially prepared clones that contain two sites a long distance apart on the chromosome – ingenious tricks are used to delete the DNA in between. It has also recently become possible to clone very long DNA fragments up to 500 000 kilobases in yeast cells.

Until recently it was impossible to separate DNA corresponding to individual chromosomes. New methods of electrophoretic separation are now making this possible, and DNA libraries composed of cloned fragments of a single chromosome make the task of assigning genes to particular chromosomes much easier.

Sequencing the human genome

Ambitious projects to sequence a complete human genome are now in the assessment stage in several countries and will undoubtedly eventually go ahead. Current estimates for the US project put costs at around $200 million annually over the next 15 years for the 3000 million base pairs of the human genome. When the project was first proposed a few years ago, the timescale was calculated at around a hundred years, but the foreseeable development of fully-automated sequencing techniques now make it more feasible.

A complete human DNA sequence would make work on all aspects of human genetics much less technically onerous. It would be an invaluable aid in locating and isolating genes involved in human disease, as well as providing definitive data for basic research into gene organization, expression and evolution. Complete genome sequences for representative organisms would allow the most exhaustive comparison possible of the changes that take place at the molecular level during evolution.

However, the usefulness of such a project, given its cost, has been questioned by some biologists. Large portions of the genome are repetitive and apparently have no coding or gene regulatory function and are unlikely to repay detailed sequencing with much useful information. Regions containing genes of interest are already being rapidly sequenced by individual research groups, and even here much of the information remains uninterpretable, as the underlying biochemistry and physiology on the function of the encoded proteins *in vivo* have not yet been done. There is also the question of how representative a single genome can be, given the variation between one individual and another – whose genome is to be sequenced?

REGULATION OF RECOMBINANT DNA WORK

When recombinant DNA techniques were first developed in the USA in the early 1970s they were not unreservedly welcomed. The ability to create entirely novel combinations of genes from different organisms raised questions of possible health and environmental hazards from genetically engineered microorganisms that might escape from the laboratory. There was also disquiet about the very idea of manipulating living organisms in this way, and its possible abuse. In 1974, a group of leading American molecular biologists, including some of those who had pioneered recombinant DNA techniques, called for a temporary halt to the work to allow these speculative risks to be properly assessed. Fears over practical risks focused on the use of the bacterium *Escherichia coli* as the host for gene transfer, and whether genetically engineered *E. coli* might colonize the human gut with harmful effects, especially if genes for lethal toxins or genes with the potential to cause tumours had been introduced.

The moratorium was observed for almost 2 years within which time regulatory bodies and experimental guidelines were established by governments and major scientific funding agencies worldwide. In the USA influential regulatory bodies are the Recombinant DNA Advisory Committee (RAC) of the National Institutes of Health, the Environmental Protection Agency (EPA) and the Food and Drug Administration (FDA); in the UK, the Advisory Committee for Genetic Manipulation (ACGM) together with the Health and Safety Commission (HSC) is responsible for formulating (voluntary) guidelines, monitoring their observance, and also, in more recent years, overseeing the deliberate release of recombinant organisms to the environment.

In the early days of recombinant DNA work, a small number of possible experiments were banned. The remainder were categorized according to their possible risks as perceived at the time (scientific opinion differed considerably on this point, dealing as it did with purely speculative risks) and conditions were laid down under which each type of experiment should be carried out. These conditions specified the required degree of both **physical** and **biological containment**. Physical containment refers to the security of the laboratory and the equipment needed to protect the operator and prevent microorganisms escaping to the outside world. Biological containment deals with the microorganism used as host for the recombinant DNA and the DNA carriers (vectors) into which the foreign DNA is spliced. For some experiments specially 'enfeebled' strains of *E.coli* or other bacteria that are quite unable to live outside the laboratory or colonize the human body must be used.

As experience with recombinant DNA has accumulated, and results from the research itself have been able to dispel some of the original fears, guidelines have been relaxed for many categories of experiment. Laboratory research and commercial production of genetically engineered products from microorganisms are generally conducted under the safety conditions one would normally use for dealing with the most hazardous of the organisms involved. The regulatory bodies now have to address the new question of the release of genetically engineered microorganisms into the environment, as for example, pest control agents or live vaccines. Most scientific opinion sees nothing inherently unsafe about genetically engineered microorganisms and favours dealing with each case on its merits, applying the criteria that a naturally produced biological agent would have to meet before it could be released. For biological pest-control agents, these include non-toxicity to humans and animals, strict host specificity and inability to survive without its host.

Research on human embryos

Questions of safety aside, the new techniques raise various ethical questions, primarily arising from the ability they bring to alter the genetic make-up of plants, animals, and even humans in ways not previously possible. *In vitro* fertilization and its associated reproductive technology (embryo storage, etc.) combined with the potential of *in vitro* genetic modification mean that one day it may be possible to produce and raise a human embryo that has been genetically modified.

In the UK, the problems raised by the new reproductive techniques and research on human embryos in general were considered by the Warnock Committee, which reported in 1983. Among the committee's many recommendations was that some research should be allowed on embryos cultured in the laboratory for up to 14 days after *in vitro* fertilization. At this stage a human embryo is still a tiny ball of cells – the blastocyst – which *in vivo* would have just implanted into the uterine wall. The cells that form the embryo proper have just become differentiated from those that form the extra-embryonic membranes (chorion, amnion, etc.). Proposed legislation in the UK has made no firm recommendation on this particular issue. Members of Parliament are to be given the choice of voting either to ban all such experimentation completely, or to allow a limited range of experiments under licence on cultured embryos up to 14 days after fertilization, subject to the consent of the donors of eggs and sperm. Use beyond this time would be a criminal offence.

The limited experimentation allowed would be aimed chiefly at overcoming problems of infertility and improving diagnosis and discovering the origins of congenital defects. Whatever Parliament decides, manipulations such as combining human embryos with those of other species, introducing new genes, cloning embryos to produce identical offspring or putting a human embryo into the womb of another species (or vice versa) will not be allowed.

Rapid technical advances mean that clear guidance on what is or is not permitted is now urgently needed. Working under current voluntary guidelines, developmental biologists have recently been able to determine the stage at which transcription starts in human embryos using spare fertilized eggs from an *in vitro* fertilization anti-infertility programme. It is now also possible to determine the sex of an *in vitro* fertilized human embryo by removing a few cells from the embryo soon after fertilization for culture and analysis. Similar techniques might also be used to determine whether an *in vitro* fertilized embryo from a mother at risk was carrying a genetic defect. It is not only experimentation on early embryos that poses immediate questions to be resolved. Brain tissue from aborted fetuses has recently been transplanted into the brains of patients with Parkinson's disease, in the hope that the foetal brain cells will supply the dopamine lacking in this condition (see Chapter 13: Repair and regeneration in the nervous system).

Patenting life

Recombinant DNA has also brought in its train the question of patent rights to living organisms. Until recently it was impossible to patent any living organism, even one so highly bred that it would be unlikely to have arisen without deliberate selection. (Named varieties of conventionally bred crop plants have for some years, however, been protected to a certain extent by 'plant breeders' rights' – a form of patent.) Precedent was first broken in 1980 when a genetically manipulated microorganism was successfully patented, and in 1987 the US Patent and Trademark Office announced that they would now accept applications for 'nonnaturally occurring' genetically altered non-human multicellular living organisms – that is, animals or plants. Applicants are required to prove that the animal constitutes a manufacture or composition not found in nature. The first such patent was granted recently to Harvard University, for an 'oncomouse', a strain of transgenic mice carrying an oncogene.

Opposition to patenting of animals so far has focused largely on the moral issue, with its implications that humans can create and claim ownership for a new living thing. The question of plant breeders' rights and the possibility of being able to patent genetically engineered seeds and the means by which they have been obtained, has taken on a more political dimension. It has sharpened the long-standing grievance felt by Third World countries that although most plant genetic resources are located in the Third World, they are collected and exploited to produce new commercial varieties mainly by the developed countries.

FURTHER READING

S.S. HALL, *Invisible frontiers: the race to synthesize a human gene*, Atlantic Monthly Press, New York, 1987. A revealing and often entertaining account of the race between rival research teams to be the first to clone and express a human gene.

J.D. WATSON and J. TOOZE, *The DNA story: a documentary history of gene cloning*, W.H. Freeman, San Francisco, 1981. The course of the controversy over recombinant DNA charted in extracts from contemporary newspapers, magazines and journals.

E.L. WINNACKER, *From genes to clones: introduction to gene technology*, VCH, Basel, 1987. For students with some background in biochemistry.

A.L. JEFFREYS, J.F.Y. BROOKFIELD and R. SEMEONOFF, Positive identification of an immigration test-case using human DNA fingerprints, *Nature*, 317 (1985) 818–19.

R.L. CANN, M. STONEKING and A.C. WILSON, Mitochondrial DNA and human evolution, *Nature*, 325 (1987) 31–6. Mitochondrial 'Eve'.

F. SANGER, Sequences, sequences, sequences, *Annual Review of Biochemistry*, 57 (1986) 1–28.

J.H. DODDS (ed.) *Plant genetic engineering*, Cambridge University Press, Cambridge, 1985. Principles and techniques.

S.B. PRIMROSE, *Modern biotechnology*, Blackwell Scientific, Oxford, 1987.

Genetic engineering of plants: agricultural research opportunities and policy concerns, National Academy Press, Washington, DC, 1984. A review of prospects for plant genetic engineering by the National Research Council Board of Agriculture.

R.D. PALMITER and R.L. BRINSTER, Transgenic mice, *Cell*, 41 (1985) 343–5.

P.D. VIZE et al., Introduction of porcine growth hormone fusion gene into transgenic pigs promotes growth, *Journal of Cell Science*, 90 (1988) 295–300, and references therein.

D.J. WEATHERALL, *The new genetics and clinical practice*, 2nd edn, Oxford Medical Publications (OUP), Oxford, 1985. An authoritative and readable account of the application of recombinant DNA technology to medicine, including the diagnosis and possible treatment of genetic disease. For a wide readership.

D.J. WEATHERALL, The slow road to gene therapy, *Nature*, **331** (1988) 13–14, and references therein. A short commentary on recent advances.

K.G. BEAM, Localizing the gene product, *Nature*, **333** (1988) 798–9, and references therein. A way in to the extensive literature on the identification of the Duchenne muscular dystrophy locus and its product dystrophin.

R.A. SCHOUMACHER et al., Phosphorylation fails to activate chloride channels from cystic fibrosis airway cells, *Nature*, **330** (1987) 752–4. An entrance to the literature on the underlying biochemical defect in cystic fibrosis.

P.N. GOODFELLOW, Classical and reverse genetics, *Nature*, **326** (1986) 824–5, and references therein. A commentary on developments in the search for the cystic fibrosis gene.

N.S. WEXLER et al., Homozygotes for Huntington's disease, *Nature*, **326** (1987) 194–7, and references therein, provides an entrance to the literature on Huntington's chorea, its inheritance and the use of DNA probes to detect individuals at risk. The exchange between D.C. Watt et al. and Gusella et al., Scientific Correspondence, *Nature*, **320** (1986) 21–2 (see also p.11, same issue) illustrates the ethical dilemmas that this new technology can pose. See *Nature*, **323** (1986) 118 for a follow-up on this subject.

M. ROBERTSON, Molecular genetics of the mind, *Nature*, **325** (1987) 755, and references therein. A short commentary on the recent discovery of genes linked to manic-depressive illness.

D.L. PRICE, New perspectives on Alzheimer's disease, *Annual Reviews of Neuroscience*, 9 (1986) 489–512.

R. WHITE and J-M. LALOUEL, Chromosome mapping with DNA markers, *Scientific American*, **258** (1988) 20–8.

Science, **236** (1987) 1223–1263. Frontiers in recombinant DNA. Includes articles on diagnosis of genetic disease; enzyme engineering; and gene transfer in cereals.

S.J. O'BRIEN (ed.), *Genetic maps 1987*, 4th edn, Cold Spring Harbor Laboratory, New York, 1987. An up-to-date compilation of more than 100 linkage and restriction maps from phage ΦX174 to petunia to humans.

CHAPTER 5

THE ORIGIN OF LIFE AND ITS EVOLUTION

THE EVOLUTION OF LIVING CELLS

The earliest signs of life on Earth, in the form of microfossils of cells resembling present-day bacteria, have been found in rocks 3500 million years old. If we assume, as most scientists do, that life originated on Earth and was not seeded here from another source, the origins of these early cells must lie further back, in the thousand million or so years between the time these rocks were laid down and the time the Earth condensed from a cloud of gas and interstellar dust around 4600 million years ago.

The most generally accepted theory of the origin of life proposes that chemical reactions on the early Earth gave rise to the essential building blocks of life (e.g. purine and pyrimidine bases, amino acids and sugars), which then combined to form nucleotides, primitive nucleic acids and primitive proteins. For life to arise in this 'primeval soup' requires first of all the appearance of a molecule that can replicate, that can make new generations of molecules like itself. The only 'biological' molecules that can do this are the nucleic acids RNA and DNA through which present-day organisms pass on genetic information from generation to generation. For this reason it is assumed that the first 'living' molecules must have been nucleic acids. (An alternative theory proposes minerals as the first 'replicators', but eventually nucleic acids have to enter the picture.)

This brings us straight up against a major difficulty. Life today depends crucially on proteins, both for energy metabolism and to replicate nucleic acids. These proteins are themselves encoded by nucleic acids, creating a circle from which there seems no escape. The recent discovery that RNA can have enzymatic properties that could even include its capacity to replicate itself without protein intermediaries (see below, Ribozymes) at last provides hope of breaking the circle and has led to speculation on the existence of an **RNA world** composed of

riboorganisms that had not yet developed protein synthesis or DNA, and that preceded the DNA-containing ancestor of the present living world. Ribozymes and certain other intriguing RNA-based processes in present-day cells and viruses may be relics of this RNA world.

The RNA world

RNA has long been held to be the evolutionary precursor of DNA as genetic material. In experiments to simulate prebiotic syntheses, ribose but not deoxyribose can be synthesized from formaldehyde (the most likely prebiotic source of sugars). In present-day organisms all deoxyribonucleotides (the components of DNA) are synthesized from RNA precursors (ribonucleotides) by ribonucleoside diphosphate reductase which indicates that pathways for RNA synthesis evolved before the need for DNA. Many present-day coenzymes (e.g. nicotinamide adenine dinucleotide (NAD)) are ribonucleotide-based, as are adenosine triphosphate (ATP) and guanosine triphosphate (GTP), the universal energy currencies. They might therefore be 'fossils' of primitive nucleotide 'enzymes'. Single-stranded RNA, unlike double-helical DNA, can also fold into a variety of three-dimensional structures by base-pairing, which can provide catalytic active sites analogous to those of protein enzymes.

A popular (but obviously still highly speculative) scenario for the origin of living cells suggests that replicating molecules (in this particular scheme primitive single-stranded RNAs) arising in the primeval soup became enclosed in self-assembling lipid membranes. A primitive nucleotide-based 'metabolism' provided the energy for replication. In these early 'cells', RNA replication would not have been very accurate, providing ample room for variation and improvement. Eventually, possibly as a result of the chance appearance of primitive tRNA-like structures, these 'cells' acquired the capacity to synthesize proteins. Once they began to do this, the primitive cell could begin to resemble the cells we are familiar with today, with a genetic apparatus (in this case based on RNA), a lipid/protein membrane through which relations with the external environment could be subtly controlled, and a protein-based intermediary metabolism. This has been designated the **RNP (ribonucleoprotein) world.**

Because of proteins' greater versatility and potential as enzymes they gradually supplanted nucleotide-based enzymes and became essential to the process of RNA replication itself. Eventually, RNA was replaced by the more chemically stable double-stranded DNA as the permanent information store of the cell and the **DNA world** we live in today had arrived.

There are naturally enormous leaps of imagination required even in this attractive scheme. One gap that must be filled is the demonstration of an RNA that can copy a single-stranded RNA template. (The technicalities of enzymatic action require that in the primeval soup two RNA replicases must have arisen simultaneously each propagating the other. A single RNA, even if it could use itself as a template, could not copy its own enzymatic active site and would thus only produce a partial and inactive copy.) All known RNA replicases, encoded chiefly by RNA viruses, are proteins and it is perhaps unlikely that ribozymes with RNA replicase activity still exist in our DNA world. No ribozyme has yet been found that can catalyse its own replication although the capacity to add ribonucleotides to an already formed ribonucleotide polymer has been shown. It may be possible, however, to use the chemical clues in surviving ribozymes to construct a replicating ribozyme in the laboratory. It would not have to be very large – RNAs composed of only 50 nucleotides can have enzymatic activity.

The second problem that has to be solved is the mystery of the evolution of the genetic code and nucleic-acid directed protein synthesis. This has posed a problem for biologists ever since its discovery. There is first the logical difficulty mentioned above, that protein synthesis in present-day cells requires proteins. There is also the puzzling fact that the genetic code – the correspondence between nucleic acid codons and amino acids – appears to have no compelling basis in structural chemistry. The correspondence between a particular amino acid and a particular codon may simply be historical accident, and the fact that the genetic code is virtually universal in present-day organisms therefore strongly argues (along with much other evidence of a basic biochemical similarity) for the origin of the whole of the living world as we know it from a single ancestral cell lineage (the **progenote**) that had developed that particular code. As far as we know, and admittedly that is not far at all, there is no reason why alternative forms of the code would not have been equally workable and could not have also been present in the early RNP world. Small divergences from the present code work perfectly well in mitochondria and certain protozoans.

Why only one code survived we may never know, but clues to its origin may lie in the RNA components of the translation apparatus – the ribosomal and transfer RNAs.

Ribozymes

Biochemical history was made in 1981 with the discovery of the first **ribozyme** – an RNA with enzymatic activity. Ribozymes catalyse a variety of reactions involving ribonucleotide polymers and this has opened up a new field of RNA biochemistry and sparked off renewed debate about the origin of life and the nature of the earliest living organisms (see above). The first ribozyme to be discovered was a **self-splicing intron** in the ribosomal rRNA genes of the protozoan *Tetrahymena thermophila*.

The *Tetrahymena* intron (or intervening sequence (IVS) as it is usually called) is a sequence of 400 nucleotides that can excise itself from the purified pre-rRNA *in vitro* without any additional protein components, leaving a correctly spliced ribosomal RNA behind. All that is required to initiate the reaction is the nucleoside guanosine, or its nucleotide derivatives and the presence of a suitable cation such as Mg^{2+} (Fig. 5.1).

Self-splicing depends on the ability of a single-stranded RNA molecule to form regions of base-paired secondary structure. A short stretch of nucleotides within the intron – the 'guide' sequence – pairs with bases at the splice sites to bring them into the correct relationship for splicing. The newly formed structure binds a guanosine nucleotide and then the RNA is simultaneously cleaved at the splice sites and the exons rejoined, cutting out the intron as a linear RNA molecule with

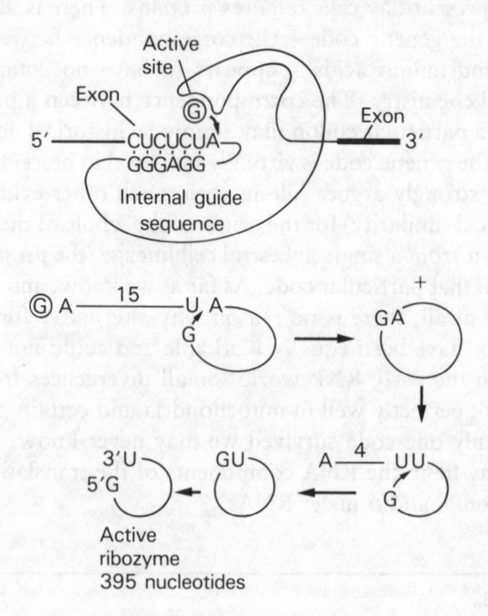

Fig. 5.1 Self-splicing of the *Tetrahymena* rRNA intervening sequence and generation of an active ribozyme. (After G. Garriga *et al. Nature*, 322 (1986) 86.)

the guanosine added at the 5′ end. Although not an enzymatic reaction in the technical sense, as there is no net use of energy, the discovery of this activity was exciting enough. But the behaviour of the excised intron proved even more fascinating.

The excised intron turns on itself like a snake swallowing its tail. The OH group on the 3′-terminal guanosine of the free intervening sequence attacks a phosphate between the fifteenth and sixteenth nucleotide from the 5′ end and forms a circular molecule, releasing a 15-base fragment from the 5′ end. If the pH is raised, the circle opens and then recircularizes, releasing a four-nucleotide fragment from the 5′ end.

The circle can be induced to reopen in mildly alkaline conditions and this truncated version of the original intervening sequence has a range of true enzymatic activity: it can add cytosine residues to an oligo(C) (Fig. 5.2), it has ribonuclease activity, phosphotransferase activity, and has acid phosphatase activity also on oligo(C) substrates.

It can also act as an 'RNA restriction enzyme', specifically cutting an RNA molecule at a site corresponding to the original 5′ splice site in the pre-rRNA from which it is derived. For this reaction it requires GTP. The specificity of the reaction is determined by base-pairing between the CUCU site (after which the ribozyme cuts) and a complementary stretch of nucleotides in the IVS, and closely resembles the first stage of the self-splicing reaction. The site-specificity of the ribozyme can be altered by substituting one base for another at the appropriate site. The potential of RNA restriction enzymes has hardly been explored yet. If they can be engineered to have a good range of specificities and to become more efficient, they may prove useful tools for analysing large RNA molecules.

Other potentially self-splicing introns have since been discovered, and other ribozymes include the bacterial enzyme **ribonuclease P**, which unusually for an enzyme contains RNA as well as protein. The RNA component has the enzymatic activity and not the protein as previously supposed. Small infectious particles of RNA found in plants – virusoids and virus satellites – are also capable of enzymatic activity *in vitro*. Certain repetitive DNA sequences in salamanders and newts are transcribed into long RNAs which are subsequently cleaved into smaller repeats. The DNA sequence shows cleavage sites resembling those in the plant virusoids so it is probable that these RNAs are also self-cleaving.

Attention is now also concentrated on the ribosome, which carries the peptidyl transferase activity that joins amino acids together during protein synthesis. The site of peptidyl synthetase activity has never been located but, for no real reason except precedent, was generally assumed

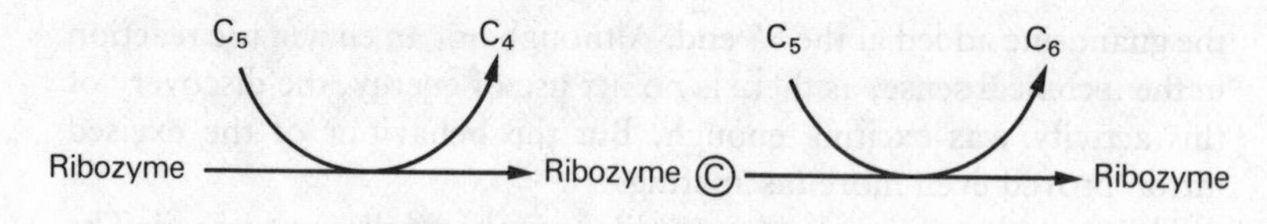

Fig. 5.2 Oligo (C) polymerase activity of the *Tetrahymena* ribozyme.

to be one of the protein components, with the ribosomal RNAs serving a purely structural role. The role of the ribosomal RNAs is now being reinvestigated. The real nature of the ribosome and the mechanism of its action is central to unravelling the mystery of how protein synthesis evolved.

Ribozymes from present-day cells represent relics of primeval enzymatic functions of RNA which have in most cases been usurped by proteins. They are, however, probably the best surviving clues to the nature of life in the first stages of its evolution.

Origins of eukaryotes

Eukaryotic cells represent a significant advance in size and internal complexity over prokaryotes and their evolution from a presumed 'bacterial'-type ancestor is one of the major events in evolutionary history. Without the structural versatility of the eukaryotic cell and its capacity to reproduce by sexual reproduction (with the consequent shuffling of genes into new combinations) it is doubtful whether multicellular organisms of any complexity could ever have evolved.

Recognizably eukaryotic cells first appear in the fossil record about 850 million years ago, and, to allow for a period of evolution, their origin is usually given as around 1500 million years ago. The characteristics that mark out eukaryotic cells from prokaryotes are their extensive internal membranes, a cytoskeleton of protein fibres and filaments, the possession of a distinct nucleus, and the compartmentalization of some aspects of energy metabolism in distinctive organelles bounded by double membranes – the mitochondria and chloroplasts (see Chapters 1, 7 and 8).

Mitochondria and chloroplasts contain their own DNA, replicate by division and show several other characteristics suspiciously like those of bacteria. They are now generally accepted to be relics of primeval bacteria that were engulfed by the eukaryotic ancestral cell. As they conferred useful functions such as aerobic respiration and photosynthetic capacity they were retained as **endosymbionts**, gradually losing their autonomy.

What that ancestral cell might have been like, and in what order and when it acquired its useful inhabitants is still much debated. The theory of **serial endosymbiosis** suggests that mitochondria were first acquired by a non-nucleated, essentially prokaryotic cell that subsequently developed (or possibly acquired) a nucleus, and then, much later, acquired chloroplasts. An opposing view is that a cell that had already developed a nucleus, internal membranes and a cytoskeleton, acquired mitochondria and chloroplasts at much the same time. Now that detailed comparative evidence from DNA and RNA sequences is becoming available and can be used to calculate evolutionary relationships more precisely some of these questions may eventually be settled.

The stimulus for the acquisition of aerobic endosymbionts may have been the rise in the oxygen content of the atmosphere, which started with the appearance of oxygen-producing photosynthesis about 200 thousand million years ago. Anaerobic cells would have found it increasingly difficult to survive. Some retreated to inhospitable and relatively anaerobic environments; others, it is suggested, acquired resident aerobic bacteria – the ancestors of mitochondria.

There is ample evidence from present-day organisms that endosymbiosis is possible. Some protozoans that lack mitochondria carry endosymbiotic aerobic bacteria instead, and similar associations between non-photosynthetic and photosynthetic cells are known.

Both mitochondria and chloroplasts show distinct evidence of an independent past. Their DNA encodes much of their own transcriptional and translational machinery, and they are self-replicating. In present-day eukaryotic cells they rely on the cell's nuclear genes to make most of their constituent proteins. Some of these nuclear genes probably derive from the original endosymbiont, having been transferred to the nucleus early in evolution.

Mitochondria are believed to originate from an ancestor of the present-day **purple photosynthetic bacteria** that had lost its capacity for photosynthesis, chloroplasts from an ancestral **cyanobacterium** (blue-green alga) or from another type of photosynthetic prokaryote, a **prochlorophyte**. These assignments are mainly based on biochemical similarities in characteristic proteins – for mitochondria much of the evidence comes from the amino acid sequence and three-dimensional structure of the cytochromes, essential components of the respiratory chains in both mitochondria and aerobic bacteria. The availability of DNA sequences of entire mitochondrial and chloroplast genomes should also aid evolutionary studies.

Mitochondria from all species are very similar in their biochemistry, suggesting that they were acquired before the plant–animal divide and that a cell (or cells) that subsequently acquired chloroplasts then gave rise to the green plants, and possibly by a separate acquisition to some branches of the algae.

Although the prokaryotic cell predates the eukaryotic cell in general evolutionary terms, the eukaryotes are of very ancient origin, and the first eukaryotic microorganisms are believed to have evolved from their prokaryote ancestors long before the majority of present-day bacteria diversified into the species we see today. The identity of the last common ancestor of prokaryotes and eukaryotes is at present a matter of much speculation (Fig. 5.3).

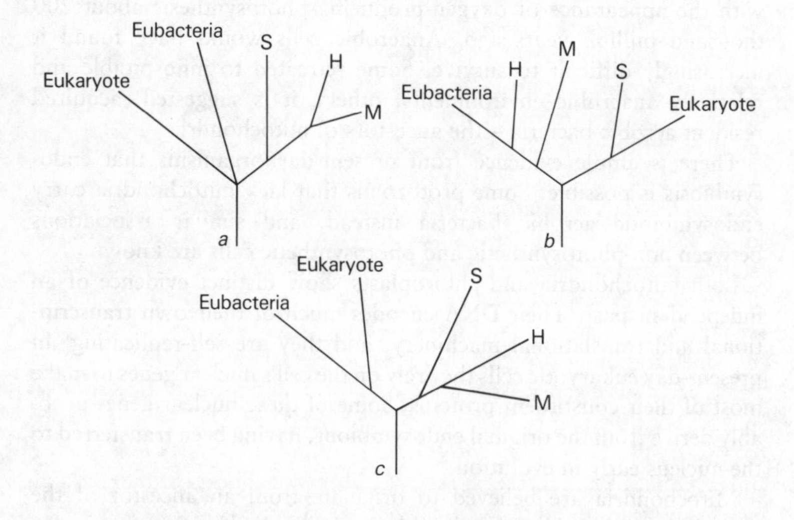

Fig. 5.3 Alternative schemes currently proposed for the relationship of the eukaryotic nucleus to the prokaryotic kingdoms. M, methanogens; H, halobacteria; S, thermophilic sulphur bacteria. *a*, The 'eukaryote–eubacteria–archaebacteria' tree; *b*, tree that proposes an ancestral thermophilic sulphur bacterium as the last common ancestor of prokaryotes and eukaryotes; *c*, derivation of the eukaryote cell nucleus from an ancestral heterotrophic bacterium. *a* and *b* are based on different interpretations of rRNA sequence differences. (*a*, C.R. Woese and G.E. Fox, *Proceedings of the National Academy of Sciences U.S.A.*, 74 (1977) 5088–5090; *b*, J.A. Lake, *Nature*, 331 (1988) 184–186; *c*, T. Cavalier-Smith, *Annals of the New York Academy of Sciences*, 503 (1987) 17–54.)

MICROEVOLUTION AND MACROEVOLUTION

The arrival of eukaryotic cells set the scene for the evolution of multicellular life, and the present living world represents only a small part of the great diversity of form and mode of life that resulted. The first fossils of multicellular animals appear just before the beginning of the Palaeozoic era, around 600 million years ago, and within a short space of time (geologically speaking) the Cambrian oceans were teeming with the ancestors of modern segmented worms, molluscs and arthropods.

Classical evolutionary biology is concerned with the questions of how, why, and in what circumstances evolutionary change in multicellular organisms occurs, and what those changes have been. The evidence of the fossil record and the study of the comparative anatomy and physiology of extant organisms provides the raw material for the study of **macroevolution**, the great changes in form, complexity of bodily organization and mode of life that have occurred over the broad sweep of evolutionary history. On a different scale are studies of **microevolution** and of speciation. Microevolution can be seen in action today as the permanent, genetically-based changes that occur within species as they differentiate into recognizably different races or subspecies living in different geographical areas, or under different climatic conditions. The study of microevolution traditionally employs the disciplines of cytogenetics, quantitative and population genetics, and ecology, supplemented in recent years by molecular studies of gene variation. There is now a third field of evolutionary study, that of **molecular evolution** – the evolution of individual genes and gene systems and the enzymes and other proteins they encode (see Chapter 3).

An account of general evolutionary history and the many questions it poses about the nature of the evolutionary process is outside the scope of this book, and readers are referred to the reading list at the end of this chapter. The remaining sections in this chapter deal mainly with some applications of new molecular biological techniques to evolutionary studies – in particular the use of DNA sequences to trace evolutionary relationships. The rest of this section briefly outlines the theory of evolution by natural selection, which in its most general interpretation is virtually universally accepted by biologists as the explanation of much (if not all) evolutionary change.

Darwinian evolution

The variety of life and the ways in which individual types of animals and plants have become adapted in very precise ways to the environments in which they live has excited the interest of natural historians for thousands of years.

The turning point in modern biological thought came with the theory of evolution put forward by Charles Darwin and, independently, by his contemporary Alfred Russel Wallace, in the middle of the nineteenth century. Darwin's book, *The origin of species by means of natural selection*, was published in 1859. It presented a mass of observation and reasoned argument supporting 'descent with modification', rather than divine creation of each individual species, as an explanation of the richness and diversity of life. Even more important, Darwin provided in the **theory of natural selection**, a coherent, and in some ways testable, explanation of how and why evolution occurred, invoking only natural biological processes.

Darwin and Wallace proposed that all species arise by descent from pre-existing ones chiefly by the agency of natural selection acting on the genetic variation normally present within a plant or animal population (see Chapter 2). Today we take this idea even further back in time than Darwin was able to, to derive the descent of all life on Earth from self-replicating molecules that were generated in the conditions present soon after the Earth was formed some 4600 million years ago.

In much simplified terms the original darwinian thesis is as follows. All organisms produce more progeny than can be supported by available resources. This leads to the 'struggle for existence' in which individuals compete with other individuals for scarce resources in the context of prevailing environmental conditions. A variant heritable trait that gives an individual a better chance of surviving to reproductive age and/or producing more surviving offspring will therefore be selected and become more common in the next generation. If it continues to be selected, the character will eventually spread throughout the population.

Selected traits will in the nature of things tend to be adaptive, fitting the organism better for life in its particular environmental niche or enabling it to survive better than its fellows in changing conditions. Having observed the ability of domestic animals to undergo considerable change and differentiation into distinct breeds under selective breeding, Darwin proposed that natural selection, acting over enormously long periods of time, could give rise to the very great morphological and functional changes that we see in the fossil record and to the differences between present-day organisms.

When the theory of natural selection was first proposed the mechanism of inheritance was still quite unknown. The rapid advances in genetics in the first half of this century led to the **neodarwinian syn-**

thesis. This was a reinterpretation of Darwin's theory in the light of the newly-discovered principles of genetics, especially of population genetics.

Darwinian theory interpreted in its most general sense is still virtually universally accepted as the best explanation so far for the sort of evolutionary change that, for example, produces organisms specifically adapted to a particular way of life or specialized environment.

From time to time the imminent demise of 'darwinism' as an explanation of evolution is announced in the popular scientific literature. But as one leading evolutionary biologist has remarked, reports of its death, like those of Mark Twain, are greatly exaggerated. There is indeed much debate at present amongst evolutionary biologists about many aspects of the evolutionary process, including the exact contribution of natural selection to driving evolution, and whether evolutionary change is always as gradual as orthodox neodarwinians have generally assumed. But darwinian evolution as a cornerstone of present-day biology is far from being discarded.

THE MOLECULAR RECORD

Running in parallel with mainstream evolutionary biology is the study of evolution at the molecular level, charting the changes in DNA and protein sequences that have occurred as lineages diverged.

Heritable (genetic) change lies at the heart of the evolutionary process, and a record of the changes that have occurred in DNA as generation succeeds generation and diversifies throughout evolutionary time is at least partially preserved in the genes of each of the millions of present-day species. Like a medieval parchment the record has been continually overwritten and some of it lost for ever, but by piecing together fragments from different sources the history of the genes we carry today can in some cases be reconstructed.

As described in Chapter 3 (Gene families and superfamilies) the ability to look directly at DNA sequences now makes it possible to trace in some detail the ways in which individual genes and gene families have evolved, and to investigate how complex physiological systems such as the immune system have developed from their primitive beginnings. One important point should be made. In molecular evolution, we are not, in general, dealing with individual genes (and their protein 'phenotypes') becoming 'better' over evolutionary time. The differences between the proteins that carry out the same basic metabolic functions in our cells and in those of bacteria are in general trivial. The message coming out of studies of gene evolution is of evolution through the generation of additional new genes by duplication and divergence of existing ones (see Chapter 3: Gene families and superfamilies for more on this).

At present we understand far more about the genes that specify and regulate cellular function and physiology than those that specify morphological structure. Molecular evolutionary studies have, therefore, been confined up to now to genes encoding common enzymes and other well-studied proteins such as the globins and cytochrome *c*. Potentially the most exciting application of molecular biology to evolutionary studies still lies in the future. The great differences between the various classes of living organism do not lie primarily in the basic biochemistry of their cells, which is constrained by the requirements of life itself. Many enzymes and other essential proteins have remained unchanged in function for thousands of millions of years. The great differences lie in the way cells have, over evolutionary time, become specialized to perform various functions and become organized into particular structures. In the foreseeable future, the rapid progress now being made in developmental biology and developmental genetics may be able to provide some clues to the genetic changes that underlie morphological evolution.

Tracing lines of descent

One practical contribution of molecular biology to general evolutionary studies so far has been to provide a means of assessing the true degree of evolutionary relatedness of one species to another, independent of sometimes subjective and disputed classification on the basis of morphological features.

Comparisons of protein and DNA sequences give, if interpreted with great care, measures of how long ago (in relative terms) the ancestors of two present-day species diverged. Over long periods of time, DNA inevitably accumulates sequence changes (see Chapter 2, p. 42) by the random process of mutation. If mutations occur at crucial sites that destroy the function of a vital gene, they are usually lost from the population. But those changes that create new and useful genes, or those that occur in 'silent' DNA not in use for coding or gene regulation, will accumulate. As mutation is random, the descendants of two diverging species will accumulate different changes, especially in DNA that is not under selective pressure, and the overall sequence of their DNA will become more and more different over evolutionary time.

The same selective pressure acting on two organisms of different lineages may, of course, produce two rather similar proteins that have evolved independently from quite different ancestral types. Convergent evolution often confuses attempts to trace true lines of descent. At the molecular level, the difference between similarity by descent and similarity by convergence as a result of selection, is sometimes difficult to determine from protein structure alone. However the nature of most eukaryotic genes, with their extensive regions of non-coding DNA, means that DNA sequences are usually more informative. If the non-coding DNA in two species is still very similar, they must be related by descent and have diverged only recently.

The difference between one species and another at the DNA level can be determined in two ways. Suitable corresponding portions of the two genomes can be sequenced and the number of nucleotides at which they differ counted directly. Alternatively the whole of one genome can be matched against the other by DNA–DNA hybridization (see Chapter 4) to get an overall measure of difference.

Along with other biochemical, cytogenetic and molecular studies these techniques have been used to tackle various outstanding questions. A few years ago, for example, cytogenetic and DNA hybridization studies confirmed earlier suggestions that the giant panda (*Ailuropoda melanoleuca*) belongs to the bear family, rather than, like the lesser panda (*Ailurus fulgens*), to the racoons. Taxonomists had been divided as to whether the giant panda was a bear, a racoon, or whether it deserved a family all to itself. DNA sequence and hybridization studies also caused a flutter in some taxonomic dovecotes when they recently placed the chimpanzees as slightly nearer relatives to humans than either of us is to our next nearest relative amongst the great apes – the gorilla (Fig. 5.4).

Non-coding DNA is generally used to give unbiased estimates of rates of sequence change, and thus the passage of evolutionary time, within fairly closely related groups of organisms. Charting evolutionary relationships over very long timescales, such as the more than a thousand million years that divide some classes of prokaryotes from each other and from eukaryotes, requires a slightly different approach. Here one has to look for DNA sequences that evolve only very slowly, as a result of very strong selective constraints. However, the DNA chosen must evolve sufficiently to provide enough change to distinguish the various lines of descent. Favoured sequences for this type of work are the rRNA genes, which have been used to trace evolutionary relationships amongst prokaryotes, including the intriguing question of what the last common ancestor shared by eukaryotes and prokaryotes might have been like (see Fig. 5.3).

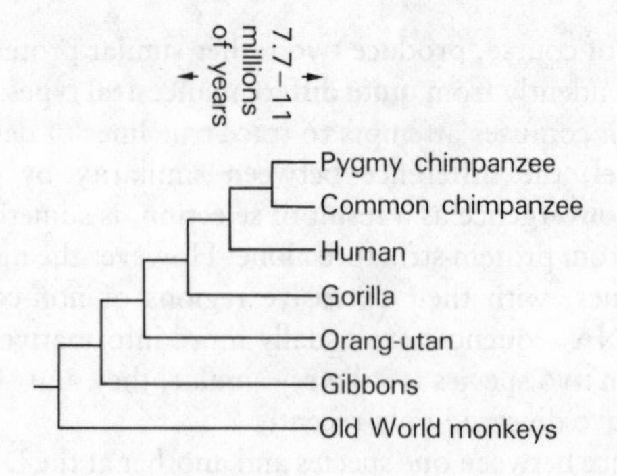

Fig. 5.4 Possible phylogeny of the apes as determined by DNA sequences and DNA–DNA hybridization studies (M.M Miyamoto *et al.*, *Science*, 238 (1987) 369–373; C.G. Sibley and J.E.Ahlquist, *Journal of Molecular Evolution*, 26 (1987) 99.)

Molecular clocks

The rate of change in nucleotide sequence can in principle be calibrated with fossil evidence to provide a 'molecular clock' counting absolute time elapsed. Molecular clocks presume that over long periods of time the number of mutations (counted as nucleotide substitutions) will be directly proportional to time elapsed (assuming that the deviations from this steady accumulation as a result of selection can be properly taken into account). However, the basic rate of change in the absence of selection turns out to differ widely from class to class for reasons that are so far unknown, but which may be due to differences in generation times, the fact that some species can repair DNA more efficiently than others, and so on. Sudden large-scale sequence changes may also be introduced by processes other than simple nucleotide substitution (see p. 72) which invalidates the basic assumption on which most molecular clocks are calculated. The hope of a universal clock that could be applied to all organisms seems to be receding.

FURTHER READING

The RNA world and gene evolution are extensively covered in Watson et al. (see general reading list). See also references in Chapter 3 for gene evolution. See general reading list for introductions to evolutionary biology.

T. Cech, RNA as an enzyme, *Scientific American*, 255 (1986) 76. A non-specialist review of ribozymes by one of their discoverers.

Evolution of catalytic function, *Cold Spring Harbor Symposium on Quantitative Biology*, **52**, 1987.

A. CAIRNS-SMITH, *Genetic takeover and the mineral origins of life*, Cambridge University Press, 1982. The theory that clay minerals were the first self-replicators.

L. MARGULIS, *Origin of eukaryotic cells*, Yale University Press, New Haven, CT, 1970. The original account of the endosymbiotic hypothesis.

D. PENNY, What was the first living cell?, *Nature*, **331** (1988) 111–12, and references therein. A short commentary on recent work using DNA sequences to trace evolutionary relationships in the prokaryotes and the problems of interpreting DNA sequences.

C. WOESE, Archaebacteria, *Scientific American*, **244**, 1981, 98–125.

C. DARWIN, *The origin of species by means of natural selection*, 6th edition, J. Murray, London, 1872.

S.J. GOULD, *Ever since Darwin*, Burnett Books/Andre Deutsch, London 1978; *The panda's thumb*, Norton, New York, 1980; *The flamingo's smile*, Norton, New York, 1985. Collections of sometimes controversial but always entertaining and thought-provoking essays on evolutionary themes, written for a wide audience, from the author's regular column in the magazine *Natural History* (published by the American Museum of Natural History).

R. DAWKINS, *The blind watchmaker*, Longman, Harlow, 1986. A defence of the role of natural selection in evolution, for a general readership.

M. KIMURA, *The neutral theory of molecular evolution*, Cambridge University Press, Cambridge, 1983. An account of the influential neutral mutation theory by its originator.

R.M. STANLEY, *Macroevolution*, W.H. Freeman, San Francisco, 1987.

C.G. SIBLEY and J.E. AHLQUIST, Reconstructing bird phylogenies by comparing DNAs, *Scientific American*, **254** (1986) 68. An introduction to the use of DNA–DNA hybridization in evolutionary biology.

PART II *The Eukaryotic Cell*

CHAPTER 6

BIOLOGICAL MEMBRANES

A fundamental property of living matter is that it actively creates within itself conditions very different from those that surround it. To do this it must isolate itself from the immediate environment by a molecular barrier, through which it can regulate chemical commerce with the environment. This barrier is the **plasma membrane** (**plasmalemma, cell membrane**) that surrounds the cytoplasm of a cell and whose integrity is essential for the cell's survival. Similar membranes also surround the internal organelles of eukaryotic cells. All biological membranes have the same basic structure, consisting of a double layer of lipid molecules in which are embedded many different proteins.

A fundamental feature of biological membranes is their selective permeability. Very few molecules can diffuse freely across them and most of the nutrients and ions a cell needs have to be transported into the cell by molecular machinery provided by proteins located in the plasma membrane. By regulating the type and activity of these transport processes, a cell is able to maintain a considerable degree of control over what enters and leaves.

In multicellular organisms, specialized cells elaborate and exploit particular components of basic membrane transport processes – as, for example, in nerve cells, where the regulated flow of ions in and out of the cell is used to generate and convey electrical signals (see Chapter 13).

The selectively permeable membranes surrounding organelles divide the interior of a eukaryotic cell into several metabolically 'sealed' compartments in which different processes can be isolated. Over the past decade or so the role of the endoplasmic reticulum and Golgi apparatus in membrane synthesis, the direction of proteins to their correct location in the cell, secretion and intracellular transport has become much clearer (see Chapter 7). Energy generation is the function of mitochondria and chloroplasts and their membranes play a crucial part in their role as powerhouses of the cell.

Transport directly across biological membranes is generally restricted to small molecules. Particulate matter, macromolecules such as proteins and large carbohydrates, and water and solutes in quantity can however, enter cells by the bulk transport process of endocytosis. In this process they are engulfed and taken into the cell in a membranous vacuole or phagosome derived from the plasma membrane, which effectively isolates them from the cytoplasm and prevents the entry of large amounts of foreign matter from upsetting the cell's delicately balanced internal environment. The functional reverse of this is exocytosis, the process by which cells secrete molecules such as antibodies, digestive enzymes, hormones and neurotransmitters.

Membranes are also important in their role as physical supports on which cellular components can be anchored and disposed in an organized way. Internal membranes provide surfaces on which enzyme assemblies can be arranged. The plasma membrane of animal cells provides anchorage points for the cell's internal cytoskeleton (see Chapter 8). The nuclear membrane is believed to provide sites at which the dispersed mass of chromosomal material can be attached in an orderly way as it is being transcribed and replicated between cell divisions.

As well as its role in regulating the selective entry and exit of material, the plasma membrane is also the distinctive face a cell presents to other cells. It bears the molecules by which one cell recognizes another and is the point at which many chemical signals such as hormones, neurotransmitters and growth factors are first intercepted (see Chapter 9). All cells, from the simplest free-living bacterium to highly specialized sensory cells such as the photoreceptors of the vertebrate eye, monitor their environment and receive physical and chemical stimuli through receptors – proteins located at the cell surface and forming an integral part of the membrane. How signals received at these receptors trigger the appropriate response by the cell is one of the foremost questions occupying biologists in many different fields from immunology to neurobiology. In recent years, several new links in the molecular chains connecting receptor stimulation and cell responses have been discovered.

Defects or alterations in receptors or in the subsequent links in the chain can have serious consequences. Such disturbances have been implicated, for example, in the development of cancer (see Chapter 11), in some forms of diabetes, and in atherosclerosis (see below, Receptor-mediated endocytosis). On the other hand, many useful drugs, from muscle relaxants to tranquillizers, act either by mimicking or blocking the action of the body's natural chemical signal molecules at particular receptors. An increasing understanding of the structure of receptors and the way drugs interact with them holds out the hope that it may be possible to design new drugs on a more rational basis to have the desired effects without unpleasant or addictive side-effects.

Although the physiological role of the cell membrane in the selective uptake of ions and nutrients has long been recognized, a full explanation of these properties in terms of its structure has only been achieved within the past 20 years. One obstacle to investigating the relationship between membrane structure and function is that biological membranes display their characteristic properties fully only while they remain intact and *in situ*.

Many technical approaches have been brought to bear on the problem – electron microscopy; novel biochemical techniques for isolating membrane proteins; immunological and other methods of identifying membrane components *in situ* by labelling them with specific antibodies and other proteins specific for cell-surface components; observations of the physical and biological properties of isolated membranes and artificial membranes; spectroscopic analysis of the electron transport chains in mitochondrial and chloroplast membranes; sophisticated electrical recording techniques to measure ion flow across membranes; and most recently, the use of recombinant DNA technology for isolating the genes for membrane proteins so that they can be produced in quantity for molecular studies.

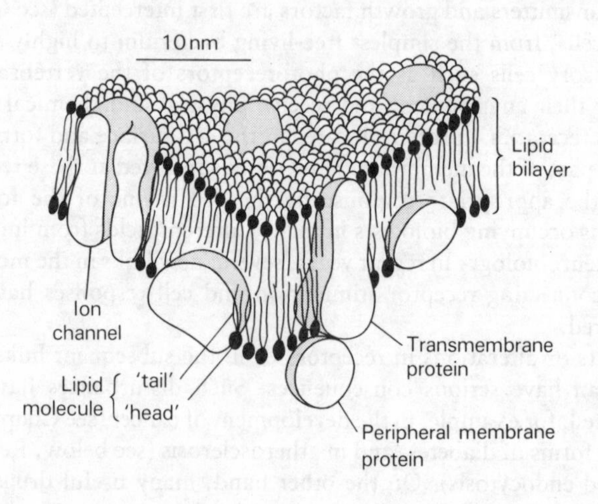

Fig. 6.1 The fluid mosaic model of membrane structure. (After S.J. Singer and G. L. Nicholson, *Science*, 175 (1972) 723).

There are few cellular processes in which membranes do not play a part. Advances in understanding membrane structure and function have been essential to the progress of cell biology towards the point at which knowledge about cell structure and biochemistry gained by many different approaches can be integrated to provide an all-round picture of cellular organization and function.

MEMBRANE STRUCTURE

All biological membranes have the same basic structure, whether they come from a bacterium, fungal, plant or animal cell. Membranes are composed chiefly of *lipids* (a class of water-insoluble materials exemplified by fats and oils) and *proteins*. Most membranes also contain appreciable amounts of *carbohydrate*, in the form of sugar side chains attached to proteins (glycoproteins) and lipids (glycolipids). The lipids form a double layer in which are embedded many different kinds of proteins (see Fig. 6.1).

Membrane lipids

The chief membrane lipids are **phospholipids** – relatively small molecules consisting of fatty acids linked to glycerol or sphingosine, which in turn are linked through a phosphate group to a headgroup such as serine, choline, ethanolamine, or the sugar alcohol inositol (see Table 6.1). The fatty acid portions, which consist of long hydrocarbon chains, form **hydrophobic** (water-repelling, 'insoluble') 'tails'; the serine, choline, etc. headgroups are polar and **hydrophilic** (water-attracting, 'soluble'). Lipids with this type of structure naturally orient themselves in an aqueous environment as a continuous closed **bilayer** with the hydrophobic groups to the centre (but see Table 6.1 for examples of 'non-bilayer-forming' lipids, which are also found in membranes in relatively large amounts).

The lipid bilayer

The lipid bilayer is the basis of all biological membranes. In itself it forms an impermeable barrier to many of the molecules and ions that cells need. The centre of the bilayer is intensely **hydrophobic**, forming a barrier to the passage of ions and most water-soluble molecules. The bilayer is permeable to lipid-soluble molecules and to small uncharged molecules such as oxygen and nitrogen and to some other small molecules such as carbon dioxide and urea. It is impassable by most water-

Table 6.1 Some lipids found in cell membranes

Phosphoglycerides

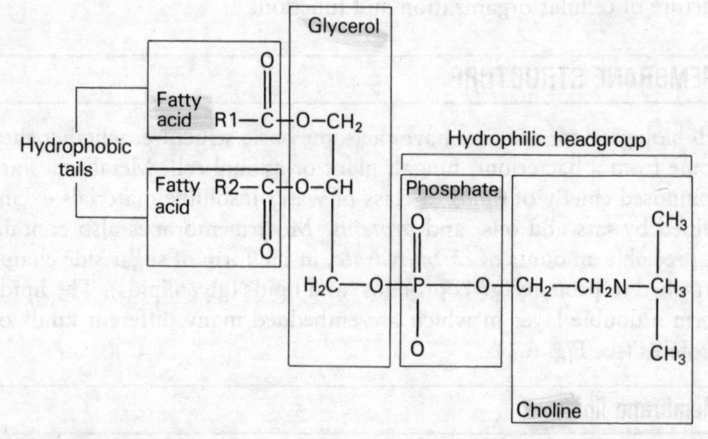

Phosphatidylcholine (lecithin)

R is a fatty acid, typically between 14 and 24 carbon atoms long. It may be saturated or unsaturated. The 16– and 18–carbon fatty acids are most common, e.g. palmitic acid and oleic acid.

Palmitate (ionized form of palmitic acid)

$$\mathrm{H_3C-CH_2-CH_2-CH_2-CH_2-CH_2-CH_2-CH_2-CH_2-CH_2-CH_2-CH_2-CH_2-CH_2-CH_2-C(=O)O^-}$$

Oleate

$$\mathrm{H_3C-CH_2-CH_2-CH_2-CH_2-CH_2-CH_2-CH_2-CH=CH-CH_2-CH_2-CH_2-CH_2-CH_2-CH_2-CH_2-C(=O)O^-}$$

Table 6.1 contd.

Phosphatidylserine	Fatty acid, Fatty acid — Glycerol-Phosphate-Serine
Phosphatidylethanolamine	Fatty acid, Fatty acid — Glycerol-Phosphate-Ethanolamine
Phosphatidylinositol (also sometimes classed as a glycolipid)	$R1-C(=O)-O-CH_2$ / $R2-C(=O)-O-CH$ / $H_2C-O-P(=O)(O^-)-O-$ inositol (OH, OH, H, H, H, H, H, HO, OH, OH, H)
Diphosphatidylglycerol	Fatty acid, Fatty acid — Glycerol—Ⓟ—$CH_2-C(H)(OH)-CH_2$—Ⓟ—Glycerol — Fatty acid, Fatty acid

Table 6.1 contd.

Sphingolipids

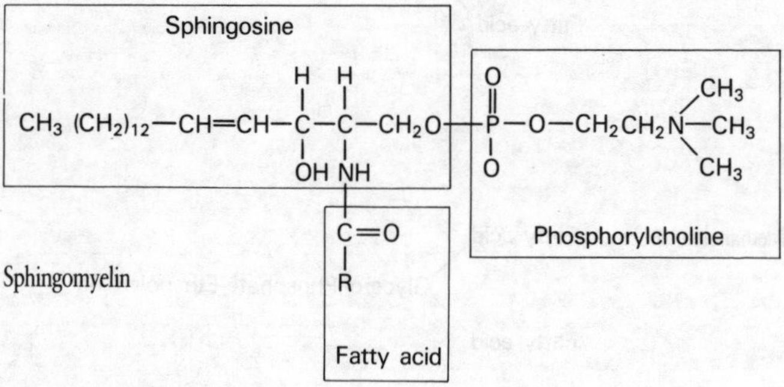

Sphingomyelin

Sphingosine: $CH_3(CH_2)_{12}-CH{=}CH-CH(OH)-CH(NH-)-CH_2O-$

Fatty acid: $-C(=O)-R$ (attached to NH)

Phosphorylcholine: $-P(=O)(O^-)-O-CH_2CH_2N(CH_3)_3$

Glycolipids

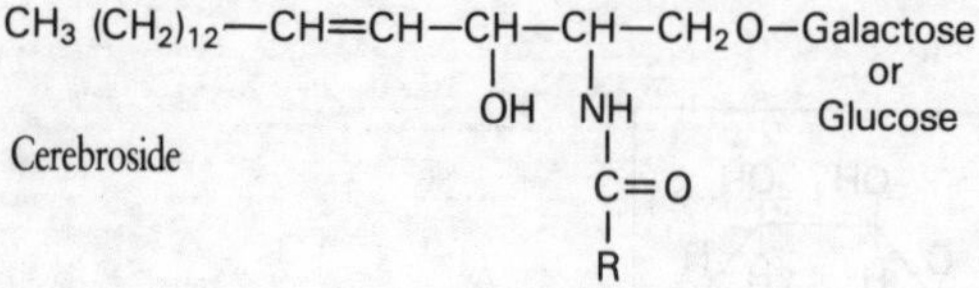

Cerebroside

$CH_3(CH_2)_{12}-CH{=}CH-CH(OH)-CH(NH-C(=O)-R)-CH_2O-$Galactose or Glucose

G_{M_1} Ganglioside

N-acyl-sphingosine—Glycerol—Glc—Gal—Gal NAc (NAN attached to Gal; Gal attached to Gal NAc)

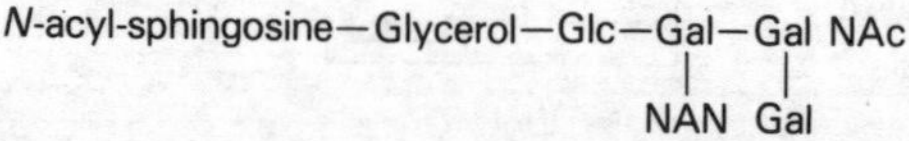

Glc, glucose; Gal, galactose; GalNAc, *N*-acetylgalactosamine; NAN, *N*–acetyl neuraminic acid

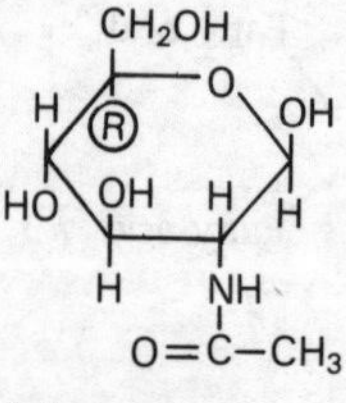

N-acetylgalactosamine (ring substituents: CH_2OH, OH, HO, OH, H, $NH-C(=O)-CH_3$, Ⓡ)

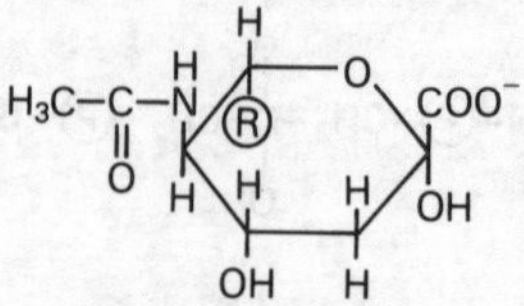

N-acetylneuraminic acid (ring substituents: $H_3C-C(=O)-NH$, COO^-, OH, OH, H, Ⓡ)

R, site of linkage to other sugars

Table 6.1 contd.

Sterols

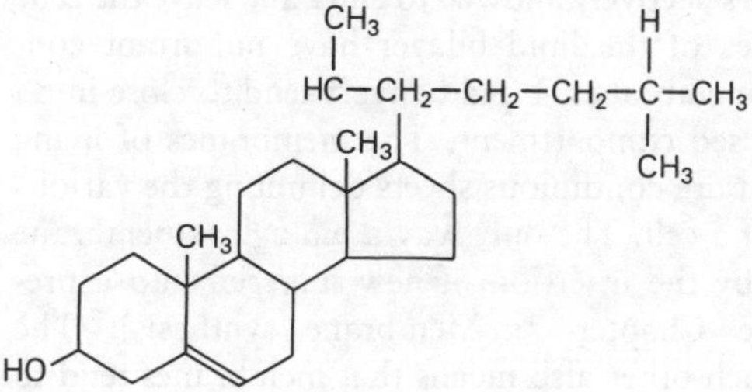

CH_3 H
$HC-CH_2-CH_2-CH_2-C-CH_3$
CH_3 CH_3
CH_3
HO

Cholesterol

More than 100 different lipids have been found in biological membranes. The most widely distributed membrane phospholipids are the **phosphoglycerides** and the **sphingolipid sphingomyelin**. Much research interest is focused on the inositol phospholipids at present as they are involved in the transduction of signals received at surface receptors for some hormones, neurotransmitters and growth factors (see Chapter 9: Second messengers) in cells as diverse as nerve, liver and the B cells of the immune system.

Glycolipids contain carbohydrate in the form of sugar residues. In one sort sugars, rather than a headgroup like phosphorylcholine, are linked to a sphingosine group. This type is found in many cell membranes but is particularly associated with the cells of the nervous system – hence their common names **cerebrosides** and **gangliosides**.

Animal membranes contain appreciable amounts of cholesterol; plants, algae and fungi contain other sterols. Cholesterol content contributes to membrane fluidity, larger amounts of cholesterol decreasing fluidity. The amount of unsaturated fatty acids and their degree of unsaturation also influences membrane fluidity, unsaturated fatty acids increasing fluidity.

The gangliosides, along with many of the other lipids found in biological membranes, do not form bilayers in pure preparations. This is a result of their 'conical' shape, compared with the bilayer-forming lipids which are roughly cylindrical. Polyprenols and lipids such as monogalactosyldiglyceride from plant chloroplasts are also of this type. They may be important in forming contact points between membranes, inducing curvature, and packing membrane proteins where they cross the bilayer at an angle.

soluble organic molecules, which include the essential nutrients a cell needs, and to ions of any sort. Somewhat surprisingly water itself seems to pass through the lipid bilayer fairly freely.

The lipid bilayer of biological membranes is, however, provided with many molecular channels, gates and pumps, formed by some of the protein components of the membrane, through which water-soluble molecules and ions can be selectively allowed to enter and leave the cell.

The physical properties of the lipid bilayer have important consequences for biological membranes. Lipid bilayers tend to close in on themselves to form a closed compartment. The membranes of living cells never have 'ends' but are continuous sheets delimiting the various internal compartments of a cell. The only way a biological membrane can grow, therefore, is by the insertion of new material into a pre-existing membrane (see Chapter 7: Membrane synthesis). The attraction of lipids for each other also means that membranes tend to be self-sealing if disturbed (within limits).

The fluid mosaic model of membrane structure

The lipid bilayer as the basis of membrane structure was first proposed more than 60 years ago. The positioning of the protein component, first introduced in Davson and Danielli's classic 1934 model, proved more intractable. Early models of membrane structure positioned the protein in ordered layers on either face, but this type of structure was not particularly successful in explaining the biological behaviour of membranes.

Figure 6.1 illustrates the generally accepted view of the organization of a biological membrane. Called the **fluid mosaic model**, it describes a lipid bilayer in which float proteins, some spanning the width of the membrane, some attached only to the interior or exterior face. This type of organization provides a sound structural basis for known transport properties of the membrane: the proteins spanning the membrane provide hydrophilic sites isolated from the hydrophobic core of the lipid bilayer through which ions and small organic molecules can either passively diffuse or be actively transported.

The fluid mosaic model of membrane structure proposed in the early 1970s is now supported by a large body of experimental evidence and its predicted properties and behaviour accommodate observed biological phenomena very satisfactorily.

The *fluid* part of the description refers to the fact that the membrane is not a rigid, static structure but a flexible assembly held together by many non-covalent interactions between lipid and lipid and protein and lipid. Both proteins and lipids can move 'sideways' in the plane of the membrane. In general, lipids move freely, as do many proteins. The observed movement of proteins on the surface of cells was an important stimulus to the development of the fluid mosaic model. Some proteins, however, are relatively immobile, and are believed to be anchored through the membrane to the cell's cytoskeleton that underlies the membrane in many places.

Membrane asymmetry

Movement is, however, limited to the plane of the membrane. Once proteins have become inserted in a membrane they cannot change their orientation, and the lipids of one half (or leaflet) of the bilayer do not exchange with the other, except by a very slow flip-over of individual molecules.

Biological membranes therefore possess 'sidedness', a physical and biochemical difference between their two faces that is determined at the time of their formation and strictly maintained thereafter. This asymmetry is of crucial importance, it defines the inside of a cell or organelle with respect to the outside and ensures, for example, that transport processes go only in the required direction.

Membrane fusion

The fluidity of biological membranes allows them to fuse with each other, a process of paramount importance in many biological processes, including the fusion of male and female gametes to form a zygote, the transport of material between organelles, delivery of new membrane to the plasma membrane, and the secretion of protein hormones, enzymes, etc. The process has been subverted by some viruses which enter cells by fusion of their lipid envelope (derived in the first place from the plasma membrane of an infected cell) with the plasma membrane. Membrane fusion is an orderly process that always preserves the topological relations of 'inside' and 'out' (Fig. 6.2).

Membrane proteins

The lipid composition of membranes is fairly similar from cell to cell although the proportion of the different lipids in different membranes varies. The type and number of membrane proteins, however, differ considerably between cell types, and between the various membranes within the cell. The protein component of the membrane determines

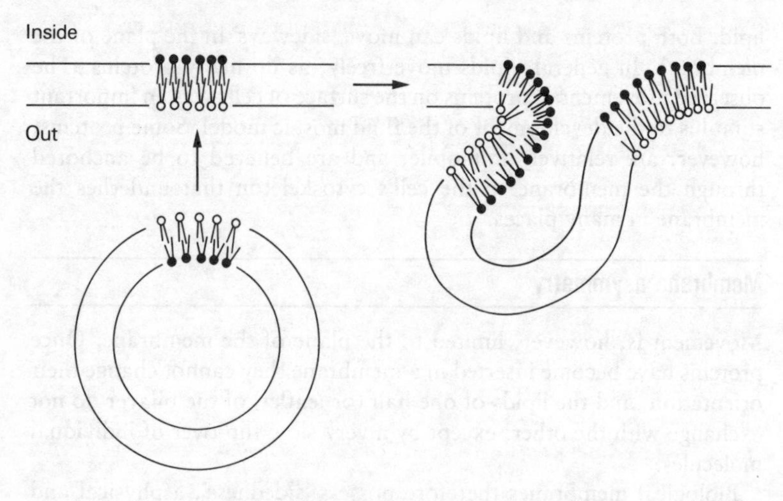

Fig. 6.2 Membrane fusion showing the preservation of membrane asymmetry.

the distinctive biological properties of the membrane and, ultimately, the behaviour of the cell or organelle it encloses.

Peripheral membrane proteins are those that can be removed from an isolated membrane simply by altering the pH or ionic strength of the surrounding medium. They are only loosely attached to the membrane, probably by weak bonds to integral membrane proteins.

Proteins that are firmly embedded in the membrane are called **integral membrane proteins**. They can only be removed by disrupting the membrane with detergent or organic solvents. The technique of **SDS-gel electrophoresis** – the electrophoretic separation of membrane proteins in the presence of the detergent sodium dodecyl sulphate – has now made it possible to isolate integral membrane proteins.

Some integral membrane proteins span the membrane entirely (**transmembrane proteins**), some are believed to be embedded mostly in either the cytoplasmic or external lipid layer (see Fig. 6.1). Membrane proteins involved in the passage of ions are generally known as **ion channels** and **ion pumps**, those involved in transporting sugars, amino acids and other metabolites to which the membrane is not intrinsically permeable are usually called **transport proteins**. They include a wide variety of specific **permeases** that transport sugars such as glucose, lactose and sucrose, and amino acids, a particular permease being specific for one or a few types of molecule.

Ion channels, ion pumps and transport proteins are found in varying types and amounts in all biological membranes, internal and external.

In addition, membranes contain many other proteins that are specific to different organelles or to the plasma membrane. The plasma membrane in particular, as the external face of the cell, contains receptor proteins for (depending on the type of cell) hormones and growth factors, antigens, neurotransmitters etc., 'recognition' proteins such as the histocompatibility antigens (see Chapter 12: The major histocompatibility antigens), and others involved in cell movement, cell adhesion, and many more specialized functions.

The membranes of mitochondria and chloroplasts are intimately involved in energy generation (see Chapter 7: Chemiosmotic theory). The mitochondrial inner membrane contains the biochemical machinery for converting the energy obtained from the controlled breakdown of glucose during cell respiration into the universal chemical energy currency of the cell – adenosine triphosphate (ATP). Chloroplasts and the cell membranes of photosynthetic bacteria contain elaborate protein complexes for harvesting the light energy trapped by photosensitive pigments (chlorophyll in green plants, bacteriochlorophyll in bacteria) and converting it into stored energy in the form of ATP or using it to drive other processes.

Because it is at present impossible to obtain high-resolution atomic structures of membrane proteins *in situ*, the three-dimensional arrangement of an integral protein in a membrane is at present largely deduced from its amino acid sequence, supplemented by a knowledge of its physiological function, and, where possible, by direct visual evidence gleaned from electron micrographs using such modern methods as cryo-electron microscopy (at very low temperatures). In regularly patterned structures, such as the outer coats of some viruses or the near-crystalline purple membrane of *Halobacterium*, the shape of the protein molecules and their arrangement in relation to each other can be retrieved in some detail from electron micrographs by modern image enhancement techniques.

The amino acid sequence of a membrane protein can be interpreted in the light of structural features common to all membrane proteins (Fig. 6.3). They all have certain peculiarities of structure that allow them to remain firmly anchored in the hydrophobic environment of the membrane – namely, one or more hydrophobic regions that are presumed to be the parts embedded in the membrane. In contrast, hydrophilic regions composed mostly of amino acids with polar or charged side groups are presumed to protrude from either the cytoplasmic or external face of the membrane. Some proteins are anchored in the membrane by covalent bonds to membrane lipids. Many membrane proteins are glycoproteins, and the hydrophilic carbohydrate side

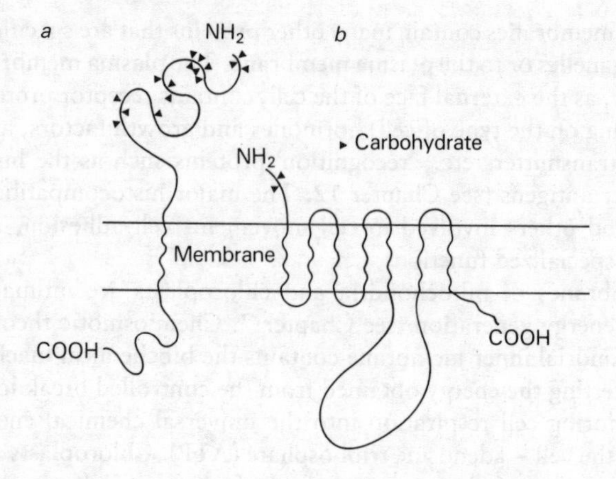

Fig. 6.3 Membrane proteins: *a*, glycophorin, an abundant protein of red cell membranes, length 131 amino acids; *b*, the ß-adrenergic receptor, the receptor for adrenaline found on several types of cell, 418 amino acids long.

chains invariably are found on the portion or portions of the protein protruding from the outer face of the membrane. This is now known to be the inevitable consequence of the way proteins are originally inserted into membranes at the time of their synthesis (see Chapter 7: Membrane synthesis).

Liposomes

Many phospholipids and glycolipids spontaneously form a bilayer in an aqueous solution. In certain conditions, this self-seals to form **liposomes** – small hollow spheres of lipid 'membrane' (see e.g. Fig. 6.2) that have proved of use as artificial model membranes in experimental studies.

As well as their usefulness in membrane research liposomes have also been tested medically as a way of delivering therapeutic drugs more efficiently to target cells. Non-lipid soluble material can be packaged into liposomes during their formation. Liposomes coalesce with cells' plasma membranes, delivering their cargo directly into the interior of the cell. This might overcome the loss of drug by degradation in the circulation before it reaches its target and also the problem of drugs that cannot by themselves gain entry to a cell. By incorporating appropriate recognition proteins into the liposome, it should be possible to target drugs more accurately to the sites where they are required.

MEMBRANE TRANSPORT

Cells need to take up from their immediate surroundings small organic molecules such as sugars and amino acids for food and fuel and to get rid of waste products. They must keep their internal pH (acidity/alkalinity), osmotic environment and ionic balance (e.g. of the ions Na^+, K^+, Ca^{2+}, Cl^-, etc.) within the limits required for maintaining the integrity and function of the delicate biochemical machinery on which the life of a cell depends. All these processes are regulated by the plasma membrane.

The lipid bilayer of a biological membrane is intrinsically freely permeable only to lipid-soluble molecules, O_2 and N_2, and small uncharged polar molecules such as carbon dioxide, urea and glycerol. The lipid bilayer is, however, provided with many molecular channels, gates and pumps, formed by some of the transmembrane protein components of the membrane, through which water-soluble molecules and ions can be selectively transported in and out of the cell.

There are a variety of transport mechanisms in biological membranes (Fig. 6.4). In general, transport is of two broad types. In **passive transport** or **facilitated diffusion** a solute or ion is conveyed across the membrane without the expenditure of metabolic energy. Passage of ions through ion channels or the uptake of nutrients by binding to certain carrier proteins is the most common form of passive transport. One example is the transport of glucose out of the side and basal membranes of epithelial cells lining the gut and into tissue fluids,

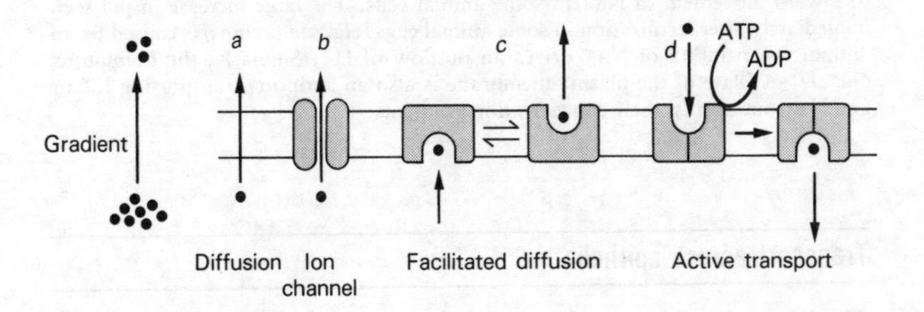

Fig. 6.4 Transport processes in biological membranes. *a*, diffusion; *b*, channel proteins allow the bidirectional flow of ions down their concentration gradients; *c*, specific carrier proteins bind small molecules such as sugars and amino acids and convey them across the membrane as a result of a change in the protein's conformation; *d*, active transport can move a solute against its concentration gradient.

mediated by a glucose carrier protein. Uptake of glucose from the gut itself, however, is an active process (see later).

Passive transport can convey material only down a concentration gradient (or for molecules carrying an electrical charge, an electrochemical gradient). To transport ions and solutes across the membrane against their concentration or electrochemical gradients, **active transport** processes are needed. These are powered by metabolic energy provided either directly by the hydrolysis of ATP, or indirectly by coupling such transport to the movement of ions down the electrochemical gradient that is permanently maintained across the plasma membrane. Coupled transport processes are either **symports**, in which both solutes are transported in the same direction, or **antiports**, in which the two solutes are transported in opposite directions (Fig. 6.5).

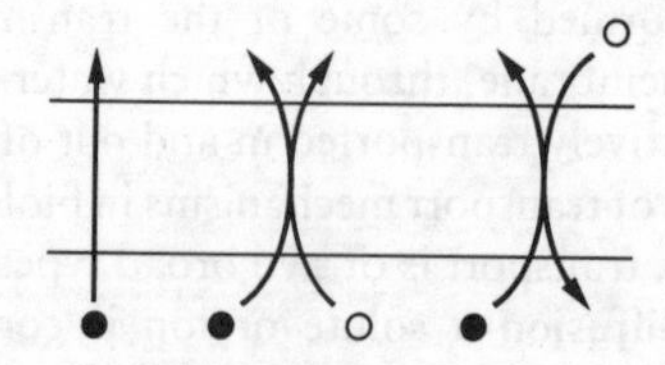

Fig. 6.5 Uniports, symports and antiports. An example of a uniport is the Ca^{2+}-ATPase of the membrane of the endoplasmic reticulum, where movement of calcium into the endoplasmic reticulum is linked to the energy-generating hydrolysis of ATP. Examples of symports are sugar uptake linked to inward movement of H^+ in bacteria (e.g. via permeases), and the uptake of some sugars and amino acids linked to inward movement of Na^+ in some animal cells. The large increase in pH seen immediately after fertilization in some animal eggs (e.g. sea urchins) is caused by an antiport – an influx of Na^+ drives an outflow of H^+. Similarly, the ubiquitous Na^+,K^+-ATPase of the plasma membrane is also an antiport, transporting K^+ in and Na^+ out against their concentration gradients.

Transport across epithelia

The internal and external cavities of the body are lined with **epithelia** (sing. **epithelium**). These are sheets of cells that effectively seal off the internal environment of tissues and organs from the exterior. Epithelial cells are joined together by **tight junctions** that form a complete seal, preventing ions or molecules, however small, penetrating between the cells. Anything entering or leaving the body therefore has to be transported across the membranes of the epithelial cell, which form a barrier that serves much the same purpose in the body as a whole as the plasma membrane does in the cell.

The way in which different transport processes localized to different regions of the cell membrane enable a cell to carry out a physiological task is illustrated by the epithelium lining the gut. Gut epithelial cells have a highly asymmetrical (*polarized*) architecture that reflects their function (see Fig. 6.6). They take up material from the apical face and deliver it into the circulation via their basal and lateral faces. The different faces of the epithelial cell represent different **membrane domains** to which particular transport proteins are restricted. The apical face is separated from the rest of the membrane by the tight junctions. Here the membranes of two adjoining cells are so closely apposed that the seal they form is a barrier to diffusion of proteins from apical to lateral and basal faces.

Nutrients such as glucose and amino acids – the products of digestion in the lumen of the gut – are taken up against a concentration gradient by active transport through the apical membrane of the epithelial cell, which faces into the gut and is specialized for this purpose with a highly infolded surface offering a large area for absorption. Uptake of glucose

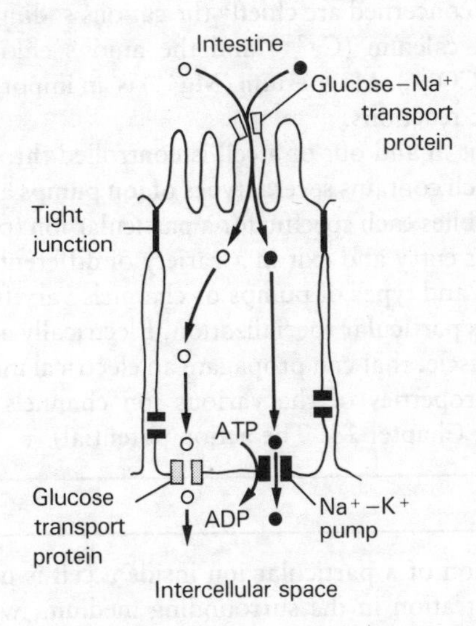

Fig. 6.6 Polarized architecture of a gut epithelial cell and the transport of glucose and Na^+. Three different transport proteins localized to different regions of the cell membrane mediate the transport of glucose across the epithelial lining of the gut.

and amino acids across the apical membranes is mediated by sodium symports. The carrier protein for glucose in the apical membrane has two separate binding sites, one for Na^+ and the other for glucose. Sodium tends to enter the cell down its concentration gradient and as it does so drags glucose in with it.

Glucose is released into the circulation on the other side of the epithelium by facilitated diffusion down its concentration gradient through the lateral and basal membranes of the cell.

ION MOVEMENT ACROSS MEMBRANES

Ions play an important part in the life of a cell in their various roles as cofactors for metabolic processes, determinants of the cell's internal osmotic environment, and as carriers of electrical charge. Hydrogen ion (proton, H^+) concentration determines the cell's internal pH, and its flow across certain membranes also drives the synthesis of ATP. Maintaining an internal environment of suitable ionic composition, pH and electrical neutrality overall means a continual flow of ions in and out of the cell, as the concentration of a particular ion inside a cell is usually very different from its concentration in the surrounding medium. The ions concerned are chiefly the cations sodium (Na^+), potassium (K^+) and calcium (Ca^{2+}) and the anions chloride (Cl^-) and bicarbonate (HCO_3^-). Magnesium (Mg^{2+}) is an important cofactor of many metabolic reactions.

Ion movement in and out of a cell is controlled through the plasma membrane, which contains several types of **ion pumps** and **ion channels** – protein assemblies each specific for a particular ion (or ions) and that regulate an ion's entry and exit in a variety of different ways.

The numbers and types of pumps or channels vary from cell to cell, depending on its particular specialization. Electrically active cells, such as nerve and muscle, that can propagate an electrical impulse, do so by virtue of the properties of the various ion channels studding their membranes (see Chapter 13: The action potential).

Homeostasis

The concentration of a particular ion inside a cell is usually different from its concentration in the surrounding medium, whether it is the intercellular fluid of the animal body, or the watery environment of free-living microorganisms. All living cells contain a much higher concentration of potassium ions (K^+) than their surroundings, and a much lower concentration of sodium ions (Na^+). Animal cells maintain a K^+ content of up to 40 times greater and a Na^+ content of up to 15 times lower than that of the fluid that bathes them. They need the potassium for essential metabolic processes; sodium, on the other hand, inhibits many of these processes. The inside of a cell must be electrically neutral. Therefore many cells also maintain a Cl^- content much lower than the external medium, because they have to get rid of excess negative electrical charge – many of their fixed constituents that are too large to leave the cell (e.g. nucleic acids, proteins, and especially the phosphate groups and carboxyl groups they carry) are negatively charged. Cells also maintain a very low internal concentration of free Ca^{2+} ions compared with the external environment. Changes in internal Ca^{2+} concentration trigger many cellular responses to external stimuli in animal cells (see Chapter 9: Calcium; and Chapter 13: Synaptic transmission).

Electrochemical gradients across membranes

Because ions carry an electrical charge their transport across membranes sets up not only a concentration gradient but also an electrical gradient. The combination is known as the **electrochemical gradient.** For example, a positive ion will not flow up an overall positive electrical gradient into a cell even if its individual concentration is higher outside than in. Ion gradients across the plasma membrane are the biochemical basis of the membrane potential (see below) because they set up a difference in electrical charge between the inner and outer faces of the membrane.

The energy stored in ionic gradients is used among other things to make ATP, to drive the transport of, for example, sugars, amino acids and other metabolites into the cell and into organelles, and to convey electrical signals.

Ion pumps

The inevitable diffusion of ions down their concentration gradient (i.e. from a region of high to a region of low concentration) and the 'leakiness' of the cell membrane to ions results, for example, in a continual slow leakage of K^+ out of and Na^+ into the cell. The cell maintains the ionic balance it requires by actively pumping ions in and out against their respective concentration gradients. This is achieved by ion pumps which move ions across the membrane by active transport. The energy required is generally derived from splitting the 'energy-rich' compound ATP by an enzyme, ATPase, which is an integral part of the pump. Ion pumps are specific for particular ions, and the best characterized is the **sodium–potassium pump** or **sodium pump** of the animal cell plasma membrane, also known as **Na^+–K^+-ATPase**. This simultaneously moves sodium out of and potassium into the cell against their concentration gradients and in doing so helps to generate the membrane potential.

The sodium pump is believed to consist of a pair of identical catalytic subunits and a pair of transmembrane glycoproteins of unknown function. The ATPase activity is located on the cytoplasmic face of the catalytic subunit. For every molecule of ATP hydrolysed, three Na^+ are transported out of the cell and two K^+ are transported inwards. Binding of Na^+ and ATP hydrolysis is believed to result in phosphorylation of the catalytic subunit and a subsequent change in its conformation which releases the Na^+ to the outside. Binding of external K^+ leads to dephosphorylation and the protein returns to its original state, releasing K^+ to the cytoplasmic side of the membrane.

A typical resting animal cell devotes a third of the total energy it produces to fuelling its sodium–potassium pumps, a measure of their importance to the cellular economy.

Calcium pumps are also important in cellular physiology. The **Ca-ATPases** located in the membranes of the endoplasmic reticulum pump Ca^{2+} out of the cytosol into the organelle, keeping cytoplasmic free Ca^{2+} low, and building up an internal store of Ca^{2+} that can be released when required.

Ion pumps in internal membranes of the cell provide the means by which organelles such as mitochondria, chloroplasts, endoplasmic reticulum and the central vacuole (in plant cells) can maintain an internal ionic environment different from that of the surrounding cytoplasm. This is vital for many of their functions (see, e.g., Chapter 7: Chemiosmotic theory).

Membrane potential

The ion gradients that cells maintain across their membranes result in a difference in electrical potential across the plasma membrane (the **membrane potential**), the interior of the cell being electrically negative with respect to the outside. Membrane potential is conventionally measured by inserting a fine glass capillary microelectrode into the cell and another electrode into the extracellular medium. The potential difference between the two is amplified and fed into an oscilloscope. It ranges between -20 and -200 mV, depending on the type of cell. The membrane potential in most animal cells is generated and maintained largely through the activity of the sodium–potassium pump (Na^+,K^+-ATPase).

Cells tend to lose K^+ and gain Na^+ and to maintain the correct balance the Na^+,K^+-ATPase in the plasma membrane is continually actively pumping Na^+ out and K^+ in against their concentration gradients. But K^+ is continually leaking out of the cell down its steep concentration gradient through a permanently open K^+ ion channel (the K^+ leak channel) (Fig. 6.7)

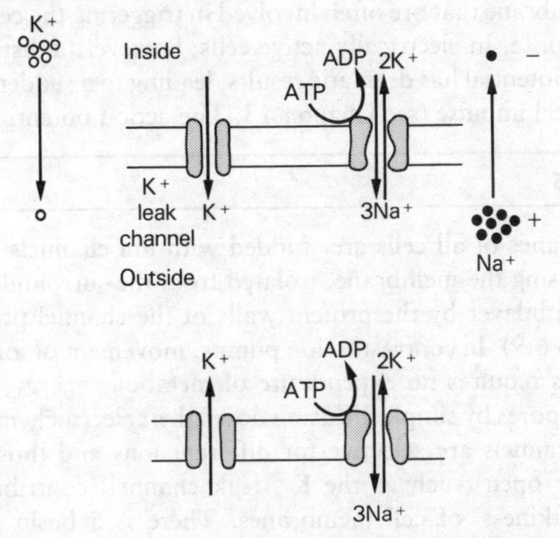

Fig. 6.7 Ion movements that generate the membrane potential (see text).

This loss of K^+ (positive charge) leads to the interior of the cell becoming electrically negative with respect to the exterior – i.e. a membrane potential is generated (Fig. 6.7 top). This then tends to retard the loss of K^+ from the cell – as K^+ is positive it will not move passively against the electrical gradient – and, instead, K^+ tends to enter the cell (Fig. 6.7 bottom). At a particular membrane potential, the tendency of K^+ to move out of the cell against its concentration gradient is exactly balanced by its tendency to enter the cell down its electrical gradient, and the membrane potential stabilizes at this value – called the **resting membrane potential.**

However, the K^+ channel is also slightly permeable to Na^+ and so there is also a continual slow flow of Na^+ into the cell. If it were not for the Na^+–K^+ pump actively maintaining the concentration gradients across the membrane, this would eventually lead to the equilibration of Na^+ and K^+ across the cell membrane and the disappearance of the membrane potential. (In many animal cells, Cl^- movement also contributes to generating the membrane potential.)

A change in the membrane potential can be experimentally forced on any cell by an internally applied small electric current. In non-excitable cells the change in membrane potential that results is usually fairly small and ion movements adjust to restore it to its normal limits. In natural circumstances small disturbances of this sort result from the actions of many extracellular signals, which lead to ion flows across the plasma membrane that are often involved in triggering the cell's physiological response. In electrically active cells, however, a disturbance of membrane potential has dramatic results, leading to a sudden discharge – an electrical impulse (see Chapter 13: The action potential).

Ion channels

The membranes of all cells are studded with **ion channels** – aqueous pores traversing the membrane, isolated from the surrounding hydrophobic lipid bilayer by the protein walls of the channel protein (Fig. 6.8; see also 6.9). In contrast to ion pumps, movement of ions through ion channels requires no expenditure of metabolic energy. Ions flow through the pores by simple diffusion down their electrochemical gradients. Ion channels are selective for different ions and those that are permanently open (such as the K^+ leak channel) contribute to the inherent 'leakiness' of cell membranes. There is a basic distinction between those that allow cations (e.g. Na^+, K^+ and Ca^{2+}) to pass and those that allow anions (e.g. Cl^- and bicarbonate HCO_3^-).

The molecular structure of ion channels has only been determined in the past few years. The amino acid sequences of several channel proteins have now been determined indirectly from cloned cDNA. The existence of many other ion channels is inferred from detectable flows of electric current in or out of a cell (conductances). The sophisticated technique of patch-clamping, in which a minute patch of membrane containing a presumed single channel is electrically sealed off from its surrounding membrane across the mouth of a hollow glass microelectrode, now allows recording of current flow through a single channel. This is helping to sort out the ion flows previously detected and their physical basis in membrane ion channels of various types.

Gated Channels

Some channels are not permanently open but open transiently only in response to a particular stimulus. They figure largely in cells' responses to many external and intracellular stimuli, the resulting ion flows in

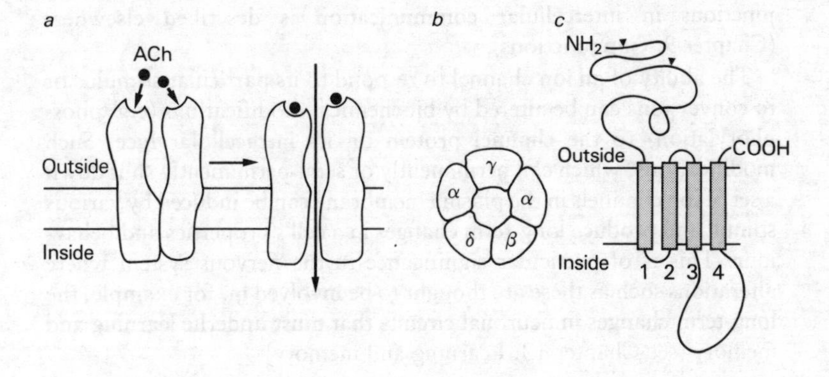

Fig. 6.8 *a*, The action of a gated ion channel, the nicotinic acetylcholine (ACh) receptor on skeletal muscle. It consists of five subunits (*b*) surrounding a central pore, the two α subunits each binding one molecule of acetylcholine. On binding ACh released from the motor nerves innervating muscle, the channel opens, allowing chiefly Na^+ to flow into the cell. This sets off a chain of events that leads to muscle contraction (see Fig. 13.3 for details). The diameter of the open channel is approx. 0.65 nm. *c*, Disposition of the polypeptide chain of the α subunit in the membrane. This arrangement is similar for the other subunits and also for the related subunits of the γ-aminobutyric acid (GABA) and glycine receptors.

and out of the cell triggering the biochemical machinery that produces the cell's typical response.

There are two functional classes of gated ion channel – those that open in response to binding of a signal molecule (**ligand-gated channels**) and those that open in response to a change in membrane potential (**voltage-gated channels**). They may stay open for times varying from a fraction of a millisecond to a few milliseconds. By analogy with the behaviour of many other proteins, it is believed that binding of another molecule (the ligand) or a voltage change in the surrounding membrane induces a conformational change in the three-dimensional structure of the channel protein that opens the aqueous pore through its centre. Selective opening of ion channels is a part of many cells' response to external stimuli, and gated ion channels probably exist in all cells. They have been best studied so far in nerve and muscle where electrical impulses are generated by the opening of voltage-gated channels, and several neurotransmitters act at ligand-gated channels.

A special type of ion channel, which also allows other small molecules to pass, is the **gap junction protein**. The function of gap junctions in intercellular communication is described elsewhere (Chapter 9: Gap junctions).

The ability of an ion channel to respond to its particular stimulus or to convey ions can be altered by biochemical modifications (e.g. phosphorylation) to the channel protein on its intracellular face. Such modifications, which can permanently or semi-permanently shut down a set of ion channels in the plasma membrane can be induced by various stimuli and produce long-term changes in a cell's properties and behaviour. This is of particular significance in the nervous system where alterations such as these are thought to be involved in, for example, the long-term changes in neuronal circuits that must underlie learning and memory (see Chapter 13: Learning and memory).

Ligand-gated channels

The best-studied ligand-gated channels are those that comprise the receptors for several neurotransmitters on the plasma membranes of nerve and muscle cells. The receptors for acetylcholine on skeletal muscle cells (Fig. 6.8), and for the neurotransmitters GABA (γ-aminobutyric acid) and glycine on neurones are ion channel proteins.

The amino acid sequences of these ion channels have been resolved and show striking similarities. This suggests that they, and possibly other neurotransmitter-activated ion channels whose sequences are not yet known, belong to a 'superfamily' of proteins descended from a common ancestor. During evolution they would have diversified to come under the control of different compounds and to convey different ions. Other ligand-gated channels have been found that are responsive to Ca^{2+} and cyclic nucleotides.

Voltage-gated channels

Voltage-gated channels open in response to a change in the membrane potential in their vicinity, opening only when the membrane potential reaches a certain 'threshold' value. Part of the channel protein is involved in 'sensing' voltage changes, presumably by a change in conformation.

The best-studied voltage-gated channels are those from the membranes of electrically excitable cells – nerve and muscle – where they are essential for generating and propagating electrical impulses, and in regulating the release of neurotransmitter from neurone terminals. The structure of the main voltage-gated Na^+ channel from the electric organ of the electric eel *Electrophorus* (a favourite material of neurophysiologists) has recently been deduced from its amino acid sequence.

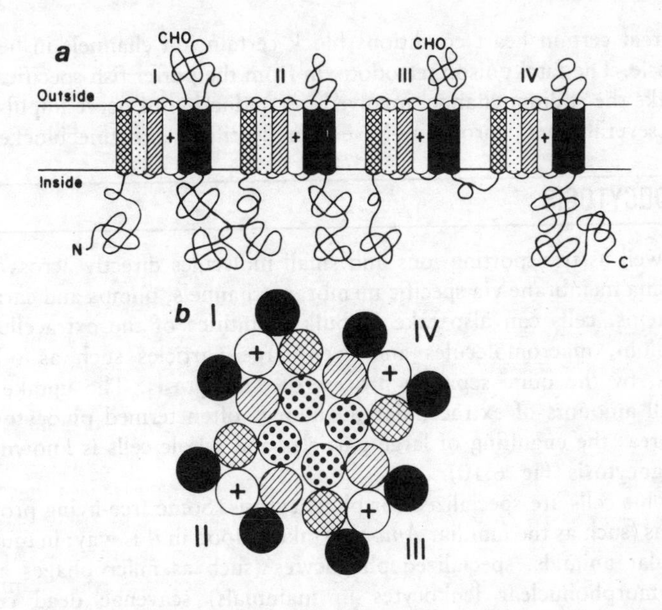

Fig. 6.9 Diagrammatic view of the voltage-sensitive sodium channel from the electric eel, *Electrophorus*. *a*, disposition of the protein chain in the membrane. Each hydrophobic region consists of six membrane-spanning segments. The segment that acts as the voltage sensor is marked + . *b*, Possible arrangement of the segments around a central pore to form the channel. (From M. Noda *et al.*, *Nature*, 320 (1986) 320.)

Unlike the ligand-gated channels described above the sodium channel is a single large polypeptide of about 1000 amino acids (Fig. 6.9). Other voltage-gated channels sequenced recently also have a similar single polypeptide structure, with the suspected voltage-sensor region in particular being very similar in each case. These similarities suggest that, like the ligand-gated channels, the voltage-gated channels also belong to an evolutionarily related superfamily of proteins descended from a primitive ion channel in the earliest cells.

Channel blockers

Many drugs and poisons are known to exert their effects by physically blocking ion channels and preventing ion flow. Local anaesthetics and 'ganglion blockers' used as anaesthetics inhibit the conduction and transmission of signals in nerve cells, and anti-dysrhythmic drugs used to treat certain heart conditions block certain ion channels in heart muscle. The fatal poison tetrodotoxin from the puffer fish specifically blocks the sodium channel involved in conduction of nerve impulses, and several other neurotoxins have been identified as channel blockers.

ENDOCYTOSIS

As well as transporting ions and small molecules directly across the plasma membrane via specific membrane channels, pumps and carrier proteins, cells can also take in bulk quantities of the extracellular medium, macromolecules and even large particles such as other cells, by the quite separate process of **endocytosis**. The uptake of small amounts of extracellular medium is often termed **pinocytosis**, whereas the engulfing of large particles and whole cells is known as **phagocytosis** (Fig. 6.10).

Some cells are specialized for phagocytosis. Some free-living protozoans (such as the familiar *Amoeba*) take in food in this way; in multicellular animals, specialized phagocytes (such as macrophages and polymorphonuclear leukocytes in mammals) scavenge dead cells and cell debris, and are also a first line of defence against invading microorganisms (see Chapter 12).

Pinocytosis, on the other hand, is a routine occurrence in virtually all animal cells – to the extent that a macrophage internalizes the equivalent of the whole of its plasma membrane in endocytotic vesicles every half hour or so, non-phagocytic cells being somewhat less active.

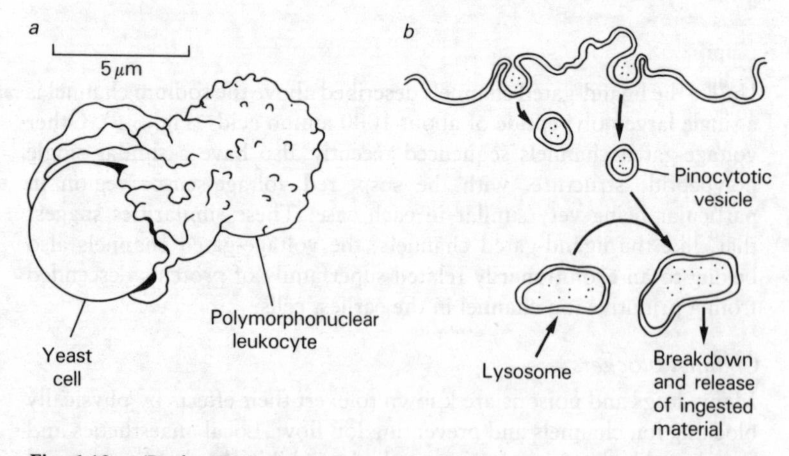

Fig. 6.10 *a*, 'Professional' phagocytes can engulf and dispose of cells approaching their own size. *b*, Pinocytosis.

Endocytosis is used by mammalian cells in various ways. Soluble material from blood, for example, is bulk transported across the endothelial cells of the walls of small blood vessels into the tissue fluid on the other side by endocytosis at one face of the cell, transport across the cell in endocytotic vesicles (**transcytosis**), and delivery into the extracellular fluid at the other face. Sometimes the endocytotic vesicles fuse to form a continuous channel through the cell.

In other cases, endocytotic vesicles (**endosomes** and phagocytic vesicles or **phagosomes**) fuse with primary lysosomes, their contents are broken down by the lysosomal enzymes, the small molecules released from the vesicle for use by enzymes in the cytosol, and the membrane is recycled, probably to the plasma membrane or to the Golgi apparatus (see Chapter 7).

Receptor-mediated endocytosis

As well as bulk transport of solute, certain molecules can be taken up by the selective process of **receptor-mediated endocytosis**. Molecules binding to a specific receptor protein at the cell surface trigger its internalization as part of the membrane of an endocytotic vesicle.

Receptor-mediated endocytosis occurs in many mammalian cells. It is, for example, the means by which they take up cholesterol. All animal cells need cholesterol as a constituent of membrane and in mammals it is manufactured in the liver and transported via the bloodstream in the form of so-called **low-density lipoprotein** (**LDL**) particles. These contain a core of cholesterol molecules esterified to fatty acids and surrounded by a lipid bilayer into which is inserted the protein component (**apolipoprotein**) of the particle. (This protein, apolipoprotein B, is incidentally one of the largest known, with a molecular weight of around 250 000. By a *tour de force* of molecular biology, the sequence of its more than 4000 amino acids has recently been determined.) Cells requiring cholesterol to manufacture new membrane display special LDL receptor proteins on their surface. When these bind the LDL protein the whole complex is then internalized and enzymes in the endocytotic vesicle release the cholesterol for the cell's use. The receptors and other membrane components are then recycled, it is presumed back to the plasma membrane or via the Golgi bodies.

Cholesterol accumulating in the blood contributes to the formation of atherosclerotic plaques on the walls of arteries. Buildup of atherosclerotic plaque narrows the artery, impeding blood flow, raising blood pressure and so predisposing to thrombosis. When it occurs in coronary arteries, it can lead to heart failure. Various genetically determined defects in the receptor mechanism for cholesterol uptake and a consequent high level of cholesterol in the blood are now known to be the cause of the inherited predisposition to severe premature atherosclerosis, called **familial atherosclerosis**. Those suffering from this condition often die at an early age from coronary artery disease.

Many of the receptors that cells possess to intercept chemical signals such as hormones at the cell surface are also 'turned-over' by receptor-mediated endocytosis once a signalling molecule has bound. Most of these signalling molecules are in fact believed to act at the cell surface and the signal transmitted rapidly via the receptor to intracellular biochemical pathways. The relatively slow process of receptor-mediated endocytosis is therefore believed to represent an internalization of used-up receptors for recycling and is also an important way in which cells can control the number of receptors on their surface, and thus their sensitivity to a particular chemical message. A decrease in the number of receptors on the surface is known as **down regulation**.

Coated pits and coated vesicles

Receptor-mediated endocytosis does not apparently occur at random on the cell surface but is confined to areas called **coated pits**, with which the receptors are associated or to which they migrate after binding their ligand. Coated pits endocytose to form **coated vesicles**. The 'coat' of coated vesicles and pits is a polyhedral lattice formed mainly of the protein **clathrin**. Coated vesicles are also involved in vesicular transport between endoplasmic reticulum and the Golgi bodies and the Golgi bodies and the plasma membrane.

The continual process of endocytosis is important in animal cells as a means of recovering 'excess' membrane added to the plasma membrane during secretion. Plant cells, however, do not routinely use endocytosis as a means of taking up material from outside the cell. (It would have little useful purpose as the cell wall prevents particles and many large molecules from reaching the plasma membrane. Endocytosis is in any case energetically unfavourable in plant cells where the plasma membrane is pressed tight against the cell wall as a result of the high internal osmotic pressure (turgor pressure) developed in the cell due to the presence of a restraining rigid cell wall.) Nevertheless, all plant cells add large amounts of membrane to the plasma membrane during secretion of cell wall material, and it is believed that, by analogy with animal cells, this excess membrane is recovered by endocytosis of coated pits that have been seen in plant cell plasma membrane preparations.

EXOCYTOSIS

Most cells manufacture and release material to the exterior – the process generally called **secretion**. Eukaryotic cells variously secrete digestive enzymes, protein, peptide and steroid hormones, polysaccharides (for plant cell walls), the materials of the extracellular matrix, the proteoglycans of cartilage, and a wide variety of different substances used as neurotransmitters in the nervous system.

Few of these substances can pass through the plasma membrane directly, and most are released by **exocytosis**. Material to be released is enclosed in small membranous vesicles which fuse with the plasma membrane delivering their contents to the exterior (Fig. 6.11).

Depending on the type of cell these vesicles are called **secretory vesicles, secretory granules, zymogen granules** (which contain inactive digestive enzyme precursors) or **synaptic vesicles** (the neurotransmitter-laden vesicles at nerve endings that are discharged in response to a nerve impulse (see Chapter 13: Synaptic transmission)). The intracellular route by which material is packaged into secretory vesicles depends on the type of material. Proteins, polysaccharides and proteoglycans (e.g. the heparan and chondroitin of extracellular matrix, cartilage and bone) are packaged via the endoplasmic reticulum and Golgi bodies. Small molecules such as the catecholamine and amino acid neurotransmitters (e.g. adrenaline, glycine and GABA), on the other hand, are synthesized in the cytoplasm and actively transported into preformed vesicles.

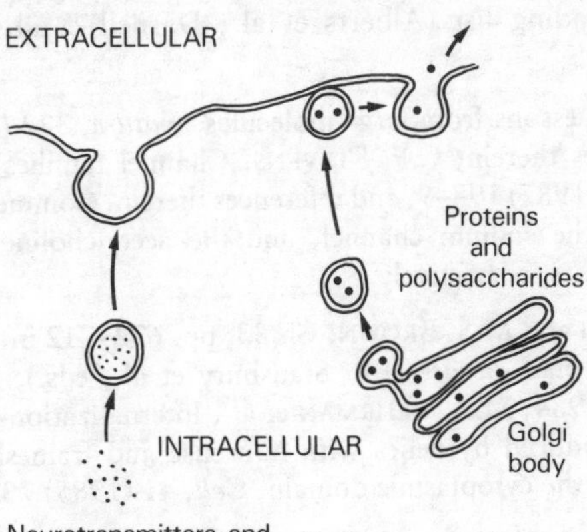

Fig. 6.11 Exocytosis (see text).

In some cases, secretion is continuous – the manufacture and release of antibodies by the plasma cells of the immune system is a case in point. In many other cases, however, material is synthesized and stored in secretory vesicles and only released when the cell is stimulated by an appropriate signal, such as a hormone or nerve impulse. The chain of intracellular events linking hormonal or nervous stimulation and exocytosis is at present receiving much attention from biochemists and cell biologists. A common factor triggering exocytosis in many cells in response to a variety of stimuli is a sudden rise in the concentration of intracellular free **calcium ions.**

The means by which Ca^{2+} triggers exocytosis is still unclear. It probably induces the fusion of transport vesicle and plasma membrane indirectly through its interaction with ubiquitous **calcium-binding proteins** such as **calmodulin**, and/or by its actions on components of the **cytoskeleton** (Chapter 8) that are believed to play a role in conveying secretory vesicles to the plasma membrane.

Like all vesicular transport processes within the cell secretory vesicles are precisely targetted to their correct destination – in this case the plasma membrane – by some means not yet clear (see, e.g., Chapter 7: Polarized secretion).

FURTHER READING

See general reading list (Alberts et al.; Darnell et al.; Stryer; and Lehninger).

W.S. AGNEW, Lessons from large molecules, *Nature*, **322** (1986) 770–1, and references therein; C.F. STEVENS, Channel families in the brain, *Nature*, **328** (1987) 198–9, and references therein. Commentaries on the isolation of the sodium channel, and the acetylcholine, glycine and GABA receptors, respectively.

J.L. GOLDSTEIN and M.S. BROWN, Ch.33, pp. 672–712 in *The metabolic basis of inherited disease* (J.B. Stansbury et al., eds.), McGraw-Hill, New York, 1984; M.A. LEHRMAN et al., Internalization-defective LDL receptors produced by genes with nonsense and frameshift mutations that truncate the cytoplasmic domain, *Cell*, **41** (1985) 735–43.

J.L. GOLDSTEIN et al., Receptor-mediated endocytosis: concepts emerging from the LDL receptor system, *Annual Review of Cell Biology*, **1** (1985) 1–39.

CHAPTER 7

THE INTERNAL MEMBRANES OF PLANT AND ANIMAL CELLS

The unit of living matter – the cell – is bounded by a surface membrane (**plasma membrane, plasmalemma**) that forms a selectively permeable barrier between the cell and its external environment. The cells of eukaryotes (protozoans, algae, fungi, plants and animals) also possess similar internal membranes that divide the cell into several compartments, represented by the intracellular organelles such as mitochondria, endoplasmic reticulum, nucleus, etc., in which different metabolic processes are segregated (Fig. 7.1). Prokaryotes (bacteria) have much simpler cells lacking internal membrane-bounded compartments (see Chapter 1).

This section focuses chiefly on the **endoplasmic reticulum** and the **Golgi apparatus** which together form an extensive membrane system within the cell, and from which the plasma membrane and several other organelles ultimately derive. The role of the endoplasmic reticulum and Golgi apparatus in the cell, unlike the better-known mitochondria and chloroplasts, remained unknown until recent years. Improved microscopic and cell biological techniques have now uncovered much of their function although many important details, especially with regard to the Golgi apparatus, still remain unresolved.

THE CYTOSOL

The **cytosol** forms the 'ground substance' of the cytoplasm and contains the 'soluble' enzymes involved in much **intermediary metabolism** – the interlocking biochemical pathways by which cells transform small organic molecules such as sugars, amino acids and nucleotides into fuel for energy generation on the one hand and into building blocks for cell components on the other.

All proteins (except for a few made within mitochondria and chloroplasts by their own synthetic machinery) are synthesized in the cytosol by the translation on ribosomes of messenger RNA exported from the

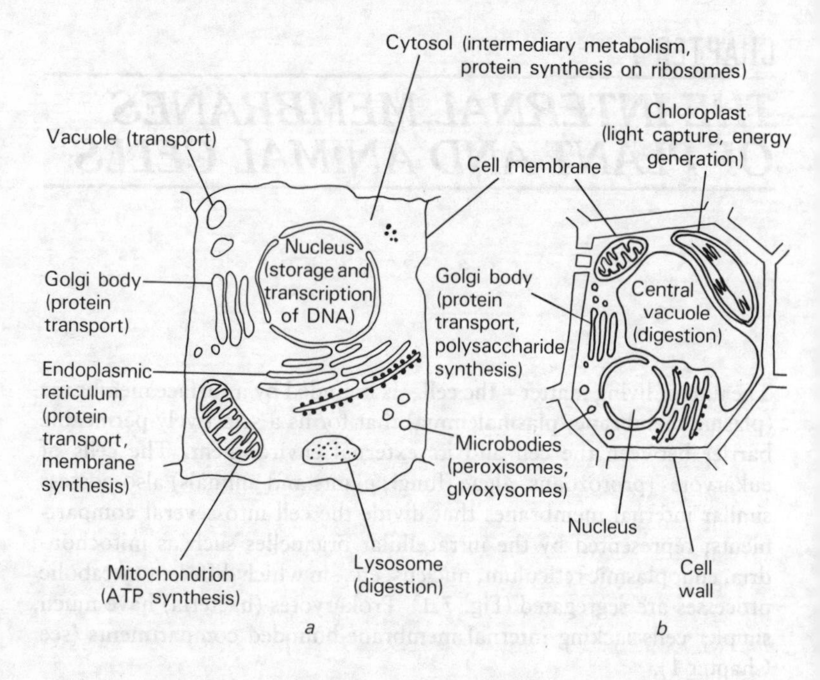

Fig. 7.1 *a*, Animal cell; *b*, cell from a terrestrial green plant, showing the metabolic compartmentation of the eukaryotic cell. The cells of vascular plants, algae and fungi differ from animal cells in many ways, not least in the possession of a rigid cell wall. In algae and vascular plants this is composed chiefly of cellulose and other carbohydrate polymers; fungi generally have chitin instead of cellulose in their cell walls. Photosynthetic eukaryotes (green plants and algae) possess chloroplasts, in which the green pigment chlorophyll captures photons of light and complex biochemical machinery converts this energy into chemical energy stored in carbohydrates – the 'light' and 'dark' reactions, respectively, of photosynthesis. Many plant and algal cells contain large internal vacuoles filled with watery, somewhat acidic cell sap. Mature plant cells usually contain a single large central vacuole taking up most of the cell and bounded by the tonoplast membrane. The central vacuole is believed to be formed from the fusion of several smaller ones that probably derive with their contents from the Golgi bodies (see below). The vacuoles contain proteolytic and other digestive enzymes, which break down and recycle cell components in safe isolation from the rest of the cell and are the counterpart of the lysosomes in animal cells. Fungi are heterotrophic and non-photosynthetic, and the multicellular fungi grow in the form of a thread-like mycelium on which fruiting bodies are borne. Unlike most other eukaryotes, the threads of a mycelium are not divided into separate cells each with a single nucleus, but often consist of a continuous multinucleated tube of cytoplasm. In some fungi this is divided into binucleate sections by cross walls (septa).

nucleus. How proteins are sorted and dispatched to their various destinations in the cell has been the topic of intensive research over the past decade or so and forms the main part of this chapter.

Electron microscopy and modern techniques of immunofluorescence microscopy have revealed a highly organized protein **cytoskeleton** running throughout the cytosol, composed of microfilaments and microtubules and associated proteins (see Chapter 8).

The many metabolic pathways of the cytosol have been largely pieced together outside the cell using cell extracts and isolated enzymes. These cytosolic enzymes are often called 'soluble' enzymes to differentiate them from enzymes that form a permanent part of cell membranes and are not easily extracted. *In vivo*, however, the cytosol is far from being simply a random mixture of enzymes in solution. Efficient intermediary metabolism implies a degree of organization within the cytosol, with enzymes of particular pathways being brought together in more or less permanent assemblies. Some of these assemblies – **multienzyme complexes** – and their characteristic location within the cytosol have already been identified.

The enzymes that synthesize and degrade glycogen, for example, are clustered together on the surface of glycogen granules. Enzymes associated with triglyceride synthesis and breakdown are similarly associated with fat droplets in adipose tissue. The internal membranes of the cell, such as the endoplasmic reticulum, provide a support for many cytoplasmic enzymes and the protein scaffolding of the cytoskeleton is also believed to provide support. It is indeed probable that there are no enzymes at all actually 'free' in the cytosol. Many 'soluble' enzymes readily bind to the protein actin *in vitro* and might therefore be expected to be associated with the ubiquitous actin filaments of the cytoskeleton. Such associations are difficult to demonstrate unambiguously *in vivo*, however, as the weak bonds that link these assemblages are broken by all but the very gentlest isolation procedures.

MEMBRANE SYNTHESIS, PROTEIN SORTING AND DELIVERY

A cell manufactures thousands of different proteins in the cytosol, many of which are destined to leave it for sites elsewhere in the cell. Some will become incorporated in membranes, some become part of organelles and some, such as hormones, antibodies and many enzymes, will leave the cell altogether. Ingenious cellular mechanisms have evolved for sorting proteins and dispatching them to their correct destinations.

These involve the **endoplasmic reticulum** and the **Golgi apparatus**, which together form an extensive, linked internal membrane system involved in the synthesis, packaging and transport of proteins for secretion, for insertion into the plasma membrane and into certain other organelles. Protein transport is intimately linked with the synthesis of new membrane at the endoplasmic reticulum, its transport to the plasma membrane and other organelles and its recycling.

The endoplasmic reticulum

The endoplasmic reticulum is the most extensive internal membrane system in most eukaryotic cells. In sections of cells viewed in the electron microscope, it appears as a series of membranous channels and vesicles (cisternae) extending throughout the cytoplasm but is believed to consist of a single convoluted membrane enclosing a continuous internal compartment. For many years the function of the endoplasmic reticulum remained unknown but it is now known to be the site of lipid synthesis, the site of new membrane synthesis, and to play an important role in the synthesis and sorting of proteins that leave the cytosol.

In most cells and especially in those actively secreting proteins, much of the endoplasmic reticulum is studded with ribosomes on its cytoplasmic face. This is the **rough endoplasmic reticulum** (RER) and the ribosomes are engaged in synthesizing proteins that are being delivered into the lumen of the organelle. Cells specialized for lipid metabolism, on the other hand, contain little RER but extensive **smooth endoplasmic reticulum** (SER) engaged in lipid synthesis and the generation of lipid storage bodies.

Together with the Golgi apparatus, lysosomes and vacuoles, the endoplasmic reticulum forms the **endomembrane** system of a eukaryotic cell, comprising those membranes that are either physically continuous or linked by membranous vesicles that shuttle between them (Fig. 7.2) delivering membrane from one to another. The plasma membrane is also part of this system, but mitochondria and chloroplasts with their double membranes are not. Proteins and membrane travel from their sites of synthesis on the face of the endoplasmic reticulum to the plasma membrane via the Golgi apparatus.

The highly compartmentalized nature of a eukaryotic cell presents an immediate difficulty in conveying proteins to their final destinations.

To reach the exterior of the cell or the interior of organelles, newly synthesized proteins must cross one of the cell's membrane systems. Proteins in general, and especially secretory proteins, tend to be bulky, water-soluble molecules, and biological membranes, with their hydrophobic lipid core, do not allow the passage of large, water-soluble molecules. In the case of secretory proteins and proteins destined for the membranes and organelles of the endomembrane system, this is overcome by a dual-purpose sorting and translocation system that initially delivers them into the lumen or the membranes of the endoplasmic reticulum. Once in place they do not have to cross another membrane to reach their destination. The lumen of the endoplasmic reticulum is therefore topologically equivalent to the outside of the cell, the interior of lysosomes and vacuoles, and the interior of the sacs (cisternae) of each Golgi body (see Fig. 7.2). The problem of conveying a bulky protein across a membrane is in this case solved by threading the unfolded protein chain through the membrane as soon as it starts to be synthesized.

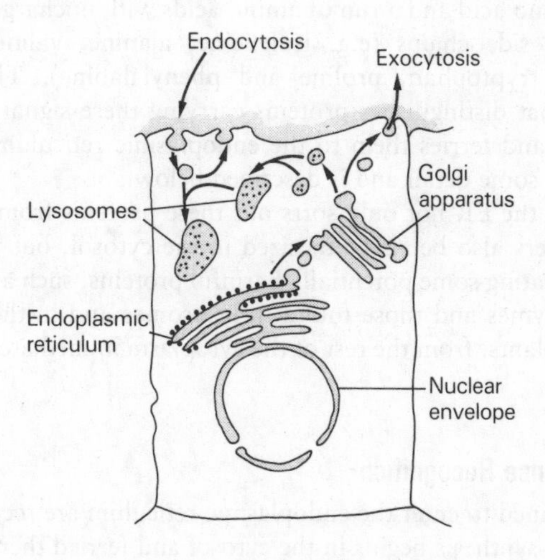

Fig. 7.2 The endomembrane system of the eukaryotic cell, and the topological relationships of the compartments (shaded) that it delimits. Arrows indicate direction of transport of membrane and lumen contents through the system.

Signal sequences and protein translocation

Most proteins that are to leave the cytosol carry molecular labels that indicate their destination. These molecular tags are part of the genetically coded protein structure and consist of stretches of amino acids that interact with recognition and transport machinery in various internal membranes.

Proteins destined for delivery into the endoplasmic reticulum carry a short sequence of 15–25 amino acids at the extreme N-terminus called the **signal** or **leader** sequence. It acts as a molecular homing device directing them out of the cytosol on the first stage of their journey. (Proteins destined for mitochondria, chloroplasts and the nucleus also possess stretches of amino acids that act as molecular localization signals – see below, Nuclear localization sequences; Mitochondrial presequences and chloroplast 'transit' sequences.)

In most proteins, the signal sequence is removed before the molecule reaches its final destination. The amino acid sequence itself varies considerably from protein to protein but always contains an N-terminal basic amino acid and a run of amino acids with uncharged, mainly hydrophobic side chains (e.g. isoleucine, alanine, valine, leucine, methionine, tryptophan, proline and phenylalanine). The sorting machinery that distinguishes proteins carrying these signal sequences from others and ferries them to the endoplasmic reticulum has been elucidated in some detail and is described below.

Entry into the ER not only sorts out these proteins from the thousands of others also being synthesized in the cytosol, but provides a means of isolating some potentially harmful proteins, such as secretory digestive enzymes and those found in lysosomes and in the digestive vacuoles of plants, from the rest of the cytoplasm immediately they are synthesized.

Signal Sequence Recognition

Proteins destined to enter the endoplasmic reticulum are recognized as soon as their synthesis begins in the cytosol and ferried there by small ribonucleoprotein particles known as **signal recognition particles** (**SRP**). Signal recognition particles bind to the signal sequence, which is the first part of the polypeptide chain to be translated, as it emerges from the ribosome (Fig. 7.3).

A signal recognition particle attached to the signal sequence arrests protein synthesis. The partially synthesized polypeptide along with its attached ribosomes is then ferried to the face of the ER where it docks at a special receptor – **the signal recognition particle receptor** or **docking protein**. Other proteins in the membrane anchor the ribosomes themselves to its cytosolic face. Once the ribosome is anchored, translation recommences and the signal sequence enters the endoplasmic reticulum membrane, with the rest of the newly synthesized protein chain threading after it (Fig. 7.3).

Entry into the endoplasmic reticulum membrane and translocation of a protein chain across it require a specific recognition protein, the **signal sequence receptor** (**SSR**), an integral membrane protein to which the signal sequence is 'handed on' from the signal recognition particle. Some kind of active transport process, as yet unidentified, is also required. One possibility is that proteins in the ER membrane form a 'pore' through which the polypeptide chain can thread.

In most cases, once the signal sequence has crossed the membrane it is enzymatically snipped off by a **signal peptidase** in the lumen (internal space) of the endoplasmic reticulum. Secretory proteins cross the membrane in their entirety, folding up into their final three-dimensional conformation in the lumen. Membrane proteins on the other hand are locked in the membrane during their transit by the hydrophobic

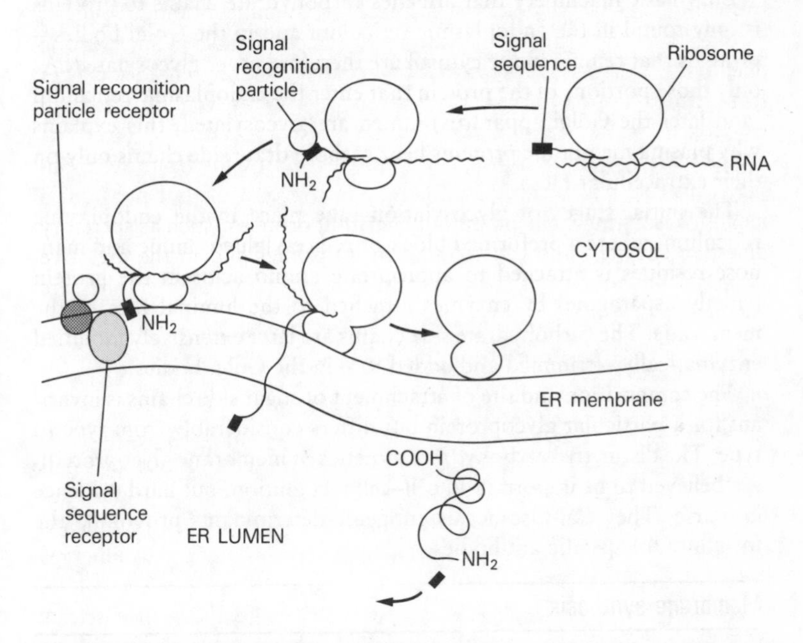

Fig. 7.3 Delivery of a protein to the endoplasmic reticulum (ER) and translocation of the protein chain into the ER lumen.

stretches of amino acids that they all carry (see below, Membrane synthesis).

Threading the newly synthesized protein chain through the membrane (**co-translational translocation**) is the usual way proteins are delivered into the endoplasmic reticulum but is not obligatory. Proteins that have already been fully synthesized and folded up can also enter. In this case they may unfold again before being transported across. 'Chaperone' proteins involved in protein folding and unfolding have recently been discovered.

Formation of glycoproteins: glycosylation

Most secretory and membrane proteins are **glycoproteins**. These are proteins that contain sometimes quite complex carbohydrate side chains, built up of branched strings of sugars and sugar derivatives, chiefly mannose, *N*-acetylglucosamine, glucosamine, fucose, galactose and sialic acid (*N*-acetylneuraminic acid) (Fig. 7.4). These are added after the polypeptide chain has been synthesized and has entered the endoplasmic reticulum.

Enzymatic machinery that attaches carbohydrate chains to proteins is only found in the endoplasmic reticulum and in the Golgi bodies – proteins that remain in the cytosol are therefore never glycosylated. As only those portions of the protein that enter the endoplasmic reticulum (and later the Golgi apparatus) lumen are glycosylated, this explains why plasma membrane proteins bear carbohydrate side chains only on their extracellular faces.

The initial stages of glycosylation take place in the endoplasmic reticulum, where a preformed block of *N*-acetylglucosamine and mannose residues is attached to appropriate amino acids in the protein (chiefly asparagine) by enzymes attached to the luminal face of the membrane. The carbohydrate side chains are later extensively modified enzymatically – trimmed and added to – in the Golgi bodies.

The composition and site of attachment of sugar side chains is invariant for a particular glycoprotein but differs considerably from type to type. The distinctive carbohydrate moieties of membrane glycoproteins are believed to be important in cell–cell recognition, but hard evidence is sparse. They can also act as antigenic determinants provoking the formation of specific antibodies.

Membrane synthesis

Protein transport is closely linked to the synthesis and transport of new membrane at the ER. The plasma membrane of most eukaryotic cells,

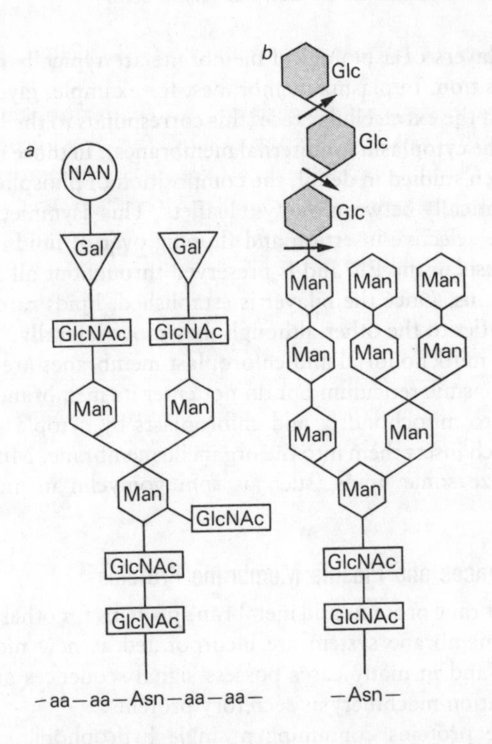

Fig. 7.4 *a*, The composition of a carbohydrate side chain of human immunoglobulin; *b*, the initial mannose rich oligosaccharide chain added to glycoproteins in the endoplasmic reticulum. Arrows indicate cleavage of the terminal glucose residues that occurs almost immediately. NAN, *N*-acetylneuraminic acid; Gal, galactose; Man, mannose; GlcNAc, *N*-acetylglucosamine.

certainly of animal cells, is continually being recycled and renewed by the complementary processes of endocytosis and exocytosis (see Fig. 7.2 and Chapter 6). New plasma membrane (as well as lysosomal membrane and the vacuole membrane of plant cells) is originally assembled at the endoplasmic reticulum. On its cytosolic face the endoplasmic reticulum carries the enzymes that synthesize the lipids that form the chief components of the lipid bilayer. Newly synthesized lipids slip into the outer leaflet of the membrane. Special proteins called 'flippases' then transfer some of them to the other leaflet, thus preserving the bilayer structure. Membrane proteins are also inserted into the lipid bilayer during its initial synthesis. Membrane is then transported to the Golgi apparatus in the form of small vesicles, for modification and onward transport to the plasma membrane.

The two layers of a biological membrane are typically of different lipid composition. In plasma membranes, for example, glycolipids are found only at the extracellular face (this corresponds to the layer facing away from the cytoplasm for internal membranes). In those membranes that have been studied in detail, the composition of phospholipids also varies dramatically between the two leaflets. This asymmetry is established by the selective insertion and flipping over of lipids during the initial synthesis in the ER and is preserved throughout all subsequent transformations. Once the bilayer is established, lipids rarely transfer from one leaflet to the other although they move laterally.

Lipids for mitochondrial and chloroplast membranes are also made at the endoplasmic reticulum but do not enter its membrane. They are transported to mitochondria and chloroplasts by cytoplasmic carrier proteins which insert them into the organelle membrane. Mitochondria can synthesize some lipids (such as sphingomyelin in animal cells) themselves.

Signal Sequences and Plasma Membrane Proteins

Plasma membrane proteins and membrane proteins for other organelles of the endomembrane system are incorporated as new membrane is synthesized, and in many cases possess signal sequences and use the same recognition machinery as secretory proteins.

Membrane proteins containing a single hydrophobic membrane-spanning segment (e.g. the red blood cell membrane glycoprotein glycophorin (see Fig. 6.3) and the ubiquitous histocompatibility antigens (see Fig. 12.20)) are thought to thread through the membrane in the same way as secretory proteins, except that the central hydrophobic segment arrests the process when it enters the membrane and remains embedded in the lipid bilayer. The signal sequence is removed from the N-terminal end as it enters the lumen.

Transmembrane proteins with a more complicated disposition of their polypeptide chain within the membrane, with several membrane spanning segments, often contain internal signal sequences, one for each hydrophobic segment, which may feed individual segments into the membrane using essentially the same machinery as for secretory proteins but which remain as a permanent part of the protein's structure. Highly hydrophobic proteins, as are many membrane proteins, may also be able to 'melt' into the lipid bilayer, their hydrophobic regions interacting with the hydrophobic tails of membrane lipids to force a particular molecular arrangement. Some membrane proteins, such as opsin (the protein component of the visual pigment rhodopsin) and the ß-adrenergic receptor, do not appear to contain a conventional cleavable N-terminal signal sequence. Some membrane proteins appear not to be anchored to the membrane via a hydrophobic domain but by a covalent linkage to a membrane lipid.

Once a protein has become oriented in the membrane with respect to 'inside' and 'outside' it does not change its orientation although it may move freely sideways in the lipid layer.

The Golgi apparatus

Most material made at the endoplasmic reticulum is transferred to its final destinations by way of the **Golgi apparatus** (**Golgi bodies, dictyosomes**). The Golgi bodies contain enzymes involved in building up and remodelling the carbohydrate chains that are added to many proteins in the endoplasmic reticulum. The Golgi apparatus also appears to be an important staging and sorting post where proteins initially sequestered into the endoplasmic reticulum are sorted and dispatched to their final destinations. In the Golgi the membrane made in the endoplasmic reticulum is also altered so that it acquires the characteristics of the various target organelles. In plant cells the Golgi apparatus is additionally the site where polysaccharides for the cell wall matrix are synthesized. They arrive at the site by the same mechanism that delivers secretory proteins to the cell's exterior (see below). Cellulose is made at the wall itself, by enzyme complexes delivered via the Golgi apparatus.

Each Golgi body comprises a stack of flattened sacs of membrane, each sac being known as a **cisterna** (pl. cisternae) (Fig. 7.5). Golgi

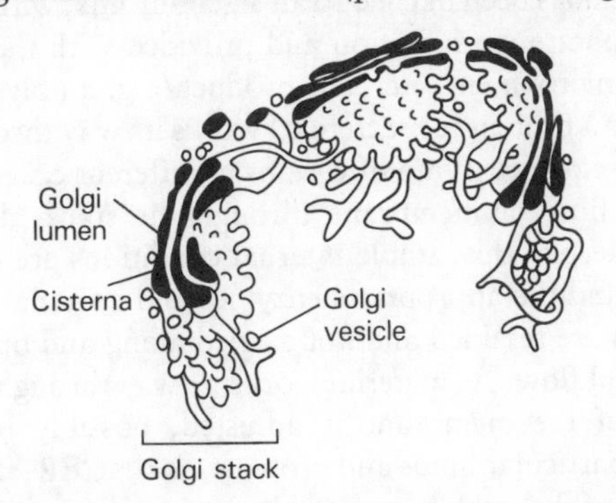

Fig. 7.5 Three-dimensional reconstruction of the Golgi apparatus of a mammalian cell. (After R.V. Krstic, *Ultrastructure of the mammalian cell*, Springer-Verlag, New York, 1979.)

stacks have a polarized structure with two quite distinct faces. In animal cells, one face always lies adjacent to a piece of endoplasmic reticulum and is known as the *cis* face, whereas the other is known as the *trans* face. The composition of the membranes, the enzymes contained in the cisternae, and the general internal cisternal environment changes as one progresses across the stack from *cis* to *trans*.

Proteins for secretion or for the plasma membrane do not go direct from the endoplasmic reticulum to their final destination but travel through a Golgi stack. Material and membrane are transported to the Golgi bodies in the form of small **transport vesicles** that pinch off the endoplasmic reticulum enclosing the contents of the lumen and subsequently fuse with the membrane of the *cis* face of the nearest Golgi stack, discharging their contents into the interior of the cisterna.

Salvage Sequences

Proteins destined to remain in the endoplasmic reticulum, such as the enzymes that carry out glycosylation, contain so-called **salvage sequences** that prevent them leaving along the normal secretory pathway. The salvage sequence identified so far is a short stretch of four amino acids (Lys–Asp–Glu–Leu) at the carboxy terminus of the protein. How this retains the protein in the endoplasmic reticulum is not yet clear: it is presumed to interact with a receptor protein on the luminal face of the membrane, which prevents it being carried along and out of the endoplasmic reticulum to the Golgi apparatus.

Progression Through the Golgi Apparatus

The Golgi stack has been likened to an assembly line, with each cisterna acting as a separate work-station and provided with the requisite enzymes, ionic environment, etc. The product (e.g. a polysaccharide for the cell wall or a protein for secretion) works its way through the stack undergoing a sequence of modifications in different cisternae. There is a continuous flow of membrane through the stack along with the product, but despite this, stable internal conditions are maintained in each cisterna and the appropriate enzymes and cisterna-specific membrane proteins are retained and not carried along and out of the stack with the general flow. As material works its way through the stack the composition of the membrane is adjusted, possibly by differential movement of particular lipids and proteins, so that 'ER-like' membrane on the *cis* face of the stack becomes 'plasma membrane-like' membrane on the *trans* face. In plants and algae, where the cisternae are not interconnected, small vesicles are thought to bud off the ends of each cisterna and fuse with the next one.

The sugar side chains on glycoproteins are extensively remodelled in the Golgi by the actions of a series of enzymes, some of which add sugars while others remove them. The enzymes are thought to be located in sequence through the stack, with each reaction providing an exclusive substrate for a subsequent enzyme further on.

In operational terms, in the absence of 'signals' to the contrary the Golgi stack processes material through to the *trans*-most cisterna, from which vesicles bud off and are conveyed to the plasma membrane with which they fuse, discharging their contents to the exterior and adding their membrane to the plasma membrane. Proteins and membrane destined for other organelles derived from the Golgi apparatus (lysosomes and their plant counterparts – vacuoles) are diverted from this default pathway at the *trans* face by mechanisms not yet well understood. A molecular signal that directs proteins into lysosomes has been identified (see below) as has a short stretch of N-terminal amino acids necessary to divert a protein into vacuoles of yeast cells. One idea, for which there is now good evidence, is that differences in the internal pH of organelles, more specifically increasing acidity, are involved in directing proteins along a particular pathway from the Golgi stacks.

Membrane recycling

The 'excess' membrane that accumulates at the plasma membrane, especially in actively secreting cells, appears to be recycled to the Golgi stack as vesicles derived from the plasma membrane by endocytosis. The extent and direction of the vesicular traffic between the endoplasmic reticulum, Golgi stacks and plasma membrane, lysosomes and other organelles, and how this traffic is controlled and the membrane composition of the various organelles maintained is not yet fully understood. Figure 7.2 illustrates present ideas of the traffic between the endoplasmic reticulum, Golgi apparatus, plasma membrane and lysosomes.

Lysosomes

The general term **lysosome** describes a diverse set of membrane-bounded vacuoles concerned with the breakdown and recycling of the cell's own components, or of material ingested by endocytosis. They contain hydrolytic digestive enzymes (e.g. proteases, nucleases, glycosidases and lipases) all of which act at a low pH and are therefore known generically as **acid hydrolases**. Small **primary lysosomes** containing these enzymes pinch off the Golgi stacks and fuse with each other and with phagocytic and endocytotic vesicles formed by uptake of material from outside the cell, delivering digestive enzymes to break down particles of ingested food, bacteria, etc. They also fuse with so-called **autophagic vesicles** probably derived directly from endoplasmic reticulum, in which cell components such as mitochondria are often seen being digested. These various types of large digestive vacuoles are often known as **secondary lysosomes.**

Enzymes destined for lysosomes have unusual mannose phosphate residues added to their carbohydrate side chains, and this apparently serves as a signal to the Golgi apparatus to sort them out of the main secretory pathway, via special mannose phosphate receptors in parts of the Golgi membrane.

Polarized secretion

The specialized secretory cells lining the ducts of exocrine glands (e.g. those that secrete digestive enzymes) secrete molecules from only one face – the apical face. The secretory vesicles produced from the *trans* face of the Golgi stack fuse only with this face and thus avoid unproductive secretion into intercellular spaces (Fig. 7.6). Polarized secretion probably involves both a specific interaction with cytoskeletal elements that guide vesicles to their correct destination and the recognition of particular 'secretory domains' in the plasma membrane with which the vesicles fuse.

The direction of secretion in such an epithelial cell is predetermined by its architecture, but non-polarized cells such as lymphocytes also achieve directed secretion. Lymphocytes secrete a range of proteins when stimulated through the variety of cell-surface receptor proteins

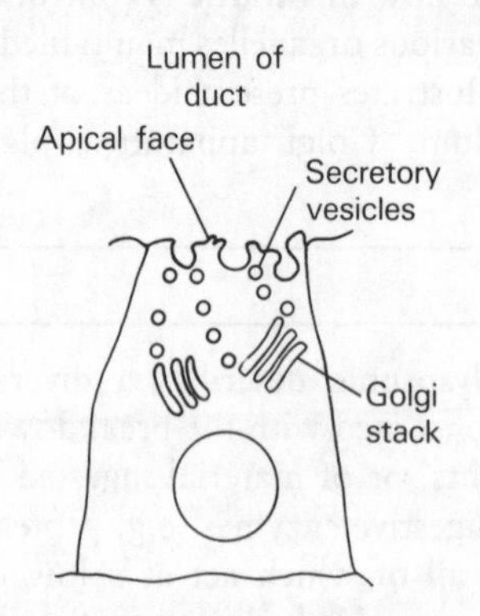

Fig. 7.6 Polarized secretion of digestive enzymes from an acinar cell of the pancreas.

they carry and sometimes secretion is directed towards the specific site on the surface at which they were stimulated. When the stimulus is a direct contact via specific antigen receptors with another cooperating lymphocyte this may ensure, for example, that the growth factors (interleukins) a lymphocyte secretes when stimulated are delivered efficiently to its antigen-specific interacting target cell (see Chapter 12).

OTHER ORGANELLES

Nucleus

The **nucleus**, which contains the cell's DNA and which is the site of DNA transcription into RNA, is bounded by a double membrane, the **nuclear envelope**, which is traversed by many **nuclear pores** through which RNA is exported and proteins imported. Inside this envelope is the **nucleoplasm** in which lie the chromosomes, the genetic material of the cell. (Chromosome structure and the replication and transcription of DNA are covered in Chapters 2 and 3.)

The outer nuclear membrane is continuous with the endoplasmic reticulum (see Fig. 7.1). In vertebrate cells (but not in all eukaryotes) the nuclear envelope fragments at each cell division and is reformed around the newly segregated chromosomes after mitosis. The fragments become associated with the chromosomes, recover nuclear pore complexes that were released into the cytoplasm and fuse to form a complete nuclear envelope. Once reformed, the nuclear envelope can expand considerably (e.g. during active gene expression) and the endoplasmic reticulum is thought to contribute material directly to it.

The ribosomes, on which protein synthesis takes place in the cytosol, are assembled in the nucleus from ribosomal RNA synthesized in the nucleolus and proteins imported from the cytosol, and then re-exported through nuclear pores. Many proteins must be imported into the nucleus at various times, including structural proteins such as the histones (see Chapter 2: Chromosome structure) and the large number of gene-regulatory proteins that must bind to DNA to exert their effects.

A nuclear pore is believed to consist of a protein complex that forms a channel about 9 nm wide through the nuclear envelope (Fig. 7.7). Small molecules (including proteins with molecular weights up to around 40 000) apparently pass freely through the pores; larger proteins and nucleoprotein complexes require some specific transport mechanism. Compared with the molecular machinery that guides proteins into the endoplasmic reticulum very little is yet known about how

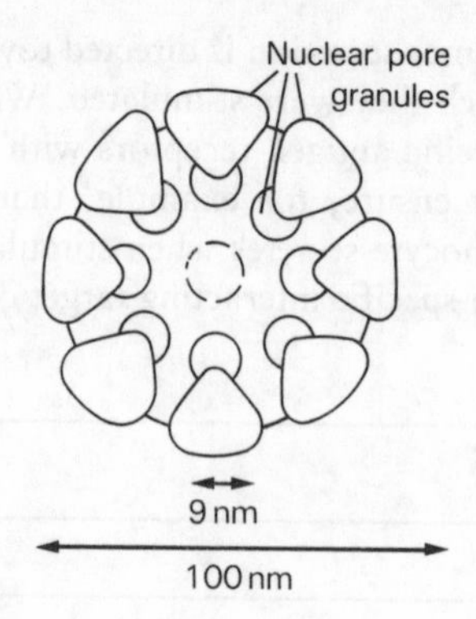

Fig. 7.7 Schematic diagram of the nuclear pore complex. The effective channel appears to be confined to a pore 9 nm in diameter.

passage of material in and out of the nucleus is controlled, despite its central importance in cell physiology.

Nuclear Localization Sequences

Molecular signals that label proteins for import into the nucleus have been found. The role of nuclear localization (or signal) sequences in proteins is not only to direct proteins into the nucleus but to ensure they remain there, as nuclear pores are large enough to let small proteins pass in and out. They are not therefore removed as are other types of signal sequence. Small proteins can pass freely through the nuclear pores and their nuclear localization signals probably interact with intranuclear receptors. For larger proteins, nuclear localization signals probably direct their active transport through the nuclear pores.

Nuclear Lamina

On the nucleoplasmic face of the nuclear membrane is the nuclear lamina, a meshwork of protein filaments that provide support to the nuclear membrane and most probably anchorage points for chromatin during the interphase period (see Chapter 8: Intermediate filaments).

Nucleolus

The nucleolus shows up in electron micrographs of interphase (non-dividing) nuclei as a dense area within the nucleus. It is composed of a mass of fibrillar material embedded in protein through which run loops of chromatin containing the rRNA genes and which are known as the **nucleolar organizer regions**. The rRNA genes are transcribed within the nucleolus to give the ribosomal RNAs that associate with protein subunits imported from the cytoplasm to form complete ribosomes. These are then re-exported to the cytosol for protein synthesis. The nucleolus disappears in the early stages of mitosis and is reformed as the chromosomes begin to decondense after nuclear division.

Mitochondria and chloroplasts

Virtually all eukaryotic cells contain **mitochondria**, organelles in which oxygen-requiring **aerobic respiration** is carried out, resulting in the synthesis of **adenosine triphosphate** (ATP), the general biological currency of stored chemical energy. The cells of the green tissues of plants and of the unicellular algae also contain **chloroplasts**, organelles in which photosynthesis takes place, when the energy of sunlight is captured and converted into stored energy in the form of carbohydrate.

Mitochondria and chloroplasts are thought to be the remnants of free-living bacteria that became symbiotic inhabitants of an early cell (see Chapter 5: The evolution of the eukaryotic cell). They replicate by division and contain small amounts of DNA, which encode their own ribosomal RNAs and ribosomal proteins, tRNAs and a few mitochondrial and chloroplast proteins. Most mitochondrial and chloroplast proteins are, however, encoded in the nucleus, made in the cytosol and subsequently transported into the organelles. In several cases, one subunit of a mitochondrial or chloroplast enzyme is encoded and synthesized in the organelle and the other by a nuclear gene.

Although very different in appearance (Fig. 7.8), in the enzymes they

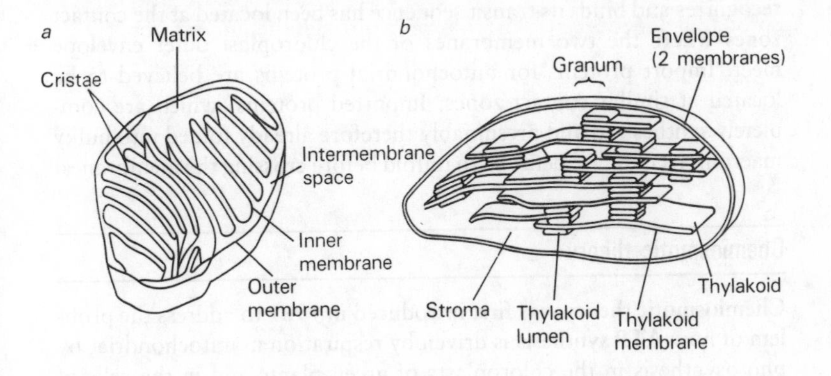

Fig. 7.8 Schematic view of the internal structure of *a*, mitochondrion from an animal cell; *b*, chloroplast. See Figs 7.9 and 7.10 for details of the metabolic reactions that take place in the various internal compartments of these organelles.

contain, and their role in the cell, mitochondria and chloroplasts are very similar in the way their structure is related to their function of energy generation. Mitochondria have an outer membrane that is generally permeable to most ions and small molecules. Within this is a highly convoluted inner membrane (each fold is called a **crista**, plu. cristae), which in contrast is highly selective and forms an inner ionically isolated compartment. This impermeable membrane is essential for ATP synthesis (see below). Enzymes and other proteins involved in ATP generation are located in the inner membrane. Chloroplasts are also bounded by a double membrane, the outer one much more permeable than the inner. In addition, there is also a separate membrane system, the **thylakoids**, which form an ionically sealed inner compartment analogous to the inner mitochondrial compartment. In the thylakoid membranes are located the chlorophyll-containing photosystems that capture light, and the enzymes and electron transport proteins that convert this captured energy into ATP, which is used to make carbohydrate in the stroma (see Fig. 7.10).

Mitochondrial Presequences and Chloroplast Transit Sequences

Proteins imported direct from the cytosol into mitochondria and chloroplasts are labelled with stretches of N-terminal amino acids called **presequences** for mitochondrial proteins and **transit** sequences for those of chloroplasts. They are in general longer than the signal sequences on secretory proteins. For one protein destined for the chloroplast stroma (the small subunit of the photosynthetic enzyme ribulose 1,5-bisphosphate carboxylase) a specific 'import' protein that recognizes and binds its transit sequence has been located at the contact zones where the two membranes of the chloroplast outer envelope meet. Import proteins for mitochondrial proteins are believed to be located at similar contact zones. Imported proteins, which are completely synthesized and presumably therefore already folded into bulky macromolecules, are believed to unfold before crossing the membranes.

Chemiosmotic theory

Chemiosmotic theory was first introduced in 1961 to address the problem of how **ATP synthesis** is driven by respiration in mitochondria, by photosynthesis in the chloroplasts of green plants and in the cells of photosynthetic bacteria. At the time it put forward an original and novel solution to what had proved an intractable problem, by proposing an answer based on biophysical processes in membranes rather than a more conventional 'biochemical' pathway. Greeted initially with disbelief and incomprehension, the basic idea is now backed up with considerable direct experimental evidence, and has proved of universal significance in explaining many types of similar processes throughout the living world. More than 25 years later, in 1978, it gained a Nobel Prize for its progenitor, the British biophysicist Peter Mitchell.

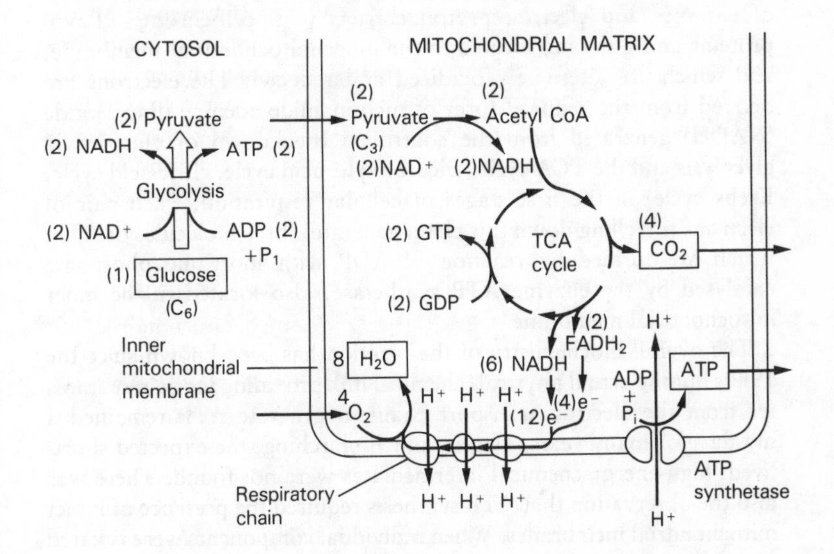

Fig. 7.9 General outline of aerobic respiration. Overall, one molecule of glucose generates a maximum of around 36 molecules of ATP formed from ADP and inorganic phosphate (P_i). Most of this ATP is synthesized during oxidative phosphorylation. The overall stoichiometry of the reaction is indicated by numbers in parentheses before each compound. Each NADH (nicotinamide adenine dinucleotide) donates two high-energy electrons to be transported along the respiratory chain (which consists of around 15 carriers, many of which are metalloproteins such as the cytochromes). Each pair of electrons provides the energy to synthesize three molecules of ATP. $FADH_2$ (flavin adenine nucleotide (reduced)) donates a pair of lower-energy electrons through which two molecules of ATP are generated. Passage of electrons along the respiratory chain leads to protons being ejected from the inner membrane into the intermembrane space, setting up a proton gradient across the inner membrane. The flow of protons back through the ATP synthetase drives ATP synthesis (see text). Some of the energy stored in this proton gradient is also used to drive transport processes in the inner membrane.

ATP is the universal currency of stored chemical energy. It is the energy source for many basic metabolic reactions as well as for more specialized energy-requiring reactions such as muscle contraction. In eukaryotes, its main site of synthesis is in mitochondria as one of the final products of aerobic respiration (Fig. 7.9).

An additional phosphate group is added to ADP to make ATP in an energy-requiring process known as **oxidative phosphorylation** from the participation of oxygen in the overall equation. ATP synthesis is powered by the flow of high-energy electrons along a respiratory 'chain' of enzymes and electron/proton carriers (e.g. cytochromes, flavoproteins and quinone) located in the inner mitochondrial membrane, and which are alternately oxidized and reduced. The electrons are derived from the reduced form of nicotinamide adenine dinucleotide (NADH) generated from the controlled breakdown of glucose via **glycolysis** and the **TCA cycle** (**tricarboxylic acid cycle, citric acid** cycle, **Krebs cycle**) in the first stages of cellular respiration. Each pair of electrons travelling down this chain generates three molecules of ATP, which are formed by reaction of ADP with inorganic phosphate catalysed by the enzyme **ATP synthetase**, also located in the inner mitochondrial membrane.

The overall biochemistry of the reaction has been known since the 1940s but the actual physical/chemical link providing for energy transfer from the electron transport chain to generate ATP remained a mystery for many years. Despite much searching, the expected short-lived, high-energy chemical intermediates were not found. There was also the observation that ATP synthesis required the presence of intact mitochondrial membranes. When individual components were isolated and the pathway reconstituted in the test tube no ATP synthesis occurred – unlike other metabolic pathways that can be reconstituted in this way.

The revolutionary idea was to link the redox reactions of the respiratory chain to ATP synthesis by ion transport (in this case of protons, H^+) across the membrane, which creates an electrochemical gradient that can drive ATP synthesis. At three points in the respiratory chain, transport of electrons from one carrier to the next leads to protons being ejected from the membrane, setting up a proton gradient across it. Protons will then tend to flow back down this gradient but the inner membrane is generally impermeable to ions. There is only one point at which they can re-enter – the ATP synthetases. These act like ion pumps in reverse. Instead of splitting ATP to provide energy to pump ions against a concentration gradient they are driven to synthesize ATP

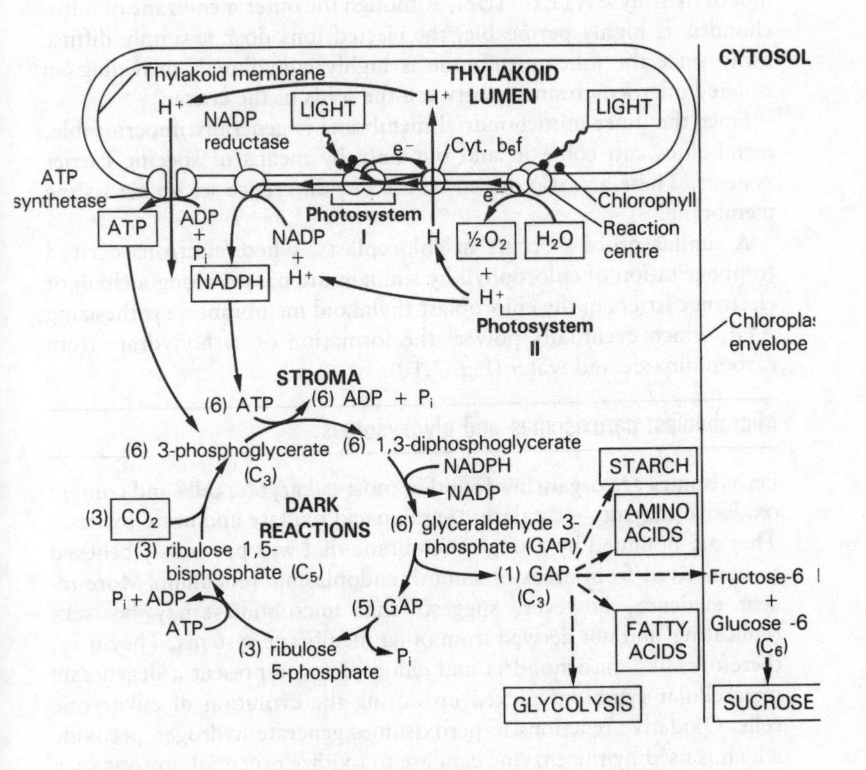

Fig. 7.10 Pathway of photosynthesis in the cell of a typical 'C3' plant (e.g. most flowering plants of temperate regions). The excitation of chlorophyll in photosystem II by light leads to the splitting of water to generate electrons, protons and oxygen. The electrons are transferred via the cytochrome b_6–f complex to photosystem I where they are used to generate the reduced form of nicotinamide adenine dinucleotide phosphate (NADPH) from NADP and H^+. Their passage through cytochrome b_6–f generates a proton gradient across the membrane, which is used to drive ATP synthesis. A separate cyclic flow of electrons from photosystem I through cytochrome b_6–f maintains the proton gradient and is the main source of power for ATP generation. NADPH and ATP are then used to power the fixation of carbon dioxide into carbohydrate in the stroma – the 'dark' reactions of photosynthesis (since they can proceed in the absence of light.) The overall stoichiometry of carbon fixation is indicated by the numbers in parentheses before each compound.

by ions (in this case protons) flowing through them in one direction. (If protons flow through ATP synthetase in the reverse direction it does indeed hydrolyse ATP to ADP.) Although the outer membrane of mitochondria is highly permeable, the ejected ions do not simply diffuse away since the inner membrane is highly convoluted, providing an isolated microenvironment between the folds of the cristae.

Since the inner mitochondrial membrane is generally impermeable, metabolites can come in and out only by means of specific carrier systems. These are also driven by the H^+ difference set up across the membrane.

A similar process occurs in chloroplasts, where electrons derived from excitation of chlorophyll by sunlight are passed along a chain of electron carriers in the chloroplast thylakoid membranes, synthesizing ATP, which eventually powers the formation of carbohydrate from carbon dioxide and water (Fig. 7.10).

Microbodies: peroxisomes and glyoxysomes

Peroxisomes are organelles found in most eukaryotic cells and contain oxidative enzymes – catalase, D-aminoacid oxidase and urate oxidase. They are bounded by a single membrane that was previously believed to be derived from areas of smooth endoplasmic reticulum. More recent evidence, however, suggests that microbodies may be 'self-replicating' and not derived from other membrane systems. They may, therefore, like mitochondria and chloroplasts, represent a degenerate intracellular symbiont picked up during the evolution of eukaryotic cells. Oxidative reactions in peroxisomes generate hydrogen peroxide which is used by the enzyme catalase to oxidize potential poisons such as ethanol (alcohol) and phenolic compounds.

In plant cells, organelles of similar appearance and presumed origin, but bearing different enzyme complements are known as **glyoxysomes** – these are abundant in certain seeds where they convert fatty acids into carbohydrate via the glyoxylate cycle (a conversion animal cells cannot do). Another type of peroxisome occurs in leaves, and is the site of **photorespiration**.

FURTHER READING

See general reading list. The endoplasmic reticulum and the Golgi apparatus and their role in protein and membrane synthesis and transport are comprehensively covered in Alberts et al. and Darnell et al. For respiration, photosynthesis and chemiosmosis also see Stryer and Lehninger.

J. LAWLOR, *Photosynthesis*, Longman, Harlow, 1987. An undergraduate-level text including a more detailed treatment of chemiosmotic theory.

M. WEIDMANN et al., A signal sequence receptor in the endoplasmic reticulum, *Nature*, 328 (1987) 830–33, and accompanying commentary by P. Walter (p. 763, same issue).

R.J. DESHAIES et al., A subfamily of stress proteins facilitates translocation of secretory and mitochondrial precursor polypeptides, *Nature*, 332 (1988) 800–5; W.J. CHIRICO, M.G. WATERS and G. BLOBEL, 70K heat shock related proteins stimulate protein translocation into microsomes, *Nature*, 332 (1988) 805–10; and accompanying commentary by H. Pelham (p. 776, same issue).

D. PAIN, Y.S. KANWAR and G. BLOBEL, Identification of a receptor for protein import into chloroplasts and its localization to envelope contact zones, *Nature*, 331 (1988) 232–7.

CHAPTER 8

THE CYTOSKELETON AND CELL MOTILITY

A definitive feature of eukaryotic cells is an internal **cytoskeleton**, a complex network of protein filaments and tubules ramifying throughout the cytoplasm. The cytoskeleton is involved in determining the distinctive shape of the cell and the changes it may undergo, cell movement, the movement of chromosomes during mitosis and meiosis, and intracellular transport of vesicles and organelles, including the processes of exocytosis and endocytosis. In muscle, cytoskeletal components present in many cell types have evolved into highly organized contractile fibres. Other specialized structures that make use of cytoskeletal elements are the cilia and flagella of eukaryotic cells. In contrast to these permanent structures, the cytoplasmic cytoskeleton of most cells is continually changing, being assembled and dismantled as the cell requires.

The cytoskeleton, by its connections with the plasma membrane, may provide an important pathway for transmitting signals from the cell's exterior, such as the nature of the extracellular substratum, to intracellular structures such as the nucleus (see, for example, Intermediate filaments, below). Cell movement depends on particular contacts between the plasma membrane and the substratum (in the body this is the extracellular matrix). The plasma membranes of many mammalian cells, for example, contain receptor proteins that bind to the protein components of the extracellular matrix, and are in turn linked to the internal cytoskeleton. Interactions with the matrix are now realized to be of great importance in embryonic development, both in relation to the extensive cell movements that occur in early embryogenesis and in the determination and differentiation of different cell types (see Chapter 10). The way in which these receptor proteins are linked to the cell's cytoskeleton, and their role in these processes are now under investigation.

The main structural elements of the cytoskeleton can be seen in electron micrographs of sectioned cells as straight filaments (**microfilaments** and **intermediate filaments**) and hollow tubules (**microtubules**) running throughout the cytoplasm. These elements can be isolated from cells and studied in the electron microscope, their protein components have been identified, and their mechanism of assembly and disassembly studied biochemically *in vitro*. Their disposition in the living cell and their relationship to cell shape and movement are less easy to determine, but the general arrangement of some parts of the cytoskeleton in a whole cell can be seen by the techniques of fluorescence microscopy, using fluorescent-tagged reagents such as specific antibodies that label filament or microtubule proteins (Fig. 8.1).

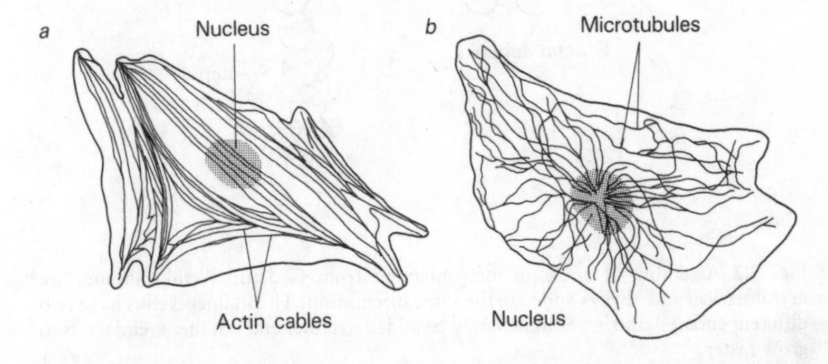

Fig. 8.1 Diagrammatic view of the arrangement of *a*, actin filaments and cables; and *b*, microtubules, in a cultured fibroblast.

STRUCTURE OF THE CYTOSKELETON

Microfilaments (actin filaments)

These are long protein threads of diameter 5–7 nm, composed of globular protein subunits of **actin** joined end to end. Actin subunits are called G **actin**, the chains are called F (**fibrous**) **actin**. A single microfilament consists of two chains wound around each other to form a helix (Fig. 8.2). Actin filaments self-assemble from a pool of G actin subunits in the cytoplasm. Energy for the reaction is supplied by the hydrolysis of ATP to ADP.

Individual microfilaments are visible only in the electron microscope, but in cultured animal cells stained for actin with specific fluorescent stains, thick **actin cables**, which consist of crosslinked bundles of actin filaments, can be seen traversing the cell from edge to edge. Several 'actin-bundling' proteins have been identified (see Table 8.1) and these may form the cross-links in actin cables.

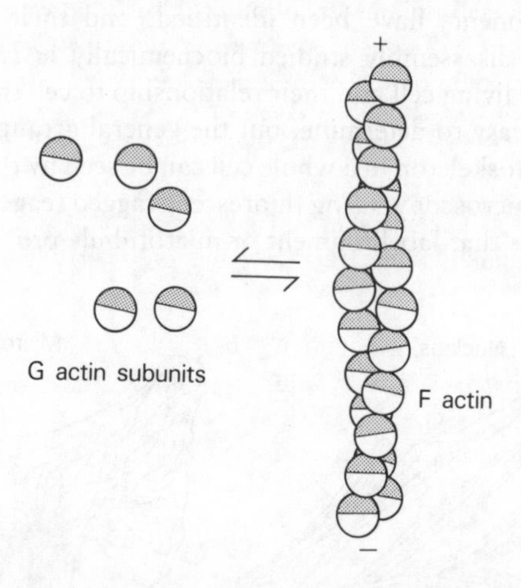

Fig. 8.2 Assembly of an actin microfilament from G actin. Actin subunits are asymmetrical and always added in the same orientation. The filaments thus have two different ends, + and – . Subunits may be added to either end but the + end tends to grow faster.

Actin filaments provide mechanical support for extensions of the cell surface, whether permanent such as the microvilli on the surface of intestinal brush border cells and the stereocilia on the hair cells of the inner ear (which detect sound vibrations), or temporary, such as the extensions cultured mammalian cells put out as they explore the surface of their culture dish. The remarkable degree of order that can be shown by a permanent assembly of actin filaments is illustrated in Fig. 8.3.

On their own actin filaments are not contractile. They can, however, cause shape changes, such as the acrosomal reaction seen in some invertebrate sperm (see below), by rapidly assembling from a cytoplasmic pool of components and equally rapidly disassembling. In most cases where actin filaments have been implicated in cell motility however, they are believed to be associated with **myosin** to form contractile assemblies resembling a primitive form of muscle fibril, but the exact relationship of actin filament and myosin is not known.

The G actin subunits are not symmetrical and are always added in the same orientation. Actin filaments thus have a distinct **polarity** – that is, one end is different from the other (see Fig. 8.2). Polarity is also

Table 8.1 Some proteins associated with the cytoskeleton

---

caldesmon Calcium-binding protein abundant in smooth muscle and in the stress fibres (actin filaments) of other cells. It binds to actin. It may be involved in the Ca^{2+}-dependent control of actomyosin contraction in smooth muscle, and possibly in non-muscle cells.

tropomyosin As well as being a component of muscle fibrils, tropomyosin is associated with stress fibres in non-muscle cells.

alpha-actinin As well as being a component of muscle fibrils, α-actinin is present in adhesion plaques and other sites of contact with the extracellular matrix.

talin Present in adhesion plaques. It binds to the membrane receptors for extracellular matrix proteins (e.g. fibronectin, vitronectin, laminin and fibrinogen) and may provide a link between these receptors and the cytoskeleton. It also binds to another cytoskeletal protein, vinculin, and to actin – it can cross-link actin filaments.

vinculin Present in adhesion plaques and in intercellular tight junctions. Its synthesis appears to be regulated in response to cell–cell and cell–substratum interactions. Binds to talin.

profilin Ubiquitous cytoplasmic protein to which unpolymerized actin is bound. It acts as an 'actin buffer' to provide a pool of actin subunits.

fimbrin Actin-binding protein forming part of the cytoskeleton of microvilli.

villin Protein component of the cytoskeleton of microvilli. Has Ca^{2+}-dependent actin-bundling and actin-severing activity.

gelsolin Fragments actin networks *in vitro* in a Ca^{2+}-dependent process by inserting between the actin monomers (see also villin); however, their role in the cell is not yet clear. It may be involved in gel–sol transitions in the cytoplasm (see below, Cell motility).

filamin Links actin filaments to form a loose network. Its addition to a solution of actin filaments changes it from a viscous fluid to a solid gel, and it may be involved in gel–sol transitions in the cytoplasm.

synapsin I Associated with small synaptic vesicles in neurones. When dephosphorylated, it bundles actin fibres. When phosphorylated, this activity is lost. Its phosphorylation when nerve endings are depolarized may be involved in the release of synaptic vesicles from their cytoskeletal restraints in preparation for exocytosis and neurotransmitter release (see Chapter 13: Synaptic transmission). It is a substrate for several intracellular phosphorylating enzymes (kinases).

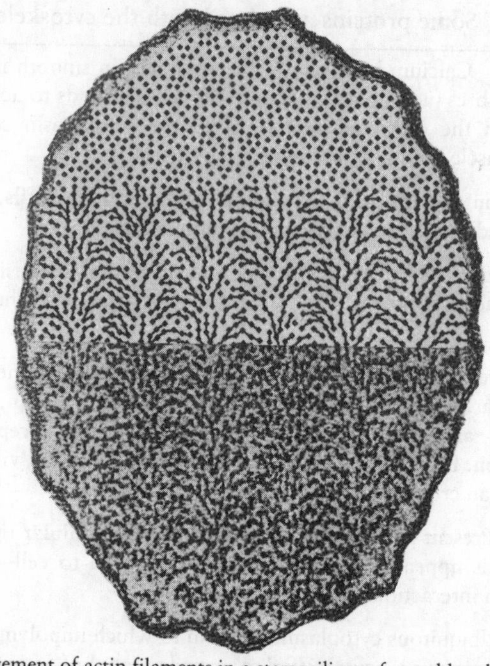

Fig. 8.3 Arrangement of actin filaments in a stereocilium of a cochlear hair cell seen in cross-section. The hair cells of the cochlea bear minute projections on their surface that are exquisitely sensitive to small movements caused by sound vibrations or movement of fluid in the semicircular canals. Each projection – a stereocilium – is stiffened by actin microfilaments, which are linked to each other by protein arms to form an ordered array. Computer-generated image from electron micrographs. (From D.J. DeRosier *et al.*, *Nature*, 287 (1980) 291–298. Figure kindly supplied by the author.)

reflected in the fact that one end (the + end) grows faster than the other, thus enabling the rapid assembly of actin filaments to drive the extrusion of a portion of a cell in one direction. A dramatic example of this is the acrosomal reaction of the sperm of the sea cucumber *Thyone* (an echinoderm). On contact with the egg, actin filaments in the sperm head assemble from monomers within a few seconds to shoot out a process, the **acrosome**, that helps the sperm to penetrate the egg.

Microtubules

Microtubules are hollow tubes formed from **tubulin** protein subunits (Fig. 8.4). The basic unit of assembly is a tubulin dimer. Microtubules assemble from a pool of tubulin subunits in the cytoplasm in a reaction requiring the hydrolysis of GTP (guanosine triphosphate) to the diphosphate form (GDP). Microtubules are also polar structures. They tend to grow at one end (the + end) and disassemble from the other (the – end). Microtubules grow out from **organizing centres in the cell**, to which the minus end of the microtubule is usually anchored. The major microtubule organizing centre in animal cells is the **cell centre** or **centrosome**, which contains the **centrioles** at its centre. Animal cells contain a pair of centrioles, each consisting of a bundle of 27 microtubules. Plant cells do not have centrioles. Microtubules grow with their + ends away from the cell centre.

Many proteins are associated in small amounts with microtubules. Few of these have yet been identified or given a role. Some probably have regulatory roles in microtubule assembly and dismantling. Others such as **dynein** and **kinesin** are ATPases powering movement involving microtubules, such as the beating of eukaryotic cilia and flagella, vesicle transport, and chromosome movement during mitosis and meiosis.

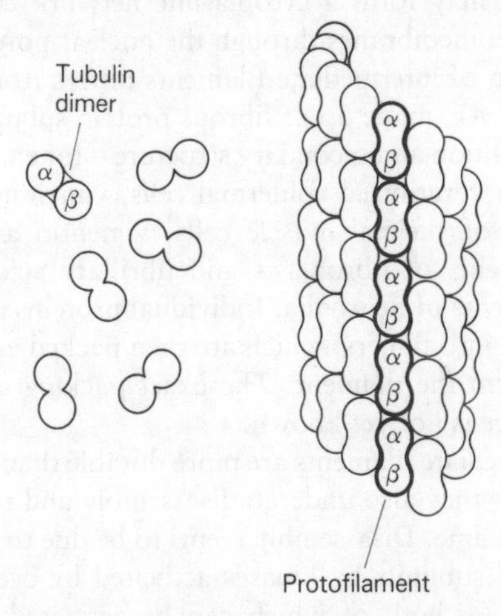

Fig. 8.4 The basic unit of assembly for microtubules is the tubulin dimer composed of two polypeptides α-tubulin and β-tubulin. Tubulin dimers are arranged in rows to form a protofilament, and a microtubule typically consists of 13 protofilaments surrounding a central space. Tubulin units can be added to or removed from either end of a microtubule. *In vivo*, microtubules tend to assemble from one end, designated the + end.

Intermediate filaments

Intermediate filaments or **10-nm filaments** as they are sometimes called, are the third structural component of the cytoskeleton of eukaryotic cells. They are present in many cell types and compose the original 'cytoskeleton' – the cell framework that remains after the rest of the cell has been removed by detergents. They differ in many respects from microtubules and actin microfilaments, being much tougher and more durable. The neurofilaments running along the axons of nerve cells are intermediate filaments, as are the keratin filaments of keratinized epidermal cells. They typically have a diameter of 10 nm and are formed from rigid rod-like protein subunits, in contrast to the globular subunits of actin filaments and microtubules.

Their role in the cell is rather obscure but has become clearer in recent years. They do not appear to be involved in cell movement or transport processes in any obvious way as are actin filaments and microtubules. If they are collapsed by antibody treatment this seems to have little effect on cell shape or growth. The most popular current view of their function is that they are 'cytoplasmic organizers', directing the spatial disposition of organelles within the cytoplasm. Recent work suggests that they may form a cytoplasmic network connecting the nuclear and plasma membranes through the nuclear pores.

The composition of intermediate filaments differs from cell to cell. They are all, however, made up of fibrous protein subunits of similar amino acid composition and secondary structure – for example, several types of keratin in keratinized epidermal cells, vimentin in the intermediate filaments of skeletal muscle cells, vimentin and desmin in intermediate filaments of fibroblasts and fibrillary acidic protein in intermediate filaments of neuroglia. Individual proteins are associated along their lengths into dimers, which are then packed end to end and side by side to form the filament. The exact packing of the protein subunits in the fibre is not yet known.

Although intermediate filaments are more durable than microtubules and microfilaments they also undergo disassembly and rearrangement during the cell's lifetime. Disassembly seems to be due to phosphorylation of the protein subunits by kinases activated by cyclic AMP and by protein kinase C, both of which can be activated indirectly by extracellular signals (see Chapter 9: Second messengers).

Intermediate Filaments in the Nucleus

Proteins called **lamins**, very similar to the components of intermediate filaments, are part of the nuclear envelope. They form the **nuclear lamina**, a durable fibrous layer on the internal (nucleoplasmic) face of the inner envelope membrane. In some cell types, it can be seen clearly in the electron microscope as a remarkably regular square meshwork of filaments of around 10 nm diameter, and it has been proposed that these filaments are a new type of intermediate filament composed of lamin subunits. (Note: lamins should not be confused with *laminin*, a large glycoprotein of extracellular matrix.)

The nuclear lamina disintegrates during prophase, soon after the chromosomes begin to condense, and reappears after mitosis. As well as providing a framework for organizing the nuclear envelope and the position of nuclear pores, the nuclear lamina may provide anchorage points for chromatin (the complex of DNA and protein of which chromosomes are made) during the interphase period of the cell cycle, when it is actively being transcribed and is not condensed into visible chromosomes. It may also provide, through the lamins, an anchorage point for cytoplasmic intermediate filaments that have binding sites for lamin at one end of their protein subunits.

Membrane cytoskeletons

Cells also possess membrane skeletons that form a layer immediately underlying the cell membrane and that are in contact with it through interactions with membrane anchor proteins. The membrane skeleton of the red blood cell (erythrocyte) is best known. A mature mammalian red blood cell lacks a nucleus or any other internal organelles and its main function is to transport the oxygen carrier protein haemoglobin around the body. It possesses a particularly stable and regular internal membrane skeleton (Fig. 8.5) immediately underlying the plasma membrane, which is, however, sufficiently flexible to allow the cell to deform to pass through small blood vessels. In other cells, less-organized membrane skeletons are likely. In some, there is a layer of actin filaments immediately underlying the membrane.

Membrane skeletons may be concerned with the organization of membrane 'domains' – that is, regions of membrane containing particular proteins – by preventing certain proteins diffusing freely over the whole cell surface.

Other cytoskeletal proteins

As well as the immediate structural components of microtubules and filaments, many other proteins are associated with the cytoskeleton. For example, many 'actin-binding proteins' have been identified. These may form structural links between individual microfilaments when they are organized into ordered arrays, or may be concerned with the dynamics of cytoskeleton assembly and disassembly.

One of the best known is **myosin** (Fig. 8.6). Myosin is the component of the thick filament in skeletal muscle (see below) where it provides the

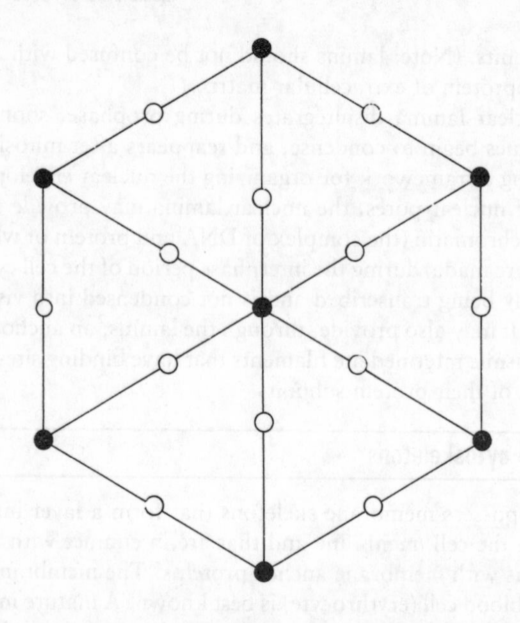

Fig. 8.5 Hexagonal arrangement of unit of membrane skeleton of a red blood cell (erythrocyte). The lines represent the fibrous protein spectrin, circles are various proteins through which it is linked to the plasma membrane. Spectrin is composed of a helix of two polypeptides. It is linked at one end to the integral membrane protein, the band III protein (an ion channel), via ankyrin. At the other end, it is linked to the membrane via actin and other membrane proteins on the cytoplasmic face. Spectrin is also present in other cells (e.g. in neurones, where it is called fodrin). Ankyrin also occurs in other cell types but its role in linking membrane proteins to cytoskeletal elements is still under investigation.

molecular motor – the ATPase activity of the myosin head – that drives muscle contraction. Myosin is also associated with actin filaments in non-muscle cells but in a far less organized fashion. Actin–myosin associations in non-muscle cells are believed to form a primitive contractile apparatus that mediates cytoplasmic streaming and possibly some forms of cell movement.

Many cytoskeletal processes are highly sensitive to Ca^{2+} and calcium-binding proteins have been found associated with the cytoskeleton. Some of the effects of calcium in eukaryotic cells may be a result of its action on the cytoskeleton (see Chapter 9: Calcium).

Some proteins associated with the cytoskeleton are listed in Table 8.1.

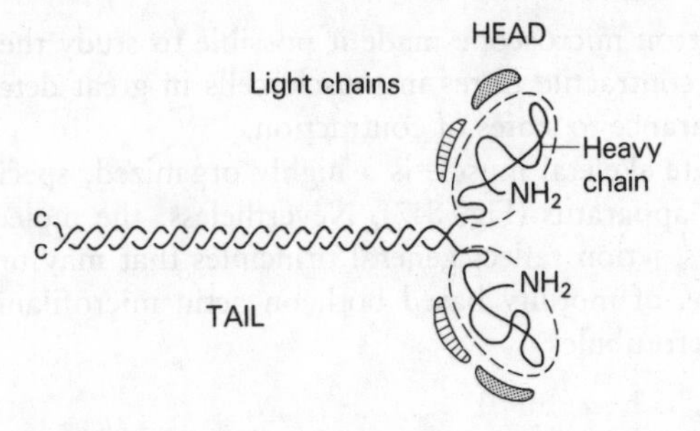

Fig. 8.6 Diagram of a myosin molecule. ATPase activity is located in the myosin heads and powers cycles of engagement and disengagement from actin filaments. Myosin can form filaments by association of the tails (see Fig. 8.8*b*).

CELL MOTILITY

To carry out many tasks cells must convert chemical energy into mechanical work. Some cells and cell structures are highly specialized to this end, such as muscle fibres, or the whip-like cilia and flagella that propel sperm and many free-living unicellular protozoans and algae along. But it has become increasingly apparent that similar molecular mechanochemical transduction mechanisms also underlie more general cellular phenomena, such as the extensive shape changes and amoeboid movement of which many animal cells are capable, the movement of segregating chromosomes during mitosis and meiosis, the constriction of animal cells at cell division, the intracellular transport of organelles and the dramatic cytoplasmic streaming seen particularly in plant cells.

The basic components of the systems that carry out mechanical work are microfilaments and microtubules (see previous sections) and proteins associated with them. Actin microfilaments are involved in muscle contraction, and are also implicated in amoeboid movement and cytoplasmic streaming. Microtubules, on the other hand, are the structures involved in the beating of cilia and flagella and the movement of chromosomes, and in some forms of intracellular transport.

Movement based on microfilaments

Skeletal Muscle

Skeletal muscle contraction is undoubtedly the best understood of all the mechanochemical transduction systems. The molecular basis of muscle contraction was uncovered in the 1950s after the introduction of the electron microscope made it possible to study the structure of individual contractile fibres in muscle cells in great detail and relate their appearance to states of contraction.

Vertebrate skeletal muscle is a highly organized, specialized force-generating apparatus (Fig. 8.7). Nevertheless, the molecular mechanisms of its action reflect general principles that may underlie many other forms of motility based both on actin microfilaments and on tubulin microtubules.

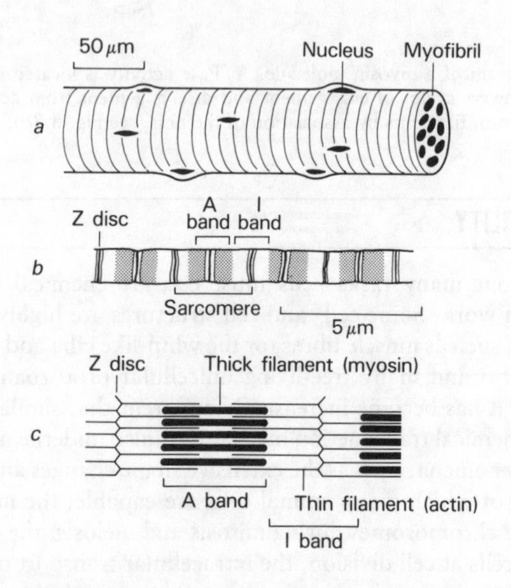

Fig. 8.7 *a*, A single fibre of skeletal muscle. *b*, Longitudinal section of a single myofibril as seen in the electron microscope. *c*, Interpretation of *b* showing the disposition of thick and thin filaments.

Skeletal muscle **myosin** is a large protein composed of four polypeptides, two 'heavy' chains that each form a tail and part of a head and two 'light' chains that form part of the heads. The head has adenosine triphosphatase (ATPase) activity which hydrolyses ATP to ADP. In skeletal muscle, myosin forms the **thick filament**, and an assembly of actin filaments and myosin is known as **actomyosin**.

Contraction is produced by sliding of thin (actin) and thick (myosin) filaments against each other. The myosin heads engage the actin filament, and are then presumed to 'bend' and pull it along in a 'rowing' movement, after which they are released to start a new cycle (Fig. 8.8). The changes in the conformation of the myosin head that drive sliding are powered by the hydrolysis of ATP. Even after nearly 30 years of muscle research at the molecular level, the actual way in which the myosin head interacts with the actin filament to generate the force that drives contraction is still not clear.

The immediate stimulus for contraction is a rise in free Ca^{2+} around the fibrils caused (at several molecular removes) by the stimulating

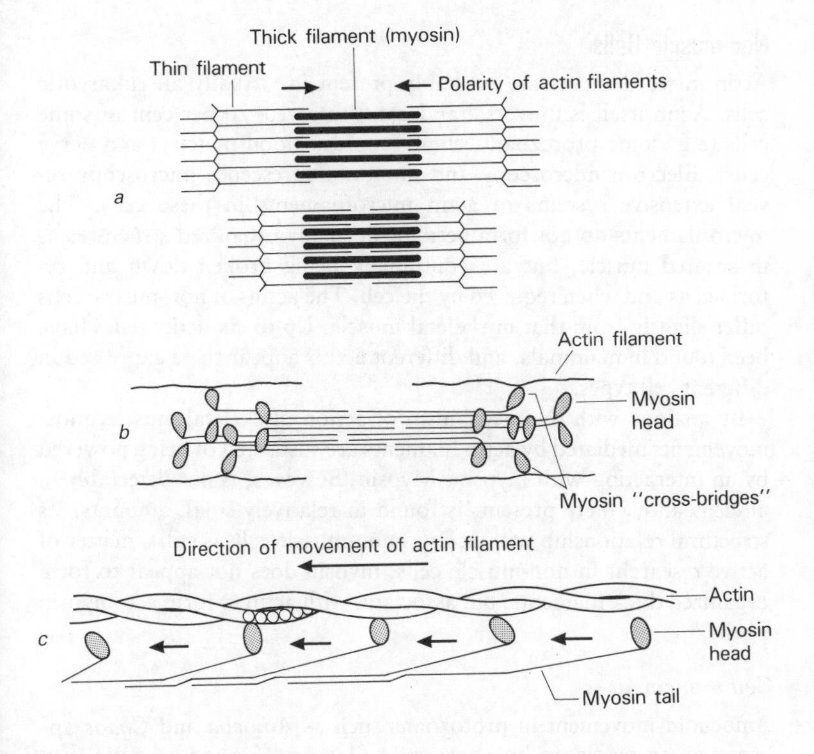

Fig. 8.8 The interaction of actin and myosin in skeletal muscle. *a*, The sliding filament model. Each sarcomere shortens by the filaments sliding against each other. The filaments themselves do not shorten. *b*, Schematic view of the arrangement of myosin in the thick filament and its relationship to the thin filament. *c*, Cycle of engagement and disengagement of the myosin head that pulls the thin filament along. The myosin head binds a molecule of ATP and is released from the thin filament. ATP is hydrolysed by the ATPase and the head resumes its original conformation. It makes contact with the thin filament and on release of ADP + inorganic phosphate (P_i) undergoes a conformational change that pulls the thin filament along. On binding another ATP molecule the cycle starts again.

nerve impulse (see Fig. 13.13). Calcium interacts with a calcium-binding protein **troponin** C in the muscle fibres. The interaction between the troponin complex and **tropomyosin** normally keeps the active sites of actin and myosin apart. When Ca^{2+} binds to troponin C, however, it causes the tropomyosin to shift position slightly allowing the myosin heads to contact the actin filament. Myosin can now interact with the actin filament to cause contraction.

Non-muscle Cells

Actin microfilaments are probably present in virtually all eukaryotic cells. Actin itself is universal and constitutes 10–20 per cent of some cells (e.g. some protozoans, slime moulds, blood platelets and nerve cells). Electron microscopy and immunofluorescence microscopy reveal extensive systems of actin microfilaments in these cells. The microfilaments do not form permanent highly organized structures as in striated muscle, but are continually being broken down and reformed as and when required by the cell. The actins of non-muscle cells differ slightly from that in skeletal muscle. Up to six actin genes have been found in mammals, and different actins appear to be expressed in different cell types.

By analogy with the mechanism of action of skeletal muscle, most movements mediated by actin filaments are suspected of being powered by an interaction with myosin. Myosin, however, is not detectable in all cells and, where present, is found in relatively small amounts. Its structural relationship with actin in non-muscle cells is still a matter of active research. In non-muscle cells, myosin does not appear to form organized thick filaments, but associates with actin to form actomyosin fibrils.

Cell movement

Amoeboid movement in protozoans such as *Amoeba* and *Chaos* appears to be mediated by contractile filaments formed of actin and myosin, but how this generates the extension of pseudopodia is still a matter of debate. During amoeboid movement fluid endoplasm streams forward until it reaches the end of the pseudopod where it becomes more gel-like and flows back along the periphery of the pseudopod as ectoplasm (Fig. 8.9). When it reaches the tail of the cell it becomes more fluid and recirculates as endoplasm.

One idea is that the formation of ectoplasm at the tip of the pseudopod involves formation and contraction of filaments within the protoplasm, and that the ectoplasm remains in the contracted state until it

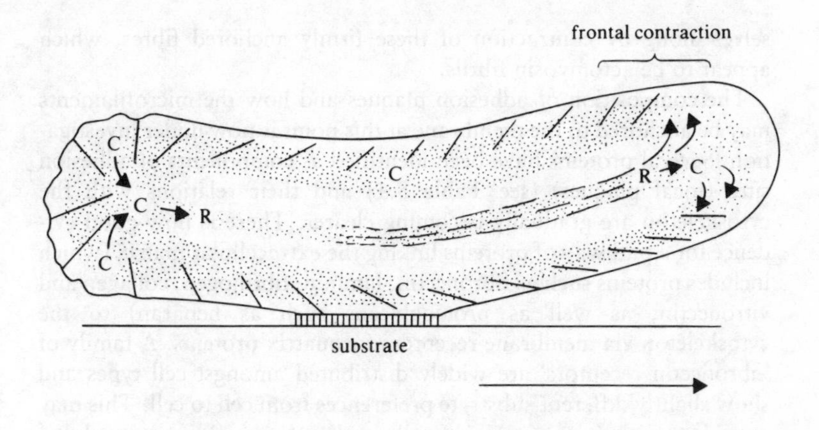

Fig. 8.9 Amoeboid movement. Schematic view of a vertical section through pseudopodium. Arrows indicate direction of cytoplasmic streaming. Transition of fluid endoplasm to gel-like ectoplasm is believed to take place by interaction of mysoin and actin filaments at the leading end of the pseudopodium. (After D.L. Taylor, J.A. Rhodes and S.A. Hammond, *Journal of Cell Biology*, 67 (1975) 427a).

reaches the tail of the cell where it 'relaxes' to become the more fluid endoplasm.

Many cells in multicellular animals are also capable of amoeba-like movement. Macrophages and polymorphonuclear leukocytes (neutrophils), for example, routinely migrate from the bloodstream into tissues. Isolated fibroblasts, if placed on a suitable surface, crawl over it (so slowly that their movement is best followed by time-lapse cinematography). Many other cells in culture explore their surroundings by extending and retracting small projections (microspikes) on the surface (Fig. 8.10).

The intracellular forces that generate cell movement have not yet been entirely clarified. The movement of a cell across a solid surface involves temporary anchorage of the leading edge to the substratum. These points of adherence can be seen if a cultured cell is made to round up suddenly by trypsin treatment. The bulk of the cell retracts leaving slender projections still anchored firmly to the substrate. The attachment sites are called **adhesion plaques** (see Fig. 8.10) and bundles of microfilaments (**stress fibres**) run back from these sites into the body of the cell. Stress fibres can normally be seen within cultured cells lying parallel to the substrate surface. The cells are believed to 'pull' themselves along by contraction of these firmly anchored fibres, which appear to be actomyosin fibrils.

The composition of adhesion plaques and how the microfilaments may be anchored to the membrane at this point is now under investigation. Several proteins have been identified as components of adhesion sites (focal plaques) (see Table 8.1) and their relations with the cytoskeleton are gradually becoming clearer. There is now good evidence for a pathway of proteins linking the extracellular matrix (which includes proteins such as fibronectin, laminin, fibrinogen, collagen and vitronectin, as well as proteoglycans such as heparan) to the cytoskeleton via membrane receptors for matrix proteins. A family of 'fibronectin receptors' are widely distributed amongst cell types and show slightly different substrate preferences from cell to cell. This may provide a basis for maintaining cells in their 'correct' tissue in adults, and the guiding and sorting of embryonic cells during development. These receptors appear to be linked to the cytoskeleton via the protein talin (see Table 8.1).

One abnormality of tumour cells is that they have far fewer regions of attachment to their substrate and fewer microfilament bundles. Whether and how this relates to their capacity for uncontrolled cell division and for migration (metastasis) is not yet clear.

Cytoplasmic streaming

Material can be transported around the cell by movement of the cyto-

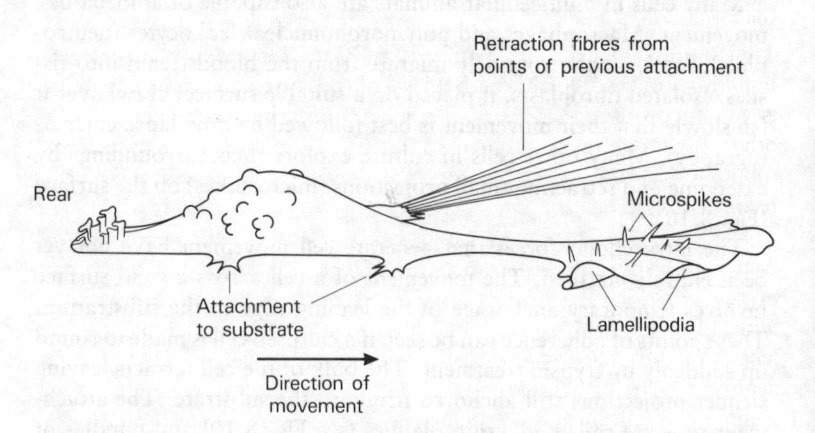

Fig. 8.10 Diagrammatic view of a cultured fibroblast moving over a surface. As it moves, the 'front' end ruffles up, it puts out lamellipodia and the back end ruffles up. The cell is temporarily anchored to the surface by small areas called adhesion sites.

plasm as a whole. The cytoplasm of large plant cells in particular is continually on the move. Rivers of cytoplasm stream around the cell, even forming ever-changing cytoplasmic bridges across the central vacuole. The outermost layer of cytoplasm, the ectoplasm just under the cell membrane, is static. There is a sharp boundary between it and the streaming endoplasm. At this boundary lies a layer of actin filaments, which together with myosin is believed to represent the mechanochemical machinery generating the streaming movement.

Protoplasmic streaming in slime moulds such as *Physarum* is also mediated by contraction of actin filaments, but in a somewhat different way. Small assemblies of actin filaments are being continually formed and disassembled at points along the stream of protoplasm. Contraction of the actin filaments (possibly in conjuction with myosin) is believed to create a pressure gradient and so push the protoplasm along.

Cell division (cytokinesis)

Animal cells divide by a process called cleavage. A contractile ring of microfilaments forms under the plasma membrane around the circumference of the cell midway between the two daughter nuclei. The contractile ring draws the membrane inwards until the two cells finally separate. In the contractile ring the actin microfilaments are associated with myosin to form a 'muscle-like' contractile assembly.

Movement based on microtubules

Cilia and Flagella

The cilia and flagella of eukaryotic cells are composed of organized bundles of microtubules (Fig. 8.11). The 'beat' of a cilium or flagellum is caused by the attempts of the microtubules to slide against each other. Because movement is constrained by radial links between microtubules, this produces a cycle of bending and straightening of the cilium resulting in the beating movement that propels the cell along.

The sliding movement that generates the force to drive the ciliary beat is powered by a molecular motor functionally analogous to the myosin ATPase of skeletal muscle (see above). It is **dynein**, a protein with ATPase activity. It is located in the side arms projecting along the length of the microtubules and produces sliding of one microtubule against another by 'walking' along the adjacent microtubule, each 'step' being powered by ATP hydrolysis. Because the microtubules are anchored in the basal membrane this leads to bending of the cilium.

Force generation by microtubules sliding against each other may also occur in the cytoplasm. Microtubules from a giant amoeba –

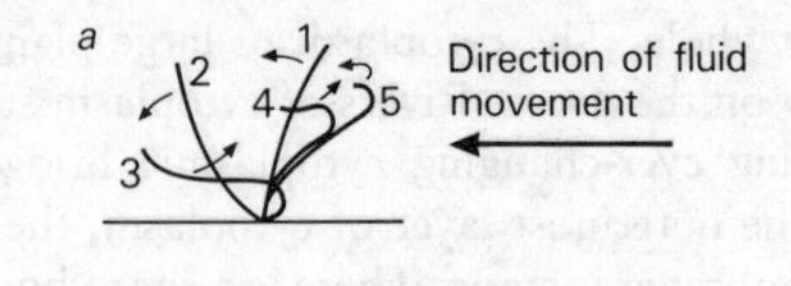

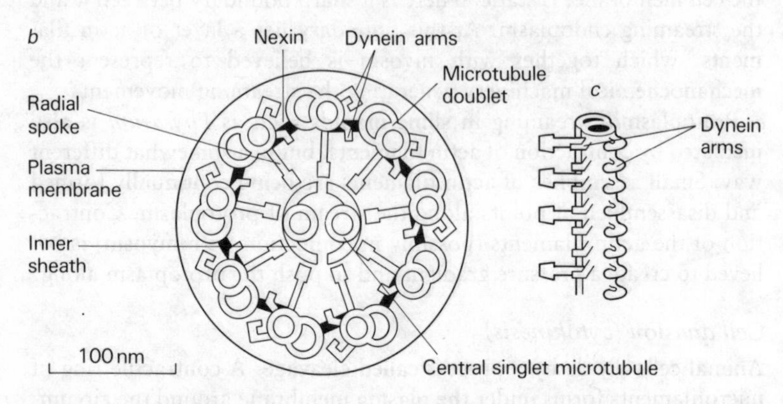

Fig. 8.11 *a*, Beating of a cilium on the surface of a eukaryotic cell. Stages 1 and 2 constitute the active stroke. *b*, Diagram of cross-sectional structure of a eukaryotic cilium or flagellum. (Note: bacterial flagella have a different structure.) *c*, Double microtubule with dynein arms and radial spokes projecting from it. (*b* and *c* redrawn with modifications from Fig. 2.10, R. McNeill Alexander, *The invertebrates*, Cambridge University Press, Cambridge, 1979.)

Reticulomyxa, which depends entirely on microtubules for its movement and intracellular transport processes – can generate force by sliding *in vitro*.

Intracellular Transport

Most cells maintain a brisk traffic in material shuttling to and from the plasma membrane and between organelles in various types of membranous vesicles (transport vesicles, secretory vesicles, lysosomes, etc.). These vesicles may be conveyed and guided to their correct destination by invisible tracks provided by the cytoskeleton. Larger organelles such as mitochondria and lysosomes, that are just visible in living cells under high-powered light microscopes, can often be seen to move jerkily backwards and forwards along a straight track that may represent an underlying cytoskeletal element.

Microtubules seem to be most involved in vesicle transport. They radiate outwards from the cell centre to the periphery and appear to be able to provide transport in either direction. The sliding of microtubules in cilia and flagella has provided the inspiration for models of how organelles may travel along cytoskeletal tracks. A molecular motor ATPase located in the crossbridges that are sometimes seen connecting vesicles and underlying microtubules, may 'walk' the organelle along the track by cycles of engagement and disengagement analogous to the action of dynein.

As microtubules have distinct polarity, growing out from the cell centre to the periphery, transport 'outwards' and 'inwards' is thought to be powered by at least two different 'motors', one of which can move material along a microtubule in the + to – direction (inwards) and the other in the – to + direction (outwards). Inward and outward transport is most clearly seen in neurones where material is transported up and down the elongated axon.

Axonal transport

Guided intracellular transport is particularly important in large free-living cells and in very elongated cells such as neurones, where the distances involved are too large for passive diffusion to convey material from one part of the cell to another rapidly enough. In neurones, proteins and other material synthesized in the cell body have to be conveyed to the axon terminal, which may be several metres away in a large animal. The axon is well supplied with neurofilaments (a type of intermediate filament) and microtubules running along its length.

As well as strengthening the axon, the microtubules appear to provide a track for two-way **axonal transport**. Newly synthesized proteins and lipids are transported from the cell body (by *anterograde transport*) and material is returned to the cell body (*retrograde transport*) for recycling. Material is shuttled along the axon in membranous vesicles of various shapes and sizes. Fast axonal transport outwards from the cell body can move a vesicle up to 400 mm per day (a passively diffusing protein would take some 20 years to travel the same distance).

There is evidence that anterograde transport occurs chiefly along microtubule tracks and is powered by an ATPase, **kinesin**. A candidate ATPase for retrograde transport has also recently been identified in one of the 'microtubule-associated proteins' (MAPs) – MAP 1C.

Chromosome Movement: The Mitotic Spindle

Before a cell divides it replicates its entire complement of DNA, and then separates the copies of each duplicated chromosome in an orderly fashion by **mitosis** to create two identical sets of chromosomes that will form the nuclei of the two new daughter cells (see Fig. 8.12). Chromosome duplication takes place in the S (DNA synthesis) phase of the

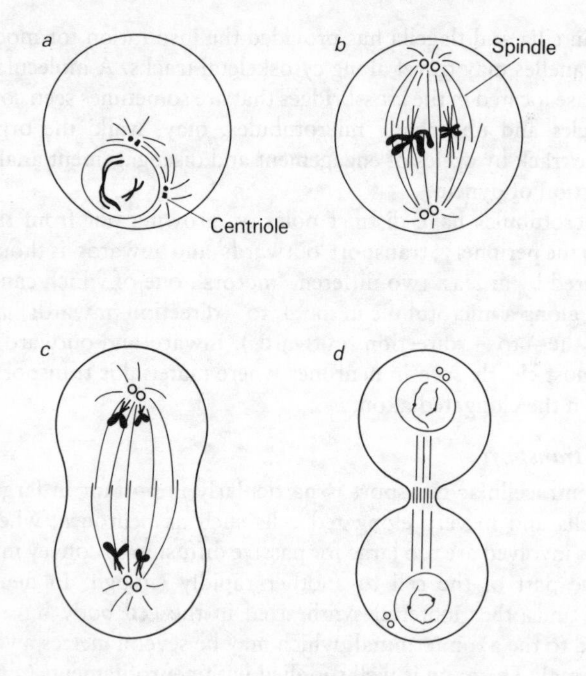

Fig. 8.12 Movement of chromosomes on the mitotic spindle: *a*, prophase; *b*, anaphase; *c*, metaphase; *d*, telophase and cytokinesis. See text for explanation.

cell cycle when the chromosomes are not visible in the light microscope. Just before mitosis, the chromosomes condense, and under appropriate illumination their movement can be followed under the light microscope in living cells.

Chromosomes are guided to their correct destination by the spindle, a highly organized system of microtubules just visible in the light microscope. (A similar spindle also guides chromosome movements through the reduction divisions of meiosis, when gametes are being formed.) How the spindle guides chromosomes to their correct destinations is now clear; how it provides the motive power for chromosome movement is, so far, less well understood. The spindle may provide an example of force generation by at least two separate mechanisms – one the differential assembly and disassembly of microtubules, the other sliding of microtubules against each other (see above, Cilia and flagella).

In animal and vascular plant cells the spindle that becomes visible under the light microscope at the end of the prophase stage of mitosis is initially formed by microtubules that grow outwards from two **organizing centres**. In animal cells, these are marked by the centrioles and the developing microtubules form structures resembling rayed stars (the **asters**). The centrioles, however, are not themselves the organizing centres from which the microtubules start to form. This function is carried out by the mass of amorphous material lying around the centrioles in animal cells and the corresponding material in plant and algal cells (which do not contain centrioles). As the microtubules extend they appear to push the organizing centres apart until they are situated opposite each other, forming the two poles of the developing spindle (Fig. 8.12*a*).

Interpolar microtubules extend from each spindle pole towards the equator, until they overlap. Because of their mode of assembly microtubules have a distinct *polarity*, that is, one end has different properties from the other. All interpolar microtubules in the mitotic spindle are oriented with their + ends at the equator. The overlapping microtubules at the equator are therefore in opposite orientations.

At the prophase stage of mitosis each chromosome consists of two identical copies – **daughter chromatids** – lying side by side and still joined at the **centromere**. On opposite sides of the centromere facing each pole a specialized region (the **kinetochore**) develops. The nucleolus disappears and the nuclear membrane starts to break down, marking the end of prophase.

Each kinetochore becomes attached to another set of microtubules, structurally distinct from the interpolar set, and known as the **kinetochore fibres**. Interactions between kinetochore and spindle microtubules align the chromosomes on the metaphase plate, a plane perpendicular to the long axis of the bipolar spindle and halfway between the poles (the metaphase stage of mitosis). After a period of oscillation the chromosomes are apparently held steady by opposing tension on each kinetochore (Fig. 8.12*b*).

The two daughter chromatids of each chromosome separate and are pulled towards the poles by their kinetochores (mitotic stage anaphase) (Fig. 8.12*c*). Chromatids move towards the poles in synchrony at a speed of about 1 μm per minute. The motive power for chromosome movement is not yet known. It may be due, at least partly, to disassembly of the kinetochore microtubules as they become visibly shorter as the chromatids move towards the poles.

At the same time, in many instances the two poles move apart. From experiments on isolated spindles from diatoms (the only mitotic spindle yet isolated that can be persuaded to elongate *in vitro*) this movement is believed to be driven by ATP-powered sliding against each other of the interdigitating anti-parallel microtubules at the equator. The molecular motor (analogous to myosin or dynein) that powers microtubule sliding has not yet been found. One strong candidate, however, is the ATPase protein kinesin, which is associated with mitotic spindles and has been implicated in vesicle transport along microtubules in neurones. *In vitro*, the interpolar microtubules also elongate on addition of tubulin units.

Once the chromatids arrive at the poles the spindle microtubules disappear and a new nuclear envelope forms around each group of daughter chromatids to make two new nuclei (the telophase stage, Fig. 8.12*d*). The chromosomes become less compact and their chromatin is dispersed throughout the nucleus.

FURTHER READING

See general reading list (e.g. Alberts et al.; Darnell et al.).

H. STEBBINGS and J.S. HYAMS, *Cell motility*, Longman, Harlow, 1979. An undergraduate-level introduction to the cytoskeleton and cell motility.

C. LLOYD, Actin in plants, *Journal of Cell Science*, 90 (1988) 185–8.

M. KIRSCHNER and T. MITCHISON, Beyond self-assembly: from microtubules to morphogenesis, *Cell*, 45 (1986) 329–42.

L.A. AMOS and W.B. AMOS, Cytoplasmic transport in axons, *Journal of Cell Science*, 87 (1987) 1–2.

J.S. HYAMS, Retrograde step for microtubules, *Nature*, 330 (1987) 106, and references therein. Commentary on the discovery of a possible retrograde molecular motor for microtubular transport in cells.

R. BURNS, Chromosome movement *in vitro*, *Nature*, 331 (1988) 479, and references therein. A short commentary on some recent work on chromosome movement and the spindle.

CHAPTER 9

HOW CELLS COMMUNICATE

The millions of cells that make up the body of a complex multicellular animal must behave in a coordinated fashion if the whole organism is to function properly. Animals send urgent messages to distant parts of the body in fractions of a second by the transmission of electrical impulses along nerves, but for short-range communication, and for less urgent long-range communication via the circulatory system they use chemical signals. Neighbouring cells can also communicate directly with each other by means of **gap junctions** in their adjoining plasma membranes that allow the passage of ions and small molecules from one cytoplasm to another.

The hormones produced by the specialized secretory cells of **endocrine** glands (such as adrenals and thyroid) enter the circulation and act on distant targets. Most cells, however, produce one or more chemical signals that are usually short-lived and only influence neighbouring cells. This is often called **paracrine** action to distinguish it from the more widely distributed actions of endocrine hormones. Some cells are also responsive to their own products, resulting in **autocrine** stimulation. Yet another type of communication is shown by nerve cells, in which neurotransmitters act at very short-range at restricted sites between one nerve cell and its immediate neighbour (see Chapter 13). Cells also communicate with each other by direct contact through mutual recognition of cell-surface proteins. Such contacts are particularly important in the immune system (see Chapter 12).

Hundreds of chemical signals in the guise of secreted hormones, neurotransmitters and neuromodulators, growth factors and mitogens, lymphokines and differentiation-inducing factors, as well as membrane-bound proteins, have already been identified and there are undoubtedly many more yet to be discovered, including those that are presumed to act transiently during embryonic development. Many different substances are used as signals by animal cells. They range from simple inorganic molecules such as nitric oxide (which has recently been identified as one of the 'endothelium-derived relaxing factors' that causes blood vessels to dilate in response to hormonal or mechanical stimuli), through small organic molecules such as the amino acid and amine neurotransmitters and the steroid hormones, to peptides and proteins (see Tables 9.1–9.6).

One important feature of the body's signalling systems is that the same signal molecule is often produced by a range of different tissues and acts at a range of different sites, and thus produces a variety of physiological effects. The polypeptide 'growth factors' are proving exceptionally versatile in this respect, with reports of new activities, sites of synthesis and sites of action appearing regularly. As just one example, the protein CSF-1, long known as a stimulant of the division and differentiation of white blood cell precursors in bone marrow, was recently reported in mouse uterine epithelium. Here its production is stimulated by oestrogen and progesterone, and where it is believed to be involved in implantation of the blastocyst and development of the placenta.

A cell's capacity to respond to any molecule usually depends on whether it carries specific **receptors** for it on the cell surface. Few of the signalling molecules used by cells can pass unaided through the plasma membrane, and most do not in fact enter their target cells to exert their effects. They are intercepted at the cell surface by receptor proteins that carry selective binding sites. The receptor proteins, in conjunction with other components of the plasma membrane, transduce the signal across the membrane to stimulate an intracellular response.

The physiological response a cell makes to a given signal is governed by its particular specialization. Apart from the steroid hormones, which act directly on DNA to switch on new gene expression, the immediate actions of most chemical signals are to activate or inhibit specific enzymes and/or to cause ion channels in the plasma membrane to open or close. The consequent intracellular changes directly or indirectly trigger the cell's response, which may be – depending on the combination of signal, receptor and cell involved – a temporary change in metabolic activity, generation of an electrical impulse, contraction or relaxation, synthesis and secretion of a particular product, or longer-term changes, such as differentiation or tissue growth.

The response is determined chiefly according to the functions a cell has become differentiated to carry out – and thus the particular enzymes, ion channel proteins, and specialized cell structures it contains. A relatively limited repertoire of signal molecules can therefore be used to produce coordinated effects in different tissues. The 'flight or fight' hormone adrenaline, for example, prepares the body for action by mobilizing carbohydrate fuel reserves in skeletal muscle and liver and speeding up the heartbeat by its action on heart muscle. Adrenaline also stimulates the breakdown of triglycerides to fatty acids in fat cells, and has effects on the smooth muscle of airways, blood vessels and other organs.

The final physiological effects of most hormones and other chemical signals are therefore bewildering in their variety. At the level of the cell, however, some unifying principles underlying signal reception and transduction have emerged during the past 30 years. The early stages of signal transduction, between receptor stimulation and the specialized cell response, proceed through one of a small number of biochemical pathways shared by many different cells (see Second messengers, G

Table 9.1 Neurotransmitters (see also Chapter 13)

Amino acids and their derivatives
Glycine, GABA (γ-aminobutyric acid, glutamate, aspartate)
Acetylcholine (ACh)
Catecholamines
Noradrenaline, adrenaline, dopamine
Serotonin (5-hydroxytryptamine, 5-HT)
Histamine

Peptides
Substance P, enkephalin

Table 9.2 The neuroendocrine system (see also Chapter 13)

Peptides released by the hypothalamus and which act in the pituitary to stimulate secretion of pituitary hormones
Corticotropin-releasing factor (CRF) (releases adrenocorticotropin (ACTH) from pituitary)
Growth-hormone releasing factor (somatoliberin)
Thyrotropin-releasing hormone (TRH)
Luteinizing hormone-releasing hormone (LHRH)
Somatostatin (inhibits release of growth hormone)
Gonadotropin-releasing hormone (GnRH) (releases follicle-stimulating hormone and luteinizing hormone)

Peptides released by the neurohypophysis (posterior pituitary) into the general circulation
Vasopressin
Oxytocin

Table 9.3 Protein hormones released from the pituitary under the influence of the neuroendocrine system

Hormone	*Main site of action*	*Effects*
Prolactin	Mammary gland	Tissue proliferation; control of milk secretion
	Corpus luteum (ovary)	Maturation of ovarian follicles
	General	Anabolic effects on metabolism
Adrenocorticotropin (ACTH)	Adrenal cortex	Production and secretion of adrenal cortical steroids (corticosterone, cortisol, aldosterone)
	Adipose tissue	Lipid breakdown
Thyroid-stimulating hormone (TSH, thyrotropin)	Thyroid gland	Secretion of thyroid hormone
	Adipose tissue	Lipid breakdown
Growth hormone (GH, somatotropin)	Liver	Production of somatomedins, intermediaries in the growth-promoting activities of GH
	General	Anabolic effects on calcium, phosphorus and nitrogen metabolism; stimulates carbohydrate and lipid metabolism; increases cardiac and skeletal muscle glycogen
Luteinizing hormone (LH)*	Ovary	Secretion of progesterone; luteinization
	Testis	Development of interstitial tissue; secretion of androgens
Follicle-stimulating hormone (FSH)*	Ovary	Development of follicles; acts with LH to stimulate secretion of oestrogen and ovulation

Table 9.3 contd.

Hormone	*Main site of action*	*Effects*
	Testis	Development of seminiferous tubules; spermatogenesis
Oxytocin	Smooth muscle, especially of uterus	Muscle contraction; parturition
	Mammary gland after birth	Milk ejection
Vasopressin (ADH)	Arterioles	Vasoconstriction, raises blood pressure
	Kidney	Water resorption

*Follicle-stimulating hormone and luteinizing hormone are also called gonadotropins as they are essential for the development and maintenance of the gonads (testis and ovaries). Other gonadotropic hormones are prolactin, and chorionic gonadotropin, a hormone produced by the placenta.

Table 9.4 Hormones produced by endocrine glands

Hormone	*Main source*	*Effects*
Amino acid derivatives		
Thyroxine (thyroid hormone, tetraiiodothyronine)	Thyroid gland	Essential for normal growth and development; wide-ranging effects
Adrenaline	Adrenal medulla (the chief source of adrenaline in mammals)	Increase in heart rate; stimulates glycogen breakdown in muscle and liver cells; stimulates triglyceride breakdown in fat cells
Steroids		
Oestrogens (oestrone, oestradiol)	Gonads	Responsible for the development of female secondary sexual

Table 9.4 contd.

Hormone	*Main source*	*Effects*
		characteristics; stimulate proliferation of the endometrium during the ovarian cycle; general anabolic effects on metabolism
Progesterone	Ovary (corpus luteum), placenta	Acts on endometrium to prepare it for implantation of the embryo; continued secretion is essential for maintenance of a pregnancy to term; has an antiovulatory effect if given at certain times during the menstrual cycle
Testosterone	Gonads	Responsible (with other androgenic hormones produced by the testis) for the development of male secondary sexual characteristics
Glucocorticoids (cortisol, corticosterone)	Adrenal cortex	Increase blood glucose by stimulating gluconeogenesis and glucose release from the liver; have anti-inflammatory action by direct effect on lymphocytes; stimulate protein synthesis in liver
Mineralocorticoids (aldosterone)		Involved in maintaining water and electrolyte balance in body

Table 9.4 contd.

Polypeptides and proteins		
Insulin*	Pancreas	Promotes uptake of glucose from the blood and synthesis of glycogen
Glucagon	Pancreas	Stimulates mobilization of carbohydrates and lipids in muscle, liver and fat cells
Parathyroid hormone (PTH)	Parathyroid glands	Involved in bone remodelling, causes increase in bone resorption and an increase in blood calcium and phosphate; also causes an increase in calcium resorption at the kidney

*Also affects ion transport and stimulates division of some cells.

Table 9.5 Some local chemical mediators

Compound	*Source*	*Effects*
Fatty acid derivatives		
Prostaglandins	Various	Wide range of actions including contraction of smooth muscle (especially of uterus), platelet aggregation and inflammation
Prostacyclin	Endothelial cells (and others)	Vasodilator by its action on smooth muscle
Small peptides		
Eosinophil chemotactic factor	Mast cells	Attracts eosinophils (a type of white blood cell)
Polypeptides and proteins		
Endothelin	Endothelial cells	Stimulates contraction of underlying smooth muscle of blood vessel wall

Table 9.6 Some biologically active polypeptides

Polypeptide growth factors (see Table 9.9)
Epidermal growth factor (EGF, urogastrone); fibroblast growth factor (FGF); platelet-derived growth factor (PDGF); insulin-like growth factor I (IGF-I, somatomedin); transforming growth factor-α (TGF-α); transforming growth factor-β (TGF-β)

Lymphokines (see Chapter 12)
Interleukins (e.g. IL-1, IL-2, IL-3); tumour necrosis factor (TNF); interferons β and γ

Factors that stimulate differentiation and proliferation of blood cell precursors (see Table 11.1)
Colony-stimulating factors (e.g. CSF-1, GM-CSF), erythropoietin

Nerve growth factors
Nerve growth factor (NGF): produced by tissues innervated by sympathetic nerves; promotes survival and growth of sympathetic neurones

Neuroleukin (identical with the enzyme glucose-6-phosphate isomerase): produced in nervous tissue; promotes growth and survival of spinal and sensory neurones, possibly because of properties unconnected with its enzymatic activity

Others
Atrial natriuretic factor (ANF, ANP, atriopeptin): produced by heart and neurones of the central nervous system; ANF produced by heart acts on kidneys to increase water flow and reduce blood pressure

Bradykinin: present in blood; causes dilatation of blood vessels and contraction of smooth muscle

Angiotensin: formed in blood from a precursor protein synthesized in the liver; increases blood pressure; stimulates release of aldosterone from adrenal cortex

---

The agents listed in this table are a few of the multifunctional polypeptides and proteins many of which were first recognized for their effects in regulating cellular proliferation. As well as stimulatory and inhibitory effects on cell division and differentiation, growth factors are increasingly being found to have other effects on cell function.

proteins and other sections below), each receptor being linked to one of these pathways.

During the past decade there has been rapid progress in this field. An important new signalling pathway, the phosphoinositide pathway, has been discovered. Also, a gap in our understanding of how receptor stimulation leads to activation of the signal transduction pathways has now been filled by the discovery of the G proteins, and their role in the cell. As the various pathways are worked out in all their biochemical detail, the points at which they may interact with each other are becoming clearer, providing the basis for a real understanding of how cells integrate the different signals they receive to produce a response appropriate to their particular circumstance.

This chapter covers some basic principles and recent discoveries in the chemical signalling systems of animal cells. In most cases the work refers to mammalian cells. Chemical signalling in multicellular plants is limited by the presence of the rigid cell wall which restricts access of large molecules to the plasma membrane. The number of plant hormones and the responses cells make are relatively limited compared with those of animals. Investigation of the molecular basis of signal transduction in plant cells is generally less advanced than in animal cells, and it is not yet clear whether exactly the same types of pathway are used.

PROTEIN PHOSPHORYLATION AND PROTEIN KINASES

One of the immediate intracellular biochemical responses to many extracellular signals is **protein phosphorylation**. Phosphorylation is one of the chief mechanisms that cells possess for rapidly activating and inactivating enzymes and other proteins, and thus very quickly altering the nature or level of their biochemical activity.

The enzymatic addition of a phosphate group (phosphoryl group) to one or more of the amino acid side chains in a protein can radically alter its activity. In certain proteins, it, and other similar covalent modifications such as methylation, results in a change in the protein's conformation, which for example, either opens up or masks an active site elsewhere on the protein (see Chapter 2: Proteins). The removal of the phosphate group by dephosphorylating enzymes restores the protein to its original state of activity. Phosphorylation and dephosphorylation regulate the activity of many important enzymes, receptor proteins, ion channels and structural proteins (Table 9.7).

Protein phosphorylation is carried out by enzymes known as **protein kinases**, which are abundant in animal cells. A particular kinase is specific both for its target protein and for the kind of amino acid it will

Table 9.7 Some proteins whose activity is altered by phosphorylation

Protein	*Kinase*	*Effect of phosphorylation*
Muscle glycogen phosphorylase	Phosphorylase kinase	Inactive *b* form converted into active *a* form
Phosphorylase kinase	Cyclic AMP-dependent protein kinase	Activated
Glycogen synthetase	Cyclic AMP-dependent protein kinase	Inactivated
β-adrenergic receptor	Various	Inactivated
Adenylate cyclase	Various	Inactivated
Myosin light chains (smooth muscle)	Myosin light chain kinase	Can interact with actin filaments

phosphorylate. The phosphoryl group is always added to the same amino acid residue. In glycogen phosphorylase, for example, phosphorylation at serine-14 converts the *b* form (usually inactive under physiological conditions) into the active *a* form.

During the past decade or so many different forms of protein kinases have been distinguished and their roles are becoming clearer. In many instances they are activated by the intracellular pathways through which external signals such as hormones, neurotransmitters and growth factors exert their effects. In the majority of cases, however, their target proteins remain to be discovered.

Tyrosine protein kinases that phosphorylate tyrosine side groups comprise part of the cell-surface receptors for some growth factors (see Table 9.9). Some of these, as well as other tyrosine kinases of unknown role, have been implicated in tumorigenesis (see Chapter 11: Table 11.4). **Cyclic AMP-dependent serine–threonine kinases** are activated by the intracellular second messenger cyclic AMP (see below, The adenylate cyclase/cyclic AMP pathway). Of particular interest at present are also the various forms of **protein kinase C**, a ubiquitous serine–threonine kinase that is activated by the second messenger diacylglycerol in response to many stimuli (see below, The phosphoinositide pathway).

CELL-SURFACE RECEPTORS

The surface of a typical animal cell bristles with many types of receptor proteins. As well as those that enable a cell to respond to chemical signals, some specialized cells – such as the rods and cones of the retina – possess proteins sensitive to light (photoreceptor proteins). Yet other receptor proteins are not concerned with signalling, but with the uptake

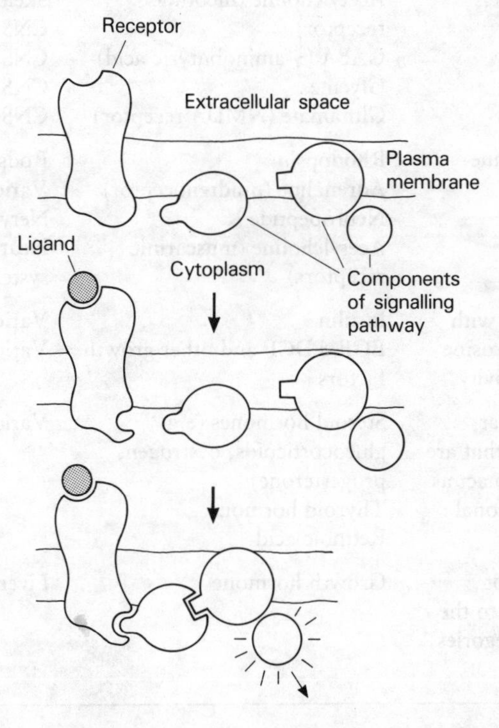

Fig. 9.1 Signal transduction across the plasma membrane. Stimulation of a receptor by its ligand leads to a change in its conformation that allows it to activate the next stage in the signalling pathway.

of molecules to meet the cell's everyday metabolic requirements (for this type of receptor see Chapter 6: Membrane transport; and Receptor-mediated endocytosis).

Receptor proteins span the plasma membrane providing a pathway for **signal transduction** across it so that intracellular changes are triggered without the signal molecule itself necessarily having to enter the

Table 9.8 Receptor classes currently distinguished on the basis of structure and function

Receptor type	*Ligand and specific receptor bound*	*Source*
Ion channels	Acetylcholine (nicotinic receptor)	Skeletal muscle, CNS
	GABA (γ-aminobutyric acid)	CNS
	Glycine	CNS
	Glutamate (NMDA receptor)	CNS
'7 membrane-spanning segments'	Rhodopsin	Rods in retina
	Adrenaline (β-adrenoceptor)	Various
	Neuropeptide K	Nervous system
	Acetylcholine (muscarinic receptors)	Heart, nervous system
Receptors with protein tyrosine kinase activity*	Insulin	Various
	EGF, PDGF and other growth factors	Various
Intracellular receptors that are thought to act as transcriptional activators	Steroid hormones (e.g. glucocorticoids, oestrogen, progesterone)	Various
	Thyroid hormone	
	Retinoic acid	
Others (not belonging to the above categories)†	Growth hormone	Liver

*These receptors are not all structurally similar.
†The receptor for growth hormone on liver cells is a single polypeptide chain of 620 amino acids with a single membrane-spanning segment.

cell. Binding of a molecule (the ligand) to a site on the extracellular face of the receptor leads to a conformational change within the protein that allows it to associate with and activate the first components of the intracellular signalling pathways (Fig. 9.1).

Over the past few years, the detailed molecular structures of some important receptor proteins have at last been elucidated with the aid of recombinant DNA techniques (membrane proteins being notoriously difficult to isolate and purify). Structural studies are revealing remarkable similarities between receptors that mediate very different responses (Table 9.8). As more receptor proteins are sequenced and their three-dimensional arrangement in the membrane resolved, it is hoped that this will throw light on the ways in which these sophisticated signal reception systems have evolved.

In many cases, the same substance will activate different kinds of receptor on different cells (or even on the same cell), so allowing a chemical signal to be interpreted in quite different ways. Receptor subtypes are at present distinguished by their different sensitivities to pharmacological *agonists* and *antagonists* that respectively bind to the receptor and activate it, mimicking the effects of the natural chemical substrate, or inactivate it, inhibiting or preventing the normal response. The neurotransmitter acetylcholine (ACh), for example, acts at two quite distinct types of receptor. The actions of one (an ion channel) are blocked by the poison nicotine, and it is therefore known as the *nicotinic* receptor. Another group of ACh receptors (with quite different structure and mode of action) are blocked by muscarine and related alkaloids and are therefore known as *muscarinic* receptors. Acetylcholine acts at nicotinic receptors on skeletal muscle to stimulate muscle contraction, and through a muscarinic-type receptor on heart muscle to slow down the heart beat. It also appears to be able to stimulate two different intracellular pathways via different types of muscarinic receptors on brain neurones.

It was by investigating the actions of serendipitously discovered poisons and drugs such as nicotine, muscarine, the opiates and the muscle relaxant atropine (belladonna), that pharmacologists were first able to find the natural chemicals they were mimicking or antagonizing, and eventually, in the past decade or so, to uncover the cell-surface receptors at which they act.

The cell-surface receptors that transduce chemical signals fall into several broad groups on the grounds of structure and mechanism of action.

Ion channels (see under Chapter 6: Ion channels, for more detail)

One group consists of ion channels that contain a binding site for a specific ligand on their extracellular face. Ligand binding opens the ion channel, and the resulting ion flow leads to the cell's response.

An example is the nicotinic acetylcholine receptor, which mediates the stimulatory effects of acetylcholine on skeletal muscle cells. Acetylcholine is released from motor nerve endings onto acetylcholine receptors in the muscle cell membrane, causing the cell to contract (Fig. 9.2).

So far, only neurotransmitters (ACh, GABA (γ-aminobutyric acid), glycine and glutamate) are known to act via ion channel receptors.

Receptors that act via second messengers

A second and much larger group of receptors acts by transducing the signal across the membrane via **G proteins**, and in many cases, stimulating the formation of intracellular **second messengers**. (G proteins and second messengers are discussed more fully later in this chapter.) To this group belong the ß-adrenergic receptor, at which adrenaline acts to stimulate glycogen breakdown in skeletal muscle and liver cells, the α1-adrenergic receptor on smooth muscle at which noradrenaline released by sympathetic nerves acts to stimulate smooth muscle contraction, and the muscarinic acetylcholine receptors on neurones (see below under Second messengers for other examples).

Rhodopsin and other opsins, the ß-adrenergic receptor, a muscarinic ACh receptor and a receptor for a neuropeptide, substance K, have been sequenced (via cDNA) and turn out to have structural features (seven membrane-spanning domains) and regions of amino acid sequence in common (see Fig. 6.3). They (and other similar receptors yet to be sequenced) are thought to have evolved from a common ancestor, acquiring their particular specificities along the way.

Receptors with protein tyrosine kinase activity

A group of receptors has been identified during the past decade that do not appear to act by any of the identified second messenger pathways. These are the receptors for certain peptide growth factors – epidermal growth factor (EGF), fibroblast growth factor (FGF), and platelet-

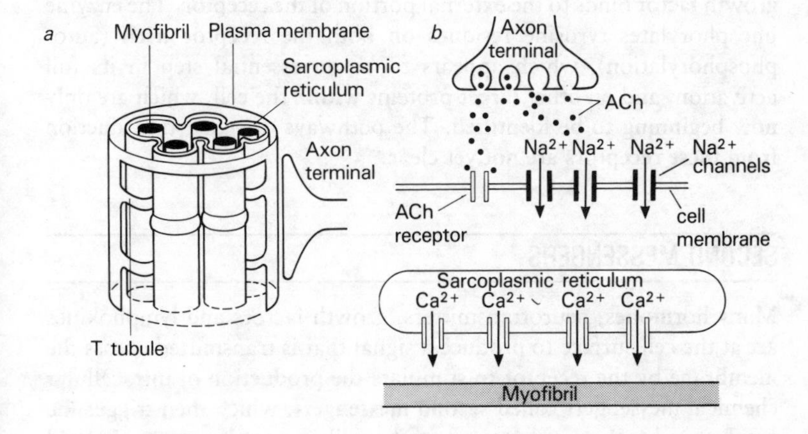

Fig. 9.2 *a*, Schematic cross-section of a skeletal muscle fibre. *b*, The sequence of events leading to muscle contraction in response to stimulation by a motor nerve. Motor neurones of the peripheral nervous system release the neurotransmitter acetylcholine (ACh) at their endings on skeletal muscle fibres when stimulated. ACh binds to its specific receptor (the nicotinic ACh receptor) in the muscle cell membrane. This is a cation channel, which temporarily opens when ACh binds, allowing sodium ions to flow into the cell down their electrochemical gradient leading to a decrease in the electrical potential difference across the membrane. (The membrane is said to depolarize as the membrane potential approaches zero from its normally negative resting value.) Membrane depolarization opens voltage-gated Na^+ channels in the plasma membrane, which results in a massive inflow of positive ions generating an electrical impulse (see Chapter 13) that is propagated throughout the muscle cell membrane. This signal is transmitted via the T-tubules to the sarcoplasmic reticulum (SR) where it leads, in some way not yet clear, to Ca^{2+} channels opening in the SR membrane releasing free Ca^{2+} into the cytosol from stores in the SR. Ca^{2+} interacts with regulatory proteins (troponin C) in the myofibrils to stimulate contraction (see Chapter 8).

derived growth factor (PDGF), and the receptors for insulin and the related insulin-like growth factor IGF-I (Table 9.9).

Although not all structurally similar, these receptors all possess an intrinsic **protein tyrosine kinase** enzymatic activity (see above, Protein phosphorylation and protein kinases), which is activated when the growth factor binds to the external portion of the receptor. The enzyme phosphorylates tyrosine residues on both the receptor itself (autophosphorylation), which appears to be an essential step in its full activation, and on other target proteins within the cell, which are only now beginning to be identified. The pathways of signal transduction from these receptors are not yet clear.

SECOND MESSENGERS

Many hormones, neurotransmitters, growth factors and lymphokines act at the cell surface to produce a signal that is transmitted across the membrane by the receptor to stimulate the production of intracellular chemical messengers called **second messengers**, which then trigger the biochemical pathways that produce the cell's eventual response. Second messengers provide a means of amplifying the signal received at surface receptors. A single activated receptor can direct the synthesis of hundreds of second-messenger molecules, each of which can in turn trigger the next step in the pathway.

A small number of second-messenger pathways transduce signals from a wide range of receptors and trigger an equally great variety of cellular responses. Each receptor is 'coupled' to a particular pathway, but there are many points at which the different pathways can intersect and influence each other, enabling a cell to integrate the various signals it is receiving.

The first second messenger to be discovered, in 1957, was the ubiquitous **cyclic AMP** (3′ ,5′ cyclic adenylic acid), which mediates the action of many hormones and neurotransmitters. Another widely used intracellular signal is provided by a rise in the cytoplasmic concentration of free **calcium ions** (Ca^{2+}) that affects many cell functions. Quite recently, an entirely new second messenger pathway has been discovered in which receptor stimulation leads to the breakdown of membrane inositol phospholipids to **inositol phosphates** and **diacylglycerol**, both of which can act as second messengers.

Table 9.9 The activities of some polypeptide and protein growth factors that act via receptors with tyrosine kinase activity

Factor	*Effects*
Epidermal growth factor (EGF, urogastrone)	Stimulates proliferation of cultured epidermal cells (keratinocytes) and fibroblasts and inhibits proliferation of hair follicle cells. Promotes precocious eyelid opening in embryonic mice. Also stimulates gastric acid secretion. Role *in vivo* and physiological site of synthesis not yet known
Fibroblast growth factor (basic FGF, bFGF)	Stimulates proliferation of many cells of mesenchymal origin (e.g. endothelial cells and fibroblasts) and inhibits growth of various tumour cells. Synthesized by endothelial cells and others. A related molecule is thought to be an embryonic morphogen in amphibians
Platelet-derived growth factor (PDGF)	Stimulates proliferation of cells of mesenchymal origin. Released from platelets during blood clotting and thought to be involved in repair of vascular system *in vivo*
Transforming growth factor-α (TGF-α)	Structurally similar to EGF, and has EGF-like activity. Can reversibly transform cultured cells to a tumour cell-like mode of growth. Produced by tumour and embryonic cells
Transforming growth factor-β (TGF-β)	Like TGF-α (to which it is totally unrelated structurally), it can reversibly transform cultured cells. Stimulates or inhibits proliferation of some cells, depending on presence of other growth factors. It and related molecules are possible embryonic developmental signals. Produced by a wide range of normal cells
Insulin	Produced by β-cells of pancreas. Stimulates uptake of glucose from blood. Also stimulates proliferation in some cultured cells
Insulin-like growth factor I (IGF-I, somatomedin C)	Produced in liver in response to growth hormone. Mediates the effects of growth hormone on cartilage and muscle. Mimics the effects of insulin in cultured cells

The adenylate cyclase/cyclic AMP pathway

Cyclic AMP (3′ ,5′ cyclic adenylic acid) (Fig. 9.3) is found throughout the living world. It is formed by the action of the enzyme adenylate cyclase on ATP. As well as being a conventional second messenger in animal cells, it is also a signalling molecule in bacteria and in slime moulds, where pulses of cyclic AMP produced by free-living amoebae cause their dramatic aggregation into a multicellular spore-forming structure. Its role in multicellular plants has not yet been established.

In animal cells, cyclic AMP synthesis is carried out by a membrane-bound adenylate cyclase. In unstimulated cells, the adenylate cyclase

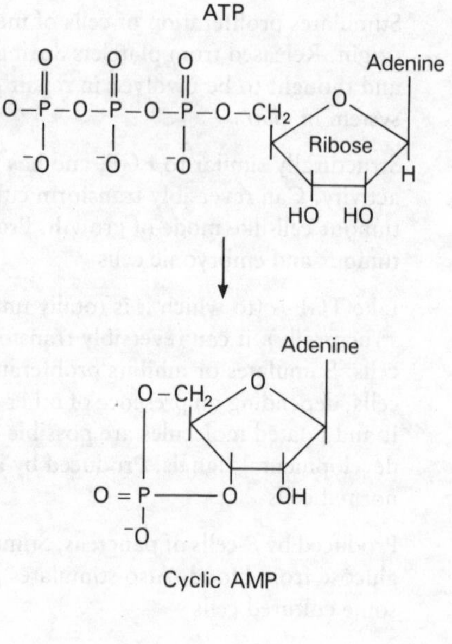

ATP
Adenine
Ribose
HO HO
Adenine
Cyclic AMP

Fig. 9.3 The formation of cyclic AMP from ATP.

converts ATP into cyclic AMP at a very slow rate. But if an appropriate receptor is activated this rapidly leads to an increase in cyclic AMP formation as the receptor in turn activates the adenylate cyclase (Fig. 9.4). The receptor and the cyclase are not physically coupled in the membrane and the nature of the link between them remained unknown for many years. This gap has now been filled by the recent discovery of the G proteins.

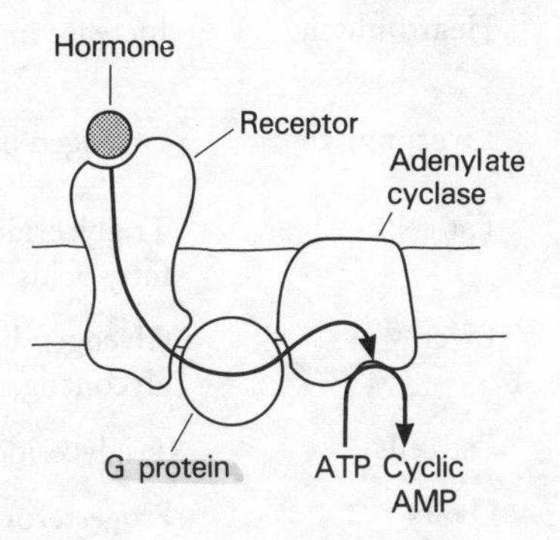

Fig. 9.4 The flow of information from activated cell-surface receptor to adenylate cyclase. Receptor, G protein and cyclase do not form a permanent complex but encounter each other as they diffuse laterally in the lipid bilayer of the membrane.

Levels of cyclic AMP in an unstimulated cell are kept very low. It is being slowly generated by adenylate cyclase and simultaneously destroyed by the enzyme **cyclic AMP phosphodiesterase** (which converts it to 5′ AMP). When the receptor is stimulated, however, the activity of the adenylate cyclase outstrips that of the phosphodiesterase and levels of cyclic AMP in the cell rapidly rise several-fold. It is this sudden change in intracellular concentration that activates the response machinery.

Action of Cyclic AMP

Cyclic AMP is a second messenger for many receptors (see Table 9. 10) and in all cases is believed to act by activating one of a number of **cyclic AMP-dependent protein kinases**. Protein kinases are enzymes that add phosphate groups to the amino acid side chains of their target proteins and in doing so alter these proteins' activity. Phosphorylation is one of the basic mechanisms that cells possess for activating and inactivating enzymes and other proteins, and thereby rapidly altering the nature and level of their biochemical activity in response to an external signal.

Table 9.10 Some agents that act via cyclic AMP

Agent	*Site of action*	*Effect*
Thyroid-stimulating hormone	Thyroid	Synthesis and secretion of thyroid hormone
Adrenaline (α-receptor)*	Heart muscle	Increase in heart rate
Adrenaline (β-receptor)	Liver; muscle	Glycogen breakdown
	Fat cells	Triglyceride breakdown to fatty acids
Glucagon†	Liver	Glycogen breakdown; glyconeogenesis; urea synthesis
	Fat cells	Triglyceride breakdown
Luteinizing hormone	Ovary	Progesterone secretion
Parathyroid hormone	Bone	Bone resorption (during remodelling and growth)
Vasopressin	Kidney	Water resorption
Odorant molecules†	Olfactory cilia	Generation of signal for transmission to brain
Histamine	CNS	Neurotransmitter
	Other tissues	e.g. acts on walls of small blood vessels making them leaky, allowing serum proteins and fluids to enter tissues
Prostaglandins	Various	See Table 9.5
Stimuli that inhibit adenylate cyclase		
Opiates and opioid peptides such as enkephalin	CNS and elsewhere	Not yet known at biochemical level; produce analgesia and mood changes
Adenosine	CNS	Neuromodulation
Somatostatin	Pituitary	Inhibition of growth hormone release

Table 9.10 contd.

Agent	*Site of action*	*Effect*
Acetylcholine (at muscarinic receptors)	CNS	Neurotransmitter

*Adrenaline also has multiple effects in other tissues, including relaxation of smooth muscle, and vasodilation of blood vessels (via the intermediary of the endothelial cells that line blood vessels).
†See also Table 9.11.

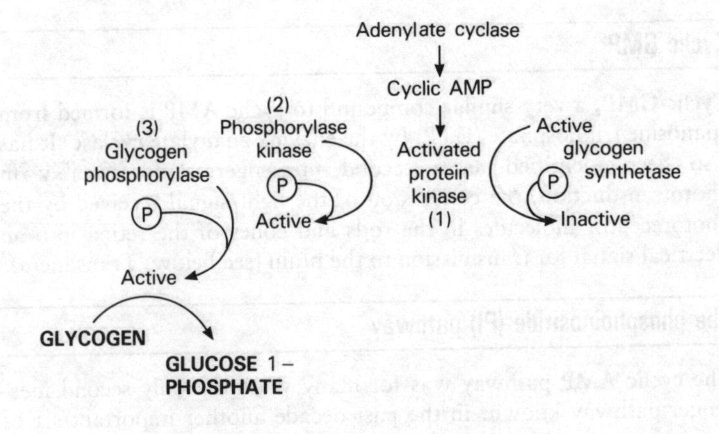

Fig. 9.5 The action of cyclic AMP produced after stimulation of the ß-adrenergic receptor of muscle by adrenaline. Cyclic AMP activates its dependent kinase (1). This phosphorylates and activates another enzyme, phosphorylase kinase (2), which in turn phosphorylates individual subunits of the multisubunit enzyme glycogen phosphorylase (3). At each stage, the reaction is amplified through this phosphorylation cascade. The phosphorylated subunits can then combine into an active enzyme that attacks the glucose polymer glycogen, hydrolysing it to glucose phosphates. Cyclic AMP activates its dependent kinase by binding to regulatory sites on the inactive form of the enzyme, causing the release of two identical protein subunits that are active enzymes in their free form. At the same time, cyclic AMP is also shutting down glycogen synthesis by activating a parallel pathway. The same cyclic AMP-dependent kinase inactivates the glycogen-synthesizing enzyme by phosphorylating it.

The targets for cyclic AMP-dependent kinases in most cells are still unknown but a complete sequence of reactions from receptor stimulation to final response has been worked out for skeletal muscle cells, in which the hormone adrenaline acting at ß-adrenergic receptors and via the cyclic AMP pathway stimulates the breakdown of glycogen to glucose phosphates that the muscle cell uses to fuel respiration (Fig. 9.5).

The action of cyclic AMP is terminated by its destruction by cyclic AMP phosphodiesterase. In the case of muscle, after the signal has ceased the status quo is restored by a series of enzymatic dephosphorylations that begin as levels of cyclic AMP fall, and that convert the enzymes back to their original states.

As well as the chemical signals that activate adenylate cyclase, there are some that act by inhibiting it and thus suppressing the formation of cyclic AMP (see below, G proteins).

Cyclic GMP

Cyclic GMP, a very similar compound to cyclic AMP is formed from guanosine triphosphate (GTP) by the enzyme **guanylate cyclase.** It has also been identified as a second messenger, most notably in **phototransduction**, the conversion of the light signal received by the photoreceptor molecules in the rods and cones of the retina into an electrical signal for transmission to the brain (see below, Transducin).

The phosphoinositide (PI) pathway

The cyclic AMP pathway was for many years the only second messenger pathway known. In the past decade another important set of second messengers has been discovered – those of the phosphoinositide (PI) pathway. Some agents known to activate this pathway are listed in Table 9.11. The ramifications of the PI pathway are so far less well understood than those of cyclic AMP. It is more complex, as two second messengers are produced on receptor stimulation, which have quite different actions in the cell. There are also many points at which the cyclic AMP and PI pathways may influence one another.

The first step in the PI pathway after receptor stimulation is the activation of a membrane-bound enzyme, a phosphodiesterase often known as **phospholipase C**, which breaks down the phospholipid **phosphatidylinositol 4,5-bisphosphate** in the plasma membrane to give **inositol 1,4,5-trisphosphate ($InsP_3$)** and **diacylglycerol** (Fig. 9.6). Receptor stimulation is believed to be linked to activation of phospholipase C via a G protein, which has not yet been identified. Both $InsP_3$ and diacylglycerol have second messenger activity.

Table 9.11 Some agents that activate the phosphoinositide (PI) pathway

Agent	*Tissue*	*Effects*
Bradykinin Bombesin α-thrombin	Various	Stimulate DNA synthesis and cell division
Glucagon*	Liver cells	Glycogen breakdown
Glutamate (via an alternative receptor to the 'NMDA' glutamate receptor, which is an ion channel)	CNS neurones	Neurotransmitter
Acetylcholine (at muscarinic receptor)	CNS neurones	Neurotransmitter
Noradrenaline (at α1b-adrenergic receptor)	Some smooth muscle (e.g. liver and spleen)	Contraction via release of Ca^{2+} from intracellular stores
Histamine (at H1 receptors)	Various, including CNS	Neurotransmitter in CNS; various effects on endothelial cells of blood vessels
Mitogen action on T cells (see Chapter 12: Lymphokines)		$InsP_3$ activates Ca^{2+} channels in plasma membrane, influx of Ca^{2+} eventually leads to cell division
Odorant molecules†	Olfactory cilia	Sensory transduction

*Glucagon used to be a textbook example of a hormone that acted via the cyclic AMP pathway. Very recently, an alternative, and possibly physiologically predominant action of glucagon via a second receptor and PI pathway has been discovered in liver cells. The gastric hormone secretin has also recently been found to stimulate secretion of pancreatic digestive enzymes via both cyclic AMP and PI pathways.

†See also Table 9.10.

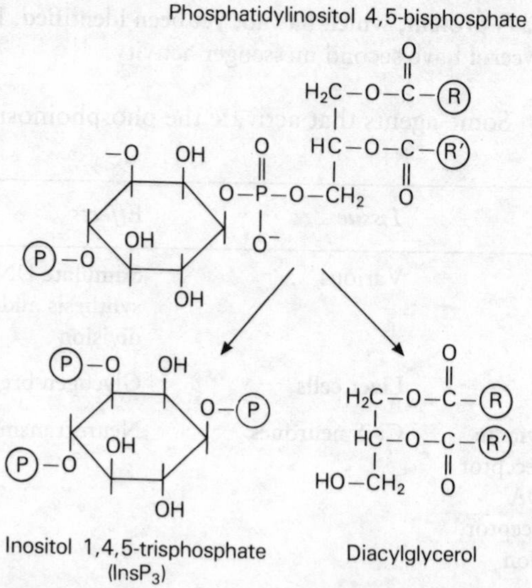

Phosphatidylinositol 4,5-bisphosphate

Inositol 1,4,5-trisphosphate ($InsP_3$)

Diacylglycerol

Fig. 9.6 The formation of inositol trisphosphate and diacylglycerol from phosphatidylinositol phosphate.

Inositol 1,4,5-trisphosphate ($InsP_3$)

The main action of $InsP_3$ so far identified in animal cells is to release Ca^{2+} from intracellular stores (see below, Calcium). Calcium is itself an intracellular messenger, stimulating a variety of cellular responses. $InsP_3$ receptors on endoplasmic reticulum have been found at which $InsP_3$ may act to open Ca^{2+} channels in the endoplasmic reticulum membrane, and $InsP_3$ has been implicated in the release of Ca^{2+} in smooth muscle in response to hormones and neurotransmitters such as noradrenaline that cause contraction of smooth muscle. Like all signalling molecules $InsP_3$ must be rapidly destroyed once it has acted. Cells that use the PI signalling pathway contain a phosphatase that converts $InsP_3$ to inositol 4,5-diphosphate ($InsP_2$). $InsP_3$ has recently been shown to undergo other conversions – to Ins 1,3,4,5-tetraphosphate ($InsP_4$), for example, which may also act as a second messenger. (Botanists will also be familiar with inositol phosphates in the form of phytate (hexaphosphoinositol), a phosphate storage compound found in some seeds.)

Diacylglycerol and Protein Kinase C

The main action of diacylglycerol identified so far is to activate the phosphorylating enzyme **protein kinase C**. Protein kinase C (a serine–threonine kinase) is a versatile enzyme and is able to phosphorylate and alter the activity of many different proteins. Very recently, the 'protein kinase C' activity studied in tissue extracts has been shown to be a mixture of at least three slightly different forms of the enzyme. This may explain the bewildering variety of targets the enzyme seems to attack. It has been demonstrated to phosphorylate receptors, adenylate kinase, the α subunit of G proteins, and many more. Unravelling the ramifications of protein kinase(s) C action in the responses of various cells to receptor stimulation is now a growth area in cell physiology. The enzyme is abundant in the brain and is believed to mediate the actions of some neurotransmitters and neuromodulators.

Calcium (Ca^{2+})

A wide variety of stimuli cause a rapid and transient increase in the level of free Ca^{2+} in the cytosol of animal cells, which acts as a second messenger to trigger many responses.

The concentration of free Ca^{2+} in the cytosol of animal cells is normally very low (less than 10^{-7} M) even though the total concentration of calcium in a cell is often equal to the extracellular concentration (on average greater than 10^{-3} M). Most calcium in a cell is bound to phosphates or macromolecules such as the calcium-binding proteins, or stored in organelles, especially mitochondria and the endoplasmic reticulum. This leads to steep calcium gradients across both the plasma membrane and organelle membranes, which are maintained by Ca^{2+}-ATPase pumps (see Chapter 6: Ion pumps) in the membranes that continually pump Ca^{2+} out of the cytosol. If a signal leads to calcium channels opening in these membranes, Ca^{2+} flows rapidly down the gradient into the cytosol.

In some cases, the intracellular increase in free Ca^{2+} is caused by extracellular stimulation of receptors that are closely coupled to Ca^{2+} channels in the plasma membrane, which open to allow Ca^{2+} to flow into the cell. In other cases, Ca^{2+} is released from the intracellular stores such as the endoplasmic reticulum by the action of intracellular second messengers – the second-messenger action of inositol trisphosphate ($InsP_3$) is chiefly the result of its release of Ca^{2+} from the endoplasmic reticulum.

Increased cytosolic Ca^{2+} has a wide variety of effects. One of the first changes seen in a fertilized egg is a massive and temporary increase in cytosolic Ca^{2+} as a result of its release from intracellular stores. This increase is the stimulus that, by means not yet known, unmasks mRNAs and activates proteins to set the egg off on its developmental programme. Calcium also triggers secretion from many cells. In nerve cells, electrical impulses arriving at the axon terminals lead to an inflow of Ca^{2+} across the plasma membrane that triggers the release of neurotransmitter (see Chapter 13: Synaptic transmission). In muscle cells, calcium released from internal stores after stimulation by a nerve impulse causes contraction by interacting with regulatory sites on the muscle fibrils. A rise in intracellular Ca^{2+} is also one of the first effects of the action of the growth factors that stimulate cell division.

How Ca^{2+} acts to trigger secretion is not yet known, but it may interact with cytoskeletal components to release secretory vesicles and allow them to fuse with the plasma membrane. Calcium is involved in the movement of spindle microtubules during mitosis and meiosis. It also acts directly to regulate enzymes and ion channels in a wide variety of cells.

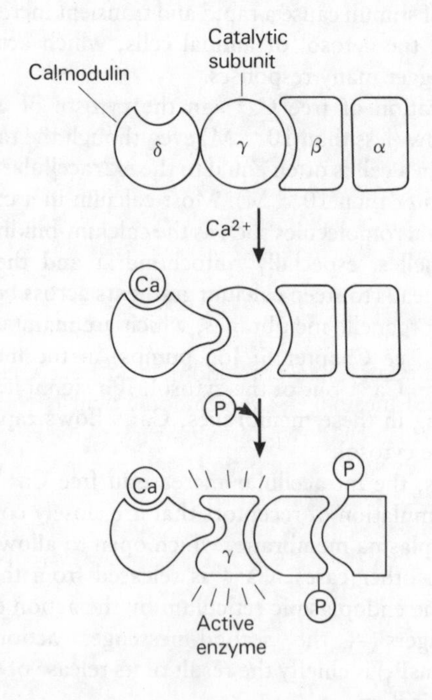

Fig. 9.7 The calcium-binding protein calmodulin forms a regulatory subunit of the multisubunit enzyme glycogen phosphorylase, which can be activated only by phosphorylation after Ca^{2+} has bound. The enzyme complex consists of four copies of each of the subunits shown schematically in the figure. Each molecule of calmodulin has four Ca^{2+}-binding sites.

The action of calcium is largely mediated through **calcium-binding proteins**. These proteins undergo an extensive change in conformation on calcium binding. In this activated state, calcium-binding proteins bind to and react with many targets, which include enzymes and elements of the cytoskeleton. The cytoskeleton is probably an important target for Ca^{2+} action, as a number of calcium-binding proteins have been found associated with it.

Calmodulin

One of the most abundant and best-studied calcium-binding proteins is **calmodulin**, and its activities illustrate the general role of calcium-binding proteins as effectors of calcium action (Fig. 9.7). It is ubiquitous throughout eukaryotic organisms and can constitute up to 1 per cent of total cell protein. It is related in structure to troponin C, the calcium-regulated component of skeletal muscle. Calmodulin forms a regulatory subunit of the enzyme phosphorylase kinase (which is involved in the cyclic AMP-regulated glycogen breakdown pathway).

Calmodulin can also regulate many other enzymes of which it is not a permanent part, including adenylate cyclase, cyclic nucleotide phosphodiesterases, membrane-bound Ca^{2+}-ATPases and the myosin light-chain kinase of both muscle and non-muscle cells. It is also associated with mitotic spindles and the actin filament bundles and intermediate filaments of the cytoskeleton.

Other second messengers

Metabolites of the fatty acid arachidonic acid have recently been implicated as second messengers. Arachidonic acid is formed from the receptor-stimulated breakdown of membrane phospholipids. The free arachidonate is rapidly metabolized to eicosanoids, of which the hydroxyeicosatetraenoic acids have been identified as second messengers mediating the action of a neuroactive peptide involved in neuromodulation in a simple molluscan nervous system.

GUANINE-NUCLEOTIDE BINDING PROTEINS

A recent discovery has been a class of membrane-associated proteins that bind guanine nucleotides (GDP and GTP) and hydrolyse GTP, and which link many cell-surface receptors to the intracellular biochemical machinery that generates second messengers such as cyclic AMP, inositol phosphates and diacylglycerol. Various forms of these proteins are involved in transducing signals as diverse as light, odours, hormones, neurotransmitters and growth factors, and, in a slightly different incarnation, they are an essential component of the protein synthetic apparatus. As well as providing a link between receptor and second messenger generation, they may also act directly in some cases to produce a response without the intervention of a second messenger. One type of G protein – the ras proteins – whose normal function is not yet known, has been implicated in human cancer (see below and Chapter 11: Oncogenes).

G proteins

The G proteins that transduce signals from cell-surface receptors are located on the inner surface of the plasma membrane and interact with other membrane components. Receptors are linked to either the adenylate cyclase or the phosphoinositide pathway via various forms of G proteins, some of which have now been isolated and purified. G proteins are also involved in mediating responses to signals (such as insulin, or acetylcholine acting at muscarinic ACh receptors in the heart) that do not appear to act through either of these pathways.

The main types of G protein isolated so far (e.g. G_s, G_i, G_o and G_T (transducin)) all consist of three subunits, α, ß and γ. Several forms of the α subunit exist for each type of protein, some encoded by separate genes (i.e. the various α_i subunits), some produced by alternative splicing of mRNA (the α_s subunits). The ß and γ subunits also exist in several forms.

The α subunit binds the guanine nucleotide and also possesses GTPase activity. The α subunit is also the target for some bacterial toxins (pertussis and cholera toxin) which irreversibly activate it by attaching an ADP-ribose derived from nicotinamide adenine dinucleotide. Depending on the cells affected this leads to the disturbances in metabolism characteristic of these diseases, such as the outflow of water from the epithelial cells of the gut that produces the often fatal diarrhoea and dehydration of cholera.

G Proteins and the Cyclic AMP Pathway

The interaction of G proteins with receptors that act via the cyclic AMP pathway is best understood at present. Hormones and other chemical signals that stimulate cyclic AMP production (see Table 9.10) have in many cases now been shown to act via a G_s (stimulatory) protein (sometimes called N_s). (Several different forms of the G_s α subunit have been identified.)

In its inactive form G_s consists of the three subunits α, ß and γ. The α

subunit carries a bound GDP molecule. In response to a signal from an activated receptor, the G protein exchanges the GDP for GTP (Fig. 9.8). The activated α subunit is then believed to dissociate from the rest of the protein and activate the enzyme adenylate cyclase which generates an increase in cyclic AMP. Eventually the GTP is hydrolysed to GDP by the GTPase activity possessed by the α subunit. The α subunit can activate several hundred molecules of adenylate cyclase before its GTPase activity hydrolyses the GTP to GDP and the subunit becomes inactive. It then becomes reunited with a β/γ subunit, which appears not to dissociate into separate subunits and which may serve to anchor the protein in the membrane.

Chemical signals that act by shutting down intracellular cyclic AMP production are transduced via a G_i (N_i) (inhibitory) protein that behaves in the same way as the G_s protein except that the activated α/GTP subunit inhibits adenylate cyclase. The β and γ subunits of G_s and G_i are similar but the α subunits are different. G_i has also been shown to stimulate the phosphoinositide pathway.

Some pathological states are now known to be caused by disturbances or defects in G protein activity. **Pseudohypoparathyroidism** is an inherited disease in which cells appear non-responsive to parathyroid hormone (PTH). The action of parathyroid hormone is mediated via a G-protein-stimulated increase in cyclic AMP. It is now known that the basic defect in pseudohypoparathyroidism is a generalized partial deficiency of G_s. This explains not only the growth defects caused by the lack of parathyroid hormone action on bone remodelling, but also the fact that many patients have a much-reduced sense of smell, G_s also being involved in the transduction of odour signals in the nasal epithelium.

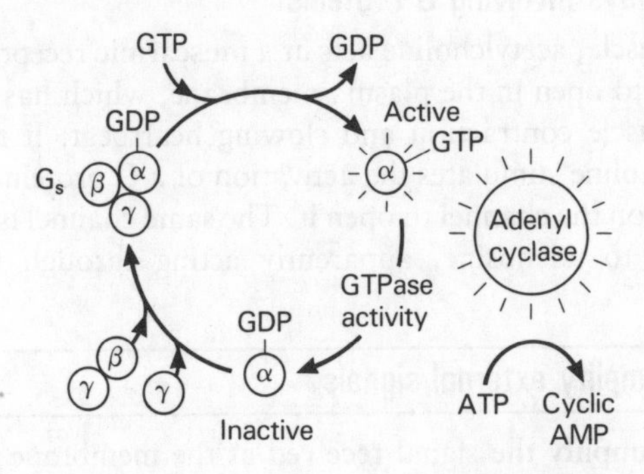

Fig. 9.8 The cycle of activation and inactivation of the guanine-nucleotide binding protein G_s.

G Proteins and Phosphoinositol and Diacylglycerol

G proteins are also involved in transducing signals from receptors that act via the inositol phosphate/diacylglycerol second messenger pathway. The G proteins concerned have not all yet been identified but include both G_i and G_o. They are presumed to act by stimulating the membrane-bound enzymes phospholipase C and phospholipase A_2 which break down membrane inositol phospholipids into inositol phosphates and diacylglycerol.

Transducin

The G protein involved in transducing light signals via the photoreceptor molecule rhodopsin in the rod and cone cells of the retina is called **transducin.** It is activated by light-activated rhodopsin. It has the same type of three-subunit structure as G_s and G_i and a similar mode of action but activates, instead of adenylate cyclase, a cyclic GMP phosphodiesterase. This enzyme destroys cyclic GMP and the sudden drop in cyclic GMP levels in the rod cell transiently closes cyclic GMP-gated cation channels in the cell membrane. When these channels are open the membrane is depolarized. As the level of cyclic GMP in the cell drops, it dissociates from the channel which then closes. The closing and reopening of this channel as cyclic GMP is replenished by the constitutive action of guanylate cyclase, and the consequent hyperpolarization and depolarization of the membrane, generates an electrical signal that is eventually transmitted via the optic nerve to the brain. There are different forms of the transducin α subunit in rod and cone cells.

Other Pathways Involving G Proteins

On heart muscle, acetylcholine acts at a muscarinic receptor to cause a K^+ channel to open in the plasma membrane, which has the effect of reducing muscle contraction and slowing heartbeat. It now appears that acetylcholine stimulates the activation of a G protein (G_K), which acts directly on the channel to open it. The same channel is also opened in response to adenosine, apparently acting through the same G protein.

G proteins amplify external signals

G proteins amplify the signal received at the membrane several hundred-fold or more since one activated receptor can activate many G proteins, and each α/GTP subunit can turn on (or turn off) many effector molecules before its active life is terminated. They also provide an additional step at which the cascade of reactions set off when the receptor is stimulated can be controlled by other cell machinery. Signals coming from two different receptors can interact at various points in the pathway, cancelling each other out or enhancing each other's effects. For example, clues to how insulin antagonizes the action of glucagon in liver cells have emerged recently with the discovery that insulin may, by some means as yet unknown, activate G_i, which inhibits the action of adenylate cyclase. Glucagon, on the other hand, stimulates adenylate cyclase, which in these cells leads to general mobilization of stored carbohydrates. Such 'cross-talk' at the level of receptors and G proteins may provide cells with a way of integrating hormonal and other signals and producing a response appropriate to the circumstances.

However, the presence of several different G proteins in most cells, the ability of a single type of G protein to be activated by many different receptors, and the ability of some types of G protein to activate several different effector pathways raises an important question. Given this apparent promiscuity amongst receptors and G proteins, how does the cell manage to make the appropriate specific response to stimulation at a given receptor? There are few answers yet.

The ras proteins

These are a group of membrane-associated proteins encoded by the *ras* family of proto-oncogenes in mammalian cells (see Chapter 11: Oncogenes). Similar proteins have been found in other eukaryotic cells, including yeasts. Ras proteins are composed of a single polypeptide chain (M_r 21 000) that bears some resemblance to the α-subunit of G proteins in amino acid sequence and its interaction with guanine nucleotides. Their role in the cell is not yet clear. In mammalian and other vertebrate cells they are implicated in the regulation of growth and cell division in response to a variety of mitogens and growth factors, including insulin, and are thought to be part of some signal transduction mechanism. They are the subject of intense research interest at present in light of the fact that altered ras proteins appear to be involved in some types of human cancer.

Like the α subunit of G proteins, ras proteins have GTPase activity. But, unlike them, it appears that in normal cells ras exists mostly in an active complex with GTP rather than GDP. Ras protein activity may be regulated by a recently discovered protein that activates the ras protein GTPase, resulting in the production of the inactive ras/GDP complex

(Fig. 9.9). In cancer cells the altered ras protein appears unable to interact with this protein (the GTPase activating protein (GAP)) and unregulated ras activity might result in uncontrolled growth. Very little is yet known about the interaction of ras proteins with GAP.

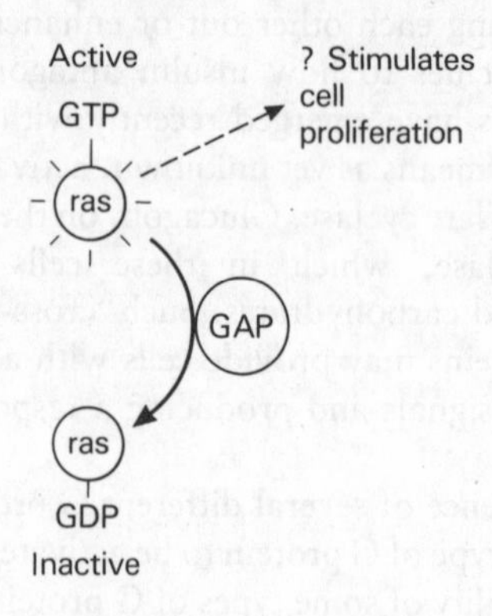

Fig. 9.9 The regulation of ras protein activity by GTPase activating protein (GAP).

Protein synthesis elongation factors

The molecular mechanism represented by the guanine-nucleotide binding proteins is of considerable evolutionary antiquity. It appears in bacterial elongation factors – essential components of the translational machinery for decoding mRNA into protein. A part of the elongation factor EF-Tu binds a GDP molecule, which is replaced by GTP. In this activated form, EF-Tu binds an aminoacyl-tRNA and delivers it to the correct site on the ribosome. As this is accomplished, GTP is hydrolysed to GDP and the EF-Tu/GDP complex is released. Another guanine-binding elongation factor EF-G powers the translocation step in which the mRNA is moved along the ribosome to present the next codon for translation (see Chapter 2: The genetic code and translation).

HORMONE ACTION VIA INTRACELLULAR RECEPTORS

Most chemical signals are intercepted at the cell surface. A few small lipid-soluble signalling molecules, however, pass through the plasma membrane by simple diffusion and are then picked up by intracellular receptor proteins. The steroid hormones produced by the adrenal cortex (the glucocorticoids cortisol and cortisone and the mineralocorticoid aldosterone) and by the reproductive organs (oestrogen, progesterone, oestradiol, etc.) and the non-steroid thyroid hormone thyroxine, behave in this way.

All these hormones have multifarious and complex physiological effects. The sex hormones oestradiol and testosterone induce complex long-term changes, such as sexual maturation and the development of secondary sexual characteristics. The glucocorticoids produced by the adrenal cortex affect carbohydrate and protein metabolism and aldosterone is involved in maintaining the water and electrolyte balance of body fluids. Thyroxine is essential for growth and development, and, in amphibian tadpoles, induces metamorphosis. Another member of this group is the insect hormone ecdysterone which is needed for pupation. Cortisol and cortisone are also potent anti-inflammatory agents, as a result of their direct and as yet unexplained cytolytic effect on lymphocytes and their induction of the body's own anti-inflammatory substances such as lipocortin.

Unlike most other hormones, whose immediate actions are to regulate enzyme or ion channel activity, thyroxine and the steroid hormones are believed to act directly on DNA to switch on the transcription of a hormone-specific set of genes. Once inside the cell, these hormones bind to receptor proteins in the cytoplasm (thyroxine binds to receptors already in the nucleus), inducing a conformational change in the protein that gives it a high affinity for DNA. The receptor-hormone complexes enter the nucleus through the nuclear pores and remain there, bound to DNA (Fig. 9.10). A few genes that are 'hormone-responsive' have been identified and it is believed that the hormone-receptor complexes preferentially bind to the control sites of these genes, switching

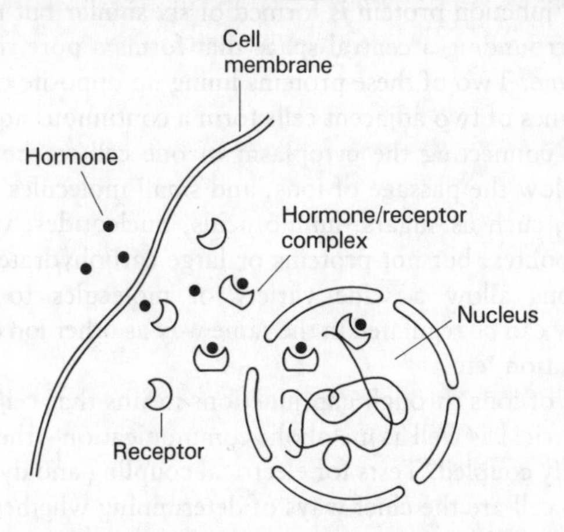

Fig. 9.10 Action of steroid hormones via their intracellular receptors.

on their transcription. It has been estimated that each hormone may switch on a set of up to 50 genes.

A few hormone-sensitive genes containing specific binding sites for the hormone/receptor complex have been identified in cultured cells (e.g. glucocorticoids can activate growth hormone and metallothionein genes), and their regulation by steroid receptor/hormone complexes *in vitro* has been intensively studied as an unusually clear-cut and easily accessible example of the action of a gene-regulatory protein in animal cells. In what circumstances the hormones activate these genes *in vivo* and how this relates to their physiological effects remains to be fully elucidated. A defective form of the thyroid hormone receptor has also been identified as an oncogene (see Chapter 11: Oncogenes).

A new and exciting addition to this group of chemical signals has been made recently. Retinoic acid, which has been identified as a vertebrate embryonic morphogen (see Chapter 10: The ZPA morphogen), binds to an intracellular receptor which has some features in common with the thyroid hormone receptor. Other members of this class still in search of a receptor are the male hormone testosterone, the poison dioxin, and the insect steroid hormone ecdysterone.

GAP JUNCTIONS

Direct communication between the cytoplasm of one cell and the next is through **gap junctions**. These are small regions of the plasma membranes of two adjacent cells in which are clustered a particular type of channel protein – the gap junction protein (Fig. 9.11).

Each gap junction protein is formed of six similar but not identical subunits surrounding a central space that forms a pore right through the membrane. Two of these proteins lining up opposite each other in the membranes of two adjacent cells form a continuous aqueous channel directly connecting the cytoplasm of one cell to the other. Gap junctions allow the passage of ions, and small molecules of up to M_r 1000–1500, such as sugars, amino acids, nucleotides, vitamins and other metabolites, but not proteins or large carbohydrates. Although gap junctions allow a wide variety of molecules to pass, their permeability can be regulated in the same way as other ion channels, by phosphorylation, etc.

The flow of ions through gap junctions means that cells are potentially in electrical as well as metabolic communication – they are said to be **electrically coupled**. Tests for electrical coupling and dye movement from cell to cell are the chief ways of determining whether cells are in direct communication.

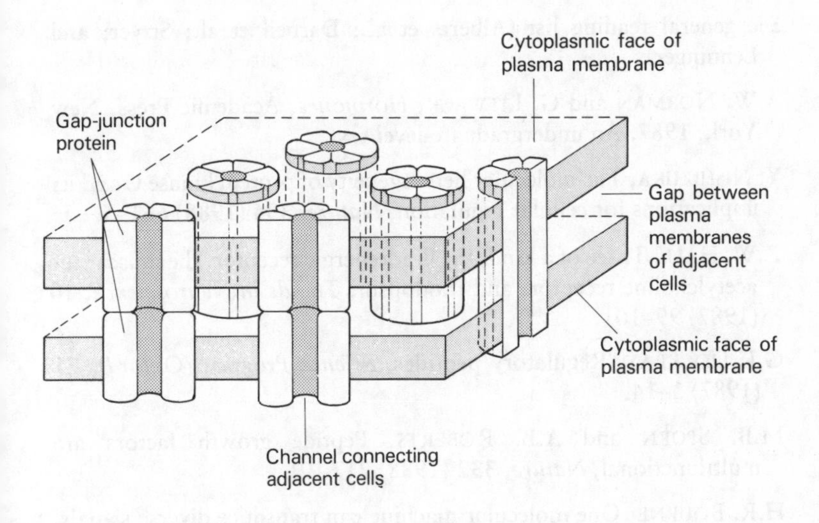

Fig. 9.11 Schematic view of gap-junction proteins in membranes of adjacent cells.

Gap junctions couple the cells of many tissues and their function is most obvious in electrically excitable cells such as muscle and nerve. Synchronous contraction of, for example, heart muscle or intestinal muscle is mediated by electrical coupling between the individual cells via gap junctions. In the nervous system, **electrical synapses** comprised of gap junctions allow a passive, rapid, localized flow of electrical signals between coupled neighbouring cells without the intervention of a chemical synapse and the consequent delay in transmission.

As a generality, gap-junctional communication in other tissues is presumed to allow an equal distribution of useful metabolites throughout a tissue or organ (such as the liver). Whether gap junctions have any more specific communication functions in cells other than muscle and nerve is still a moot point. The cells of many developing tissues are in direct communication with each other via gap junctions, and at various stages of development changes occur in the pattern of coupling. But whether this is simply a consequence of development or whether, for example, developmental signals may diffuse through gap junctions within a circumscribed area of tissue is still not clear.

FURTHER READING

See general reading list (Alberts et al.; Darnell et al.; Stryer; and Lehninger).

A.W. NORMAN and G. LITWACK, *Hormones*, Academic Press, New York, 1987. An undergraduate-level text.

Y. NISHIZUKA, The molecular heterogeneity of protein kinase C and its implications for cellular regulation, *Nature*, 334 (1988) 661–5.

Z.W. HALL, Three of a kind, the ß-adrenergic receptor, the muscarinic acetylcholine receptor, and rhodopsin, *Trends in Neuroscience*, 10 (1987) 99–101.

G.J. DOCKRAY, Regulatory peptides, *Science Progress (Oxford)*, 71 (1987) 1–14.

M.B. SPORN and A.B. ROBERTS, Peptide growth factors are multifunctional, *Nature*, 332 (1988) 217–19.

H.R. BOURNE, One molecular machine can transduce diverse signals, *Nature*, 321 (1986) 814–16; S.R. PENNINGTON, G proteins and diabetes, *Nature*, 331 (1987) 188–9; H.R. BOURNE, Discovery of a new oncogene in pituitary tumours, *Nature*, 330 (1987) 517–18; A.H. DRUMMOND, Lithium affects G-protein receptor coupling, *Nature*, 331 (1987) 388; Short reviews and commentaries on various aspects of G protein action.

E.J. NEER and D.E. CLAPHAM, Roles of G protein subunits in transmembrane signalling, *Nature*, 333 (1988) 129–34.

J. HURLEY, M. SIMON, D. TEPLOW, J. ROBISHAW and A. GILMAN, Homologies between signal transducing G proteins and ras gene products, *Science*, 226 (1984) 860–2.

M.J. BERRIDGE, Inositol trisphosphate and diacylglycerol: two interacting second messengers, *Annual Review of Biochemistry*, 56 (1987) 159–93.

M.J. BERRIDGE and R.F. IRVINE, Inositol trisphosphate, a novel second messenger in cellular signal transduction, *Nature*, 312 (1984) 315–21.

R.E. EVANS, The steroid and thyroid hormone receptor superfamily, *Science*, 240 (1988) 889–94.

PART III *Multicellular Systems*

CHAPTER 10
ANIMAL DEVELOPMENT

Understanding how a multicellular animal develops from a single cell – the fertilized egg – poses one of the greatest challenges in biology today. The dramatic but orderly changes that occur during embryonic development have now been studied for more than a century but it is only during the the past 20 years that the developmental principles uncovered by traditional embryology have begun to be translated into the universal language of cell biology, genetics and biochemistry. The aim now is to explain development in biochemical terms – to provide a molecular basis for the process that establishes form and structure.

A multicellular animal develops by cell proliferation and movement on which is superimposed the differentiation of the unspecialized 'primitive' embryonic cells into those of muscle, nerve, skin, blood, etc. The form of an animal and the events that go into its making are ultimately specified by its genes, and it is now accepted that the key to development must lie in an orderly, spatially organized and selective expression of the organism's genetic blueprint during the developmental process. A chief concern of developmental biology is to explain how differential gene activity gradually imposes pattern and generates morphological structure from the mass of potentially identical cells that arise from the single egg cell.

There is of course much more to development than changes in gene expression. Cells have inherent capacities to change shape, to move and to adhere to each other. Once set in train, cellular activity can generate considerable differences between cells without the need for further genetic intervention. Much of what we see during development is *epigenetic* in this sense, in that it is not the direct result of gene activity but of the intrinsic properties of a eukaryotic cell (see, for example, Chapter 8 on the cytoskeleton and its role in cell movement and shape changes). We are only now starting to discover how these properties can be influenced by the changes in extracellular environment and intracellular milieu that gene activity could cause. The general question of how cells recognize and communicate with each other and how they respond to signals from their local environment is also central to development (see Chapter 9).

Development from egg to adult involves the sequential expression of virtually the whole of an organism's genetic instructions, both in the mother as she lays down developmental cues in the egg and in the embryo itself. Some genes are expressed in almost all cells throughout an animal's life, others in only a selected set of cells for perhaps just an hour or two during embryogenesis, and yet others only in a single type of specialized cell after development is complete. If we could distinguish those genes whose action influences development in a precise and specific way we would have a direct route into fundamental developmental processes whose workings are difficult to elucidate by other means. In vertebrates particularly there are few clues to what to look for, and most information on the role of genes in development comes from two invertebrates, the tiny fruitfly *Drosophila melanogaster* and the even smaller nematode worm *Caenorhabditis elegans*. How regulated changes in gene expression can generate pattern and form is now being beautifully revealed in *Drosophila*, and the prospect of being able at last to work out in some detail how complex multicellular structure is genetically specified has brought the fruitfly and its esoteric genetics once again to the forefront of biology. Vertebrate development – in particular that of the embryologist's favoured amphibians – is much more intractable to genetic probing, but recombinant DNA techniques are now providing new ways into the developmental process.

There seems on the face of it little in common between the development of an insect, a worm, a frog and a mouse. However, they all have to solve the same fundamental problems, and to solve them with much the same basic biochemical machinery. Despite the diversity of developmental programmes throughout the animal kingdom, shared strategies, underlying rules and widespread developmental mechanisms are being revealed.

The first sections in this chapter provide a brief descriptive outline of the very early stages in embryonic development during which the basic form of the animal is established, using amphibia and mammals as the main examples, and introducing developmental strategies other animals deploy. Multicellular development is essentially a process of creating differences within an initially formless field – the problem of differentiation in its widest sense. One main theme of the developmental process is the changes that take place within cells as their fate becomes specified and they progress towards their final differentiated state. This is covered in the sections on determination, cell memory and differentiation. The other interwoven theme is how cells become specified for a particular fate in the right time and place – the problem of pattern formation – and this is covered in the sections on pattern formation and developmental signals. The final two sections, on the developing chick limb and the generation of segment pattern in insects, illustrate these principles in action in two very different systems.

The primitive body form

Whatever their adult appearance, most animals share a common body form, a legacy of their common evolutionary origin. At its most basic it can be represented as a cylinder with a tube – the primitive gut cavity or archenteron – running longitudinally through the centre and opening in a mouth at one end (the head or **anterior** end) and an anus at the other (the tail or **posterior** end). This cylinder is composed of three concentric layers. The innermost is the gut, around it are disposed various internal organs (the second layer), and the whole is enclosed in an outer 'skin' (the third layer), which meets the lining of the gut at mouth and anus. As well as the **anteroposterior axis** (head to tail) there is usually also a distinction between upper (**dorsal**) and under (**ventral**) side, establishing a **dorsoventral axis**. In addition many animals display **bilateral symmetry**. The first task of the developmental programme is to establish these main axes of symmetry on which body structure will later be built.

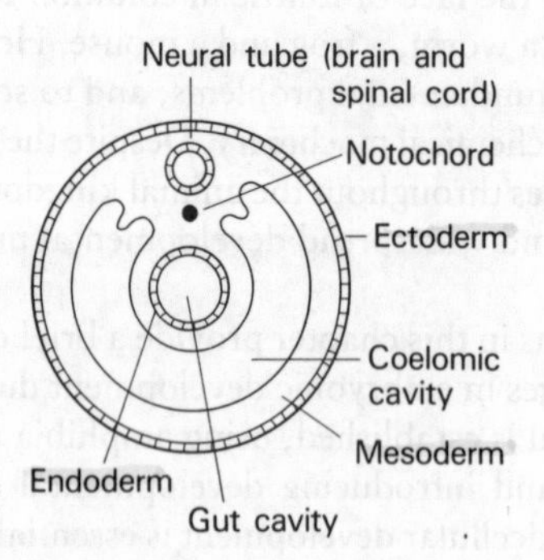

Fig. 10.1 Diagrammatic cross-section through the body of an early vertebrate embryo showing common features.

The three body layers more or less correspond to three types of tissue that arise early in embryogenesis. Working from the outside in these are the **ectoderm, mesoderm** and **endoderm** (Fig. 10.1). Each consistently gives rise to a particular set of adult tissues, and they are therefore termed the **primary germ layers.** Ectoderm develops into nervous system and the epidermis and structures derived from it; in insects, for example, the wings are epidermal in origin. Some connective tissues in the head are also of ectodermal origin. The mesoderm provides muscle, connective tissues (cartilage, bone and fibrous tissue), the circulatory system (blood vessels, blood and lymphoid tissues where present), and the urogenital system (the kidneys and the gonads). The endoderm provides the epithelial lining of the alimentary canal and the organs associated with it such as liver, pancreas, salivary glands and lungs. The smooth muscle and fibrous and elastic connective tissue associated with these organs are of mesodermal origin as is the dermis, the deep layer of the skin. The germ cells, from which new sperm and ova will be produced, are in general set aside very early, often before the differentiation of the primary germ layers.

An early episode of cell movement and rearrangement (**gastrulation**) generates the primitive body form from an initially featureless mass of cells. At gastrulation the gut cavity is created and the three primary tissues are disposed around it in more-or-less concentric layers – the outer ectoderm, an intermediate layer of mesoderm and the innermost endoderm.

Soon after the basic three-layered structure has been established, the primordial nervous system becomes distinguished from the rest of the ectoderm. In chordates (the vertebrates and their close invertebrate relatives such as amphioxus), there is a very early specialization of mesoderm into a thin rod of cells (the **notochord**) running dorsally along the midline. This structure will eventually form the core of the spinal column in vertebrates and persists as a stiff rod of tissue in primitive chordates. The ectoderm immediately overlying the notochord becomes the **neural tube** from which the brain and spinal cord develop. The disposition of invertebrate nervous systems varies but they too are derived from ectoderm.

The primary germ layers, and in vertebrates a virtually identical early embryonic form (Fig. 10.1), supply the common frame of reference in which the developmental programmes of different organisms can be compared. The transformation of a (more or less) spherical egg cell into the three-layered body proceeds through comparable stages in many animals, although the details differ considerably from class to class.

FROM OOCYTE TO EMBRYO

The oocyte

The female haploid gamete – the **ovum** – is the only cell produced by a fully developed animal that can give rise to a complete new organism. Ova are much larger than normal somatic cells and at maturity their cytoplasm contains large reserves of mRNAs, the tRNAs and ribosomes needed to translate them, and stores of membrane and pre-synthesized proteins. These are synthesized by the immature ovum or **oocyte** as it matures. This maternal survival pack supports the zygote through the first period of rapid nuclear and cell division before it starts to transcribe its own genes and produce its own metabolic machinery. Much of it is concerned simply with keeping the egg alive and dividing but in most animals the mRNAs and proteins include some that specifically direct the course of development. These are termed **maternal developmental determinants** or **cytoplasmic determinants** to distinguish them from the products of the zygote's own genes. In most animals (except, perhaps, mammals) one role of maternal determinants is to set the main body axes.

Many eggs also contain a long-term source of nutrient in the form of yolk, a dense material rich in lipids and proteins and which tends to gravitate to one end of the egg. At one end of the scale are the yolk-stuffed ova of birds; the 'yolk' of a bird's egg is the ovum, the white and the shell are laid down around it during its passage down the oviduct. At the other extreme are the eggs of mammals with almost no yolk.

The distribution of the yolk introduces an initial asymmetry into many eggs. In such eggs (e.g. those of amphibians), the yolky end is known as the **vegetal pole** and the opposite end the **animal pole**. This distinction is not simply a matter of nomenclature but describes an inbuilt polarity in the egg that is reflected in its development, the animal–vegetal axis always indicating the future anterior–posterior axis of the embryo.

Fertilization: the developmental trigger

Fertilization brings the paternal genes into the egg and restores the diploid genome. In many animals, however, this seems a lesser consideration in setting off the developmental programme than the physical act of sperm penetration. Some eggs (e.g. those of species of insects, lizards and amphibians) can develop parthenogenetically – that is, in the absence of a paternal genetic contribution; in some species parthenogenesis is a natural mode of reproduction. Even the eggs of the frog *Xenopus*, which is not naturally parthenogenetic, can be induced to divide and start developing normally simply by pricking them with a needle. Parthenogenesis has not yet been convincingly demonstrated in mammals and there is some evidence that here the presence of both maternal and paternal genes is needed for correct development.

One of the immediate consequences of sperm entry is a sharp and transient rise in intracellular Ca^{2+}. This leads to longer-lasting changes (exactly how and what are not yet known), which activate the ovum's dormant capacity for protein synthesis. Proteins are activated, stored mRNAs start to be translated, and development begins.

In many animals the site at which the sperm enters the egg is inconsequential, but in amphibians it indirectly sets the second main body axis, with the part of the egg opposite the site of sperm entry being fated to become the back of the animal (Fig. 10.2). The centriole, a bundle of microtubules brought into the egg with the sperm, is believed to induce a cytoplasmic reorganization that is crucial in establishing the future dorsal–ventral axis. Thus even before the first division several important positional references are already in place. This is true for the eggs of most animals, although the mechanisms for generating the positional cues differ. But mammalian eggs differ radically in this respect (see below).

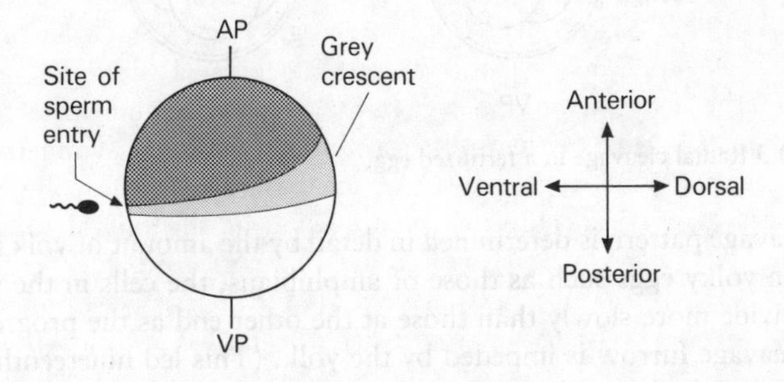

Fig. 10.2 The polarities of the fertilized amphibian egg and their relation to the tadpole body axes. AP, animal pole; VP, vegetal pole.

Cleavage

After fertilization the male and female haploid **pronuclei** combine and the first mitotic **cleavage division** occurs. Repeated cleavage initially divides the egg into a solid mass of smaller cells (the **morula**) (some exceptions are noted below). Each cleavage division is preceded by DNA replication and nuclear division, but there is at first no increase in overall cytoplasmic mass, so that the original cytoplasm of the egg is partitioned into the cleavage products or **blastomeres**. Parcelling out of maternal cytoplasm in which developmental cues have become localized to different regions is believed to guide early development in many animals, and to direct virtually the whole of development in a few (see below, Strategies of development).

In amphibian and some invertebrate eggs (e.g. sea urchins), the first cleavage runs vertically through the animal and vegetal poles, dividing the egg into symmetrical halves. A second vertical division takes place at right angles to the first, followed by a horizontal division slightly above the equator. This gives four upper blastomeres sitting directly on four lower ones, a cleavage pattern called **radial cleavage** (see Fig. 10.3). Further divisions continue more or less at right angles to each other. Frog eggs take about 4 hours to reach the 64-cell stage.

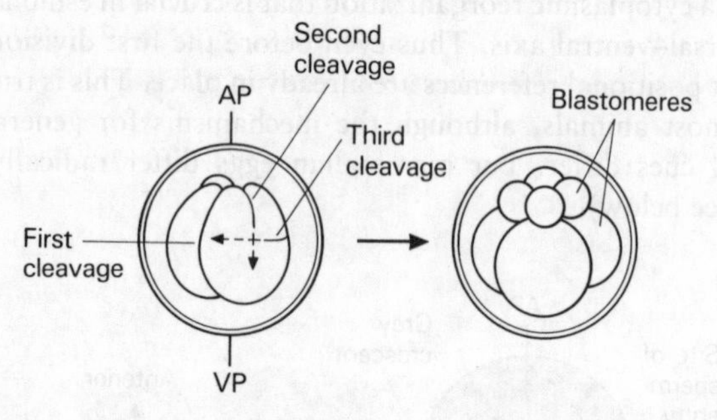

Fig. 10.3 Radial cleavage in a fertilized egg.

Cleavage pattern is determined in detail by the amount of yolk in the egg. In yolky eggs such as those of amphibians, the cells in the yolky half divide more slowly than those at the other end as the progress of the cleavage furrow is impeded by the yolk. (This led nineteenth-century embryologists to designate the more active half the 'animal' and the less active half the 'vegetal' pole.) The different rates of cleavage result in a large number of small cells in the animal half sitting on a mass of fewer, larger, yolky cells at the vegetal end. The cytoplasm of the animal half also contains more organelles (mitochondria, etc.) which have been displaced by the yolk. A cleavage pattern forced on the egg by the presence of the yolk has already produced in these asymmetrical eggs two different types of cells.

Amphibian and sea urchin eggs display regular patterns of pigmentation on their surface and these provide a visible guide to how cleavage is beginning to divide the egg into asymmetrical or apparently non-equivalent blastomeres. The animal half of amphibian eggs is heavily pigmented, and in some species a lightly pigmented 'grey crescent' appears on the side of the egg opposite sperm entry as a result of cytoplasmic reorganization. The first cleavage division runs through the centre of the grey crescent dividing the egg into symmetrical halves. It can be shown that the plane of the first division always delimits the right and left halves of the individual, and that where it runs through the grey crescent marks the future back of the animal. If the two halves of the egg are gently separated at this stage each can give rise to a new individual, showing that they are still equivalent and totipotent. The second vertical division results in non-equivalent blastomeres that have lost totipotency, and the equatorial division in effect divides the animal half from the lower yolky vegetal half. Sea urchin eggs retain totipotency up to the four-cell stage, each blastomere being able to produce a small but complete individual if separated.

Mammalian eggs with their negligible yolk divide into two, four, eight, etc. similarly sized cells. They differ from other animals in that the cells of the morula remain equivalent and totipotent until a much later stage than other animals (see the section on mammals below).

In the eggs of birds and reptiles, the massive yolk prevents the cleavage of the whole ovum. Repeated nuclear division occurs in a small space between yolk and outer membrane, followed eventually by the formation of a cell around each nucleus and the embryo develops from this small disc of cells (the **blastodisc**) sitting on top of the yolk, which does not divide at all.

In insect eggs there is also repeated nuclear division but no formation of cells until around the thirteenth nuclear division. The presence of large amounts of yolk in insect eggs restricts cell formation to the periphery. The result is a thin layer of cells (the **blastoderm**) surrounding an acellular yolky mass (see later section on *Drosophila*).

Some invertebrate eggs, exemplified by those of ascidians (sea squirts), annelids and molluscs, undergo a particular pattern of cleavage termed **spiral cleavage**, in which cleavage is successively in right and left oblique planes. Partitioning of maternal cytoplasmic determinants into different blastomeres plays a particularly important part in the development of these animals, apparently specifying the fates of most individual blastomeres almost from the start (see below, Strategies of development).

The blastula

The roughly spherical morulas of echinoderms (sea urchins), amphioxus (a primitive chordate) and amphibians are transformed from a solid ball of cells into the basic body form through a hollow **blastula** stage followed by extensive cell movements (**gastrulation**) that generate the three-layered **gastrula**. Mammalian eggs develop in a similar fashion into a hollow cellular ball – called the **blastocyst** – but their subsequent development towards the embryonic form common to all vertebrates differs considerably and is covered separately below.

In frogs, formation of the blastula starts around 4 hours after fertilization. A fluid-filled cavity – the **blastocoel** – forms inside the morula, and the outer layer of cells develops the properties of an epithelium, sealing off the interior of the blastula from the exterior. In the yolky eggs of amphibians the blastocoel is confined to the animal half (Fig. 10.4); in the less-yolky eggs of sea urchins and amphioxus the blastula comprises a single layer of cells enclosing the blastocoel. In frogs, the blastula eventually contains some 10 000 cells, but there has been little increase in absolute mass.

In the tropical frog *Xenopus*, whose large, easily manipulated eggs and robust embryos have long made it a favourite of developmental biologists, the first 12 cleavage divisions are rapid and synchronous (producing a blastula of about 4000 cells). Thereafter, synchrony breaks down and cells go on to divide at rates characteristic of their particular position in the egg. The point at which synchrony is lost is known as the **mid-blastula transition** and is also the point at which appreciable transcription starts from the embryo's own genes (the **zygotic** genes).

The mid-blastula transition illustrates an important consequence of the imbalance that develops between the amount of DNA, which is

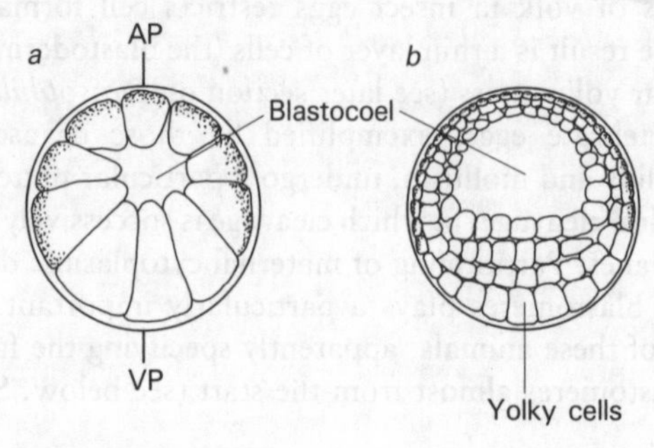

Fig. 10.4 Formation of amphibian blastula. *a*, Early; *b*, complete.

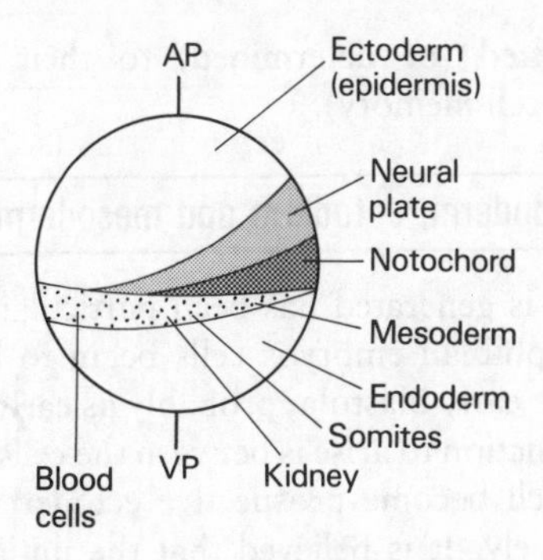

Fig. 10.5 Simplified fate map of the *Xenopus* blastula.

being replicated at each cleavage, and the mass of cytoplasm, which has until now grown little. The mid-blastula transition can be shifted to an earlier time by simply injecting the blastula with more DNA – any DNA. This suggests that the breakdown in synchrony and the start of transcription arise from the sequestering of a fixed amount of some maternally supplied DNA-binding protein by the increasing amount of DNA. The identity and mode of action of this regulatory protein are not yet known. However, there is another molecular switch in *Xenopus* blastulas that similarly relies on titrating out a fixed amount of a DNA-binding protein, and this has been analysed in exquisite detail. This is the shut-down of the 'oocyte-specific' 5S RNA genes during gastrulation (see Chapter 3: Control of gene expression).

The programmed exhaustion or dilution of a finite amount of maternal material as the number of cells grows is one possible general way of timing early developmental events.

Fate Maps

In some animals the regions of the blastula that will give rise to different tissues, organs or parts of the body can already be delimited by injecting blastomeres with a harmless stain and tracing their subsequent progress. In amphibian embryos for example, the subsequent movements of gastrulation are so orderly that quite detailed **fate maps** can be drawn on the blastula surface (Fig. 10.5). Cells that under normal circumstances go on to form a particular tissue or structure are often termed 'presumptive' or 'prospective' mesoderm, ectoderm, notochord, etc. The fact that one can predict which tissue a region of the blastula or even of the egg will normally produce should not, however, be confused with the quite separate issue of when and how those cells become irreversibly committed (or **determined**) to their fate (see below, Determination and cell memory).

Specification of endoderm, ectoderm and mesoderm

How this fate map is generated has been particularly well studied in amphibians. In amphibian embryos cells begin to be specified for a particular fate in the early blastula, probably as early as the 32/64 cell stage. The first distinction to arise is between the cells of the animal and vegetal halves, which become prospective ectoderm and prospective endoderm, respectively. It is believed that the initial specification of these two tissues is due to localized maternal cytoplasmic determinants (their identity remains unknown) that have been unequally partitioned into the animal and vegetal halves of the dividing egg.

The third primary tissue, the mesoderm, then arises from an interaction between endodermal and ectodermal cells. Where endoderm and ectoderm meet, the endoderm directs (induces) an adjacent band of ectodermal cells to become prospective mesoderm. **Induction** – in which one cell or tissue directs the development of another while itself remaining unchanged – is a widespread feature of development.

Differences are created within the mesoderm before gastrulation, in particular the specification of a small area on the 'back' of the blastula as a crucial **organizing region**. This initiates gastrulation, forms the notochord and also directs the future correct development of the dorsal region of the embryo (the generation of this primary organizing region is described in more detail in the section on pattern formation).

Gastrulation

The hollow blastula now invaginates to form the gastrula in which the primary germ layers – the ectoderm, mesoderm and endoderm – are in place. In chordates, it brings the crucial primary organizer area into a relationship with ectoderm in which it can then direct the formation of the primordial nervous system.

Chordate gastrulation is seen at its simplest in the relatively non-yolky eggs of amphioxus (Fig. 10.6). Amphioxus forms a dome-shaped blastula with a base of larger vegetal (endoderm) cells. These become drawn into the hollow dome of animal cells (ectoderm) forming a double-walled cup and obliterating the blastocoel. The new internal cavity is the archenteron or prospective gut cavity. The opening is the **blastopore**, and the rim of the cup where the two layers of cells meet is its lip. At this stage the whole mass reorientates so that the animal pole

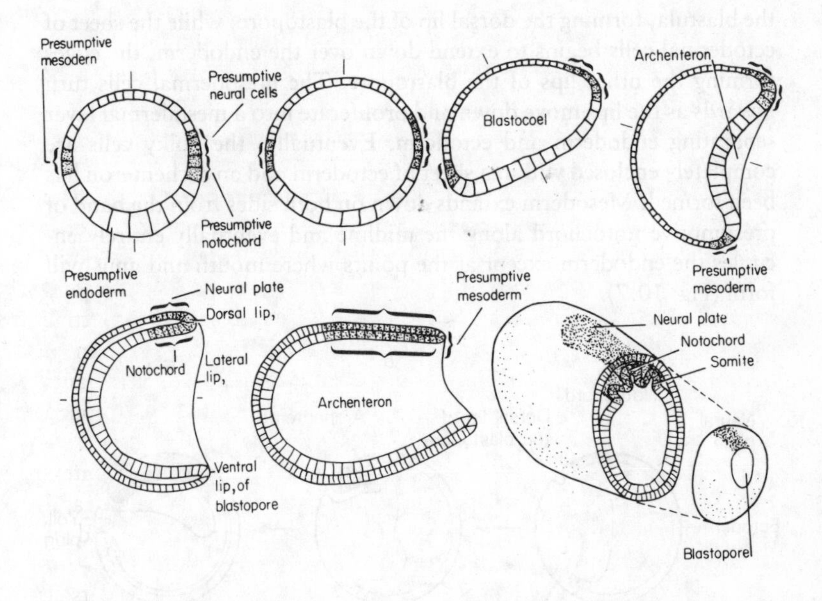

Fig. 10.6 Gastrulation in amphioxus (*Branchiostoma*), a primitive chordate. *a–d*, vertical sections along midline of embryo. *e*, Cross-section through *d*. (Fig. 5.2, J. McKenzie, *An introduction to developmental biology*, Blackwell, Oxford, 1976.)

becomes the anterior end and the vegetal pole, where the blastopore has developed, the posterior end.

Cell proliferation at the lips of the blastopore elongates the developing gastrula until it is a tube-like structure ending blindly at the head and opening at the blastopore. The mesodermal cells of the inner part of the upper or **dorsal lip** of the blastopore lay down a narrow band of cells along the midline in the roof of the archenteron. These cells will be the notochord. On either side lies a band of presumptive mesoderm that will invaginate into the plane between endoderm and ectoderm to form the intermediate mesoderm layer. Along the midline above the notochordal mesoderm ectodermal cells lay down a band of prospective neural ectoderm.

Gastrulation in amphibians is somewhat contorted because of the presence of the mass of yolky endodermal cells. This is too bulky to invaginate directly into the dome of ectoderm. Instead, the presumptive notochordal mesoderm tucks in and starts to move into the interior of the blastula, forming the dorsal lip of the blastopore, while the sheet of ectodermal cells begins to extend down over the endoderm, the edges forming the other lips of the blastopore. The mesodermal cells turn inwards as the lips move down and proliferate into a mesodermal layer separating endoderm and ectoderm. Eventually, the yolky cells are completely enclosed within a sheet of ectoderm and an archenteron has been formed. Mesoderm extends down on both sides from the band of presumptive notochord along the midline and eventually entirely encircles the endoderm except at the points where mouth and anus will form (Fig. 10.7).

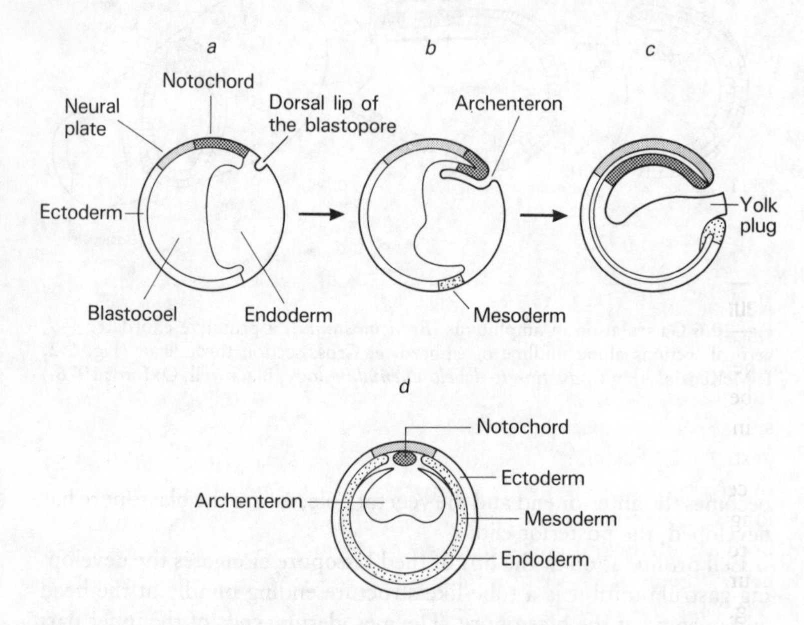

Fig. 10.7 Gastrulation in the amphibian embryo. *a–c*, Vertical sections along midline of embryo. *d*, Cross-section through *c*.

The role of the dorsal-most mesoderm as the primary organizing region in vertebrate embryos was first discovered at the beginning of the century. If the dorsal lip from an early amphibian gastrula is transplanted to a site elsewhere on another gastrula it forms a second site of gastrulation, eventually producing a 'double' embryo with two heads, neural systems and other dorsal tissues (Fig. 10.8). It is presumed to exert its dramatic effects, at least in part, by secreting a diffusible signal (a morphogen) but this has proved elusive.

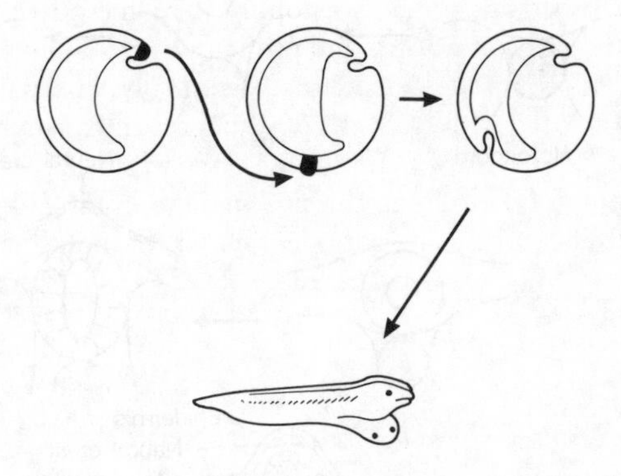

Fig. 10.8 The effects of transplanting a dorsal lip from an early amphibian gastrula.

Neurulation

The next stage in chordate development is the formation of the neural tube along the dorsal long axis of the embryo; from this the brain, spinal cord and peripheral nervous system will develop. In amphibian gastrulas, the band of notochordal mesoderm lengthens into a thin rod of cells, and under its influence the overlying ectoderm begins to form a longitudinal furrow (the **neural fold**). This starts at the head end, bordered by **neural crests** that eventually meet and fuse, forming the neural tube, which then lies beneath a continuous layer of ectoderm (Fig. 10.9). Some cells from the neural crests aggregate separately on either side of the neural tube and later migrate to form the dorsal root ganglia of the spinal cord and the peripheral nervous system.

Once the neural tube has formed the thick slab of mesoderm on either side breaks up to form a series of separate blocks of cells called **somites**. These correspond to the segmented structure (vertebrae, ribs) of the typical vertebrate body. Somite formation starts from the head end as blocks of mesoderm cells rotate to form individual somites, breaking their connections with the remainder of the mesoderm. From these somites derive the skeleton and associated muscles of the trunk, and the muscles of the limbs. The mesoderm towards the ventral side will form the remaining internal tissues and organs of mesodermal origin.

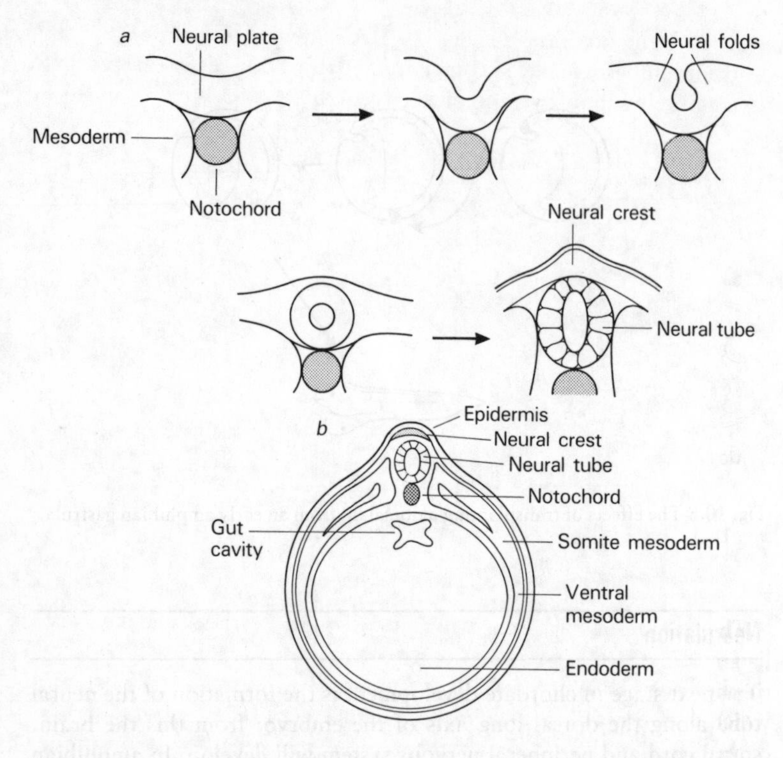

Fig. 10.9 *a*, Stages in the formation of the neural tube in vertebrates (neurulation). *b*, Transverse section through an amphibian embryo after neurulation.

Mammals are different

The development of the mammalian egg to the 'neurula' stage presents a considerable contrast to amphibians, both in its topology and in the prolonged totipotency of the cells of the morula. In mammals only a part of the morula will form the embryo proper; the rest produces the tissues that attach the embryo to the uterine wall, and continued totipotency is required until the embryonic cells are set aside.

The mammalian zygote initially cleaves into a solid morula composed of similarly sized cells. The morula develops into a hollow **blastocyst,** the stage at which the conceptus becomes implanted in the wall of the uterus. A small group of cells within the morula is set aside as the **inner cell mass,** which aggregates at one pole of the blastocyst, and the outer layer of cells – the **trophectoderm** or **trophoblast** –

attaches to the uterine wall (Fig. 10.10). The tissues that derive from the trophectoderm make no contribution to the body of the embryo and have a purely supportive and protective role.

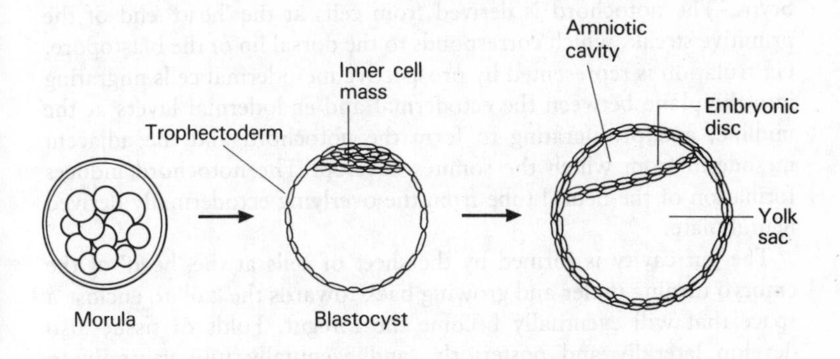

Fig. 10.10 Development of the mammalian blastocyst.

The first detectable point at which cell fates diverge in the mammalian morula is the commitment to being inner cell mass (ICM) or trophectoderm. Which cells of the morula will form the embryo and which will form trophectoderm seems to be decided solely on the basis of the effects of the local environment – those that happen to be on the outside of the morula become trophectoderm and those inside become the inner cell mass. Differential partitioning of maternal factors into different blastomeres appears to have no part to play in this choice.

Before this point, the cells of the morula are still equivalent to each other and totipotent. In the mouse, cells can be removed or added to an eight-cell morula or two morulas can even be combined to form a giant morula, and a perfectly normal animal will still be produced after replacement of the morula in the uterus. The morula is said to **regulate**, to be able to compensate for lost cells by providing replacements and to adjust to incorporate additional cells of equal developmental potential. The totipotency of the inner cell mass appears still to be retained in the early blastocyst. If an embryonic stem cell from another blastocyst is injected into the blastocoel it becomes incorporated into the inner cell mass and can be shown to give rise to all types of tissue.

After **implantation** the inner cell mass thins out into an **embryonic disc** two cells thick, spanning the blastocoel. At this stage, it closely resembles the avian blastodisc and the movements that transform these flat discs into a three-dimensional embryo are very similar. The upper and lower layers of the disc initially correspond respectively to prospective ectoderm and endoderm. Mesodermal cells that arise in the disc migrate towards the midline to produce a longitudinal thickening and furrow visible as the **primitive streak** of avian and mammalian embryos. The notochord is derived from cells at the 'head' end of the primitive streak, which corresponds to the dorsal lip of the blastopore. Gastrulation is represented by prospective mesodermal cells migrating into the plane between the ectodermal and endodermal layers at the midline, and proliferating to form the notochord and the adjacent mesoderm from which the somites develop. The notochord induces formation of the neural tube from the overlying ectodermally derived neural plate.

The gut cavity is formed by the sheet of cells at the 'head' of the embryo turning under and growing back towards the 'tail' to enclose a space that will eventually become the foregut. Folds of tissue also develop laterally and posteriorly, and eventually fuse ventrally to enclose the gut (Fig. 10.11).

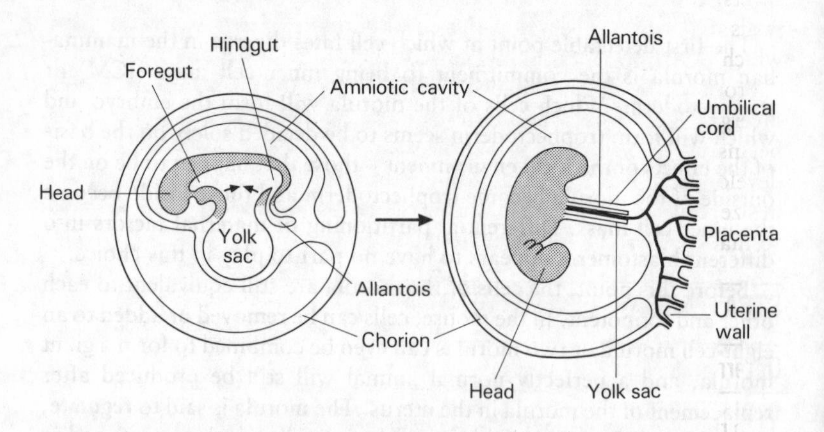

Fig. 10.11 Gastrulation and formation of the extra-embryonic membranes in a human embryo.

Embryonic development in reptiles, birds and mammals is also complicated by the production of the various extra-embryonic membranes (chorion, amnion, yolk sac membrane and allantoic membrane). The chorion is a development of the trophectoderm whereas the others are produced by outgrowths from the early embryonic disc. These are evolutionary developments of the extra-embryonic membranes that our first terrestrial reptilian ancestors needed to protect the developing embryo from dehydration and to provide a large surface area for gas exchange through the shell. Apart from the egg-laying monotremes, all mammalian embryos gain nourishment from their mother directly through modifications of trophectoderm and other extra-embryonic structures to form the various types of placenta. Marsupials form a short-lived placenta and the young are 'born' at a very immature stage after which they are suckled in the pouch.

Despite these evolutionary diversions, all vertebrate embryos converge on the basic vertebrate embryonic form at the completion of neurulation, when they are all remarkably alike in shape, size and morphology.

Soon after neurulation, different tissues begin to be distinguishable. Much of the mesoderm is initially in the form of primitive undifferentiated **mesenchyme**, a meshwork of cells bearing multiple cytoplasmic processes that contact each other to form a loose space-filling tissue. Ectoderm and endoderm are characteristically in the form of epithelial sheets, while the neural tube fills up with neuroblasts derived from its walls and starts to develop the characteristic layered structure from which the layers of the spinal cord and brain will eventually derive.

From this point, the detailed morphology of the different vertebrates diverges. The shaping and differentiation of internal organs and limbs begins, to be completed in outline while the embryo is still tiny. Later development – the **foetal** stage in mammals – is concerned with growth in size, fine detailing, and maturing of organs and physiological systems so that they will be fully functional at birth.

STRATEGIES OF DEVELOPMENT

Determinate development

Two types of development are traditionally distinguished – determinate and regulative. Many animals use both, but at either end of the spectrum are animals whose development is largely determinate, or almost entirely regulative. In highly determinate animals, a large part of the guidance system needed for development is already laid down in the cytoplasm of the egg itself and sets early blastomeres off on a predetermined developmental course. This strategy is followed by many invertebrates – nematodes, annelids, platyhelminths, molluscs and ascidians (sea squirts). In their developmental programme the fate of many cells is already set by the time they are formed. In these animals, removal of a blastomere or destruction of a particular cell at a later stage simply results in the complete absence of the tissue or structure that cell would normally give rise to. Cells cannot in general be recruited to provide replacements for each other (although there are exceptions, see below, Induction). Most blastomeres are committed almost from the start by the maternal cytoplasmic determinants that have been dealt them in the first few cleavage divisions. Once determined at this early stage, each cell lineage (i.e. the descendants of each blastomere) seems to develop without much regard to interactions with other cells or its local environment. If an ascidian blastomere is removed and its division, but not DNA replication, prevented, it will eventually undergo a form of differentiation and start to produce proteins and internal structures typical of the cells it would normally give rise to.

This type of development is often termed **mosaic development**, and such eggs as mosaic eggs, as they appear to comprise a collection of independently developing parts. Maternal cytoplasmic determinants are presumed to be precisely localized to particular regions of the cytoplasm and thus to be differentially partitioned into different blastomeres at cleavage.

An extreme example of maternal determinism is the microscopic rotifer, in which parthenogenetic development from egg to a newborn adult female containing around 1000 nuclei takes less than 30 hours and is entirely directed by translation of maternally supplied RNAs. There is no transcription at all from the developing rotifer's own genes until after hatching, which enables it to devote all its time to replicating DNA and to nuclear and cell division.

Invariant cell lineages

An animal not much larger than a rotifer has over the past decade joined *Drosophila* and amphibians as one of the main sources of information on development. The larva of the tiny nematode worm *Caenorhabditis elegans* is less than 1 mm long and has around 550 cells when newly hatched. It has a highly determinate development. Most of its cells have arisen by an invariant pattern of cell division from blastomeres which become committed at a very early stage (Fig. 10.12). This results in the same cell, as judged by its line of mitotic descent, always occupying the same position in all individuals, making them virtually exact cellular copies of each other. The developing worm is transparent and so cell division can be followed and lineages traced in the living worm under the light microscope. The larvae goes through several moults to emerge as a sexually-mature adult containing 945 somatic cells. The complete family tree for each of these cells has now been painstakingly constructed, a feat that is unlikely to be repeated for any other animal.

The study of *Caenorhabditis* was pioneered in the hope that such a simple animal with entirely predictable cell lineages could provide an insight into the genetic basis of development and behaviour that is difficult to obtain with more complex animals. Many developmental genes have now been identified. The advantage of *Caenorhabditis* is that mutations resulting in developmental abnormalities can be readily correlated with the changes they cause in the cell lineages, thus pinpointing their site of action much more easily than is usually the case with other animals.

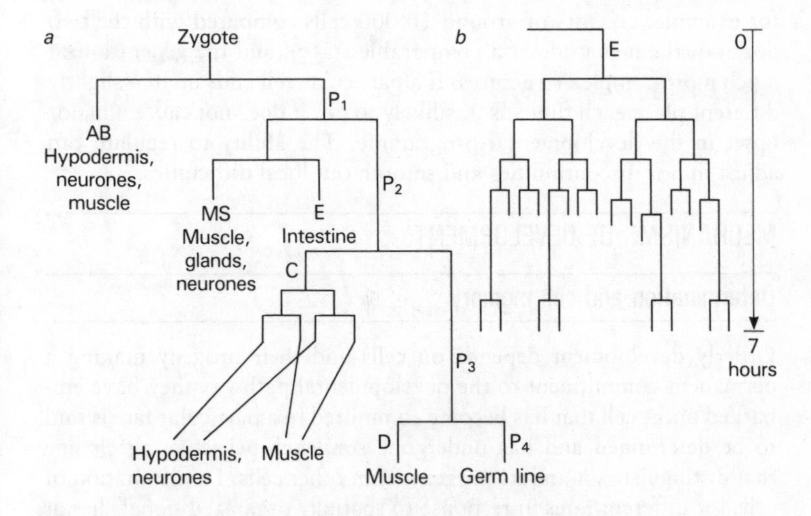

Fig. 10.12 *a*, The derivation of the cell lineages of *Caenorhabditis*. *b*, Lineage tree for the cells of the intestine. Horizontal lines represent cell divisions. (Adapted from J.E. Sulston *et al.*, *Developmental Biology*, **100** (1983) 64–119.)

Regulation

At the other end of the developmental spectrum are the highly **regulative** eggs and embryos of mammals. There is no evidence that even the earliest stages of their development depend on maternally specified differences in the cells of the morula, which remain equivalent to each other and totipotent to a later stage than other animals. If one cell of the morula is removed, another divides to replace it. If extra cells are added, development is adjusted to incorporate them.

The ability to regulate depends essentially on groups of neighbouring cells at any one time having the same developmental potential – that is forming an **equivalence group**. Any cell of the group can be recruited to replace one that has been destroyed or removed. Equally the group can adjust to the presence of additional cells. Regulation is the outward sign of a developmental programme in which cells become assigned to their fate not just as a result of their line of descent (although that also plays a part – see below, Determination and cell memory). Equally influential is the position they occupy in the embryo at any given time and in consequence the local environment (other cells and their products) they are exposed to (see below, Pattern formation and developmental signals). Regulation allows for extensive and necessarily less precise mass movements of cells (the *Xenopus* blastula at the time of gastrulation, for example, consists of around 10 000 cells compared with the two dozen of the nematode at a comparable stage), and the generation of much more complex structures. If a particular cell ends up in a slightly different place each time, as it is likely to do, it does not cause a major upset in the developmental programme. The ability to regulate can adjust minor discontinuities and smooth out local difficulties.

MECHANISMS OF DEVELOPMENT

Determination and cell memory

Orderly development depends on cells and their progeny making a permanent commitment to the developmental pathway they have embarked on. A cell that has become committed to a particular fate is said to be **determined** and has undergone some self-perpetuating change that distinguishes it and its progeny from other cells. Determination of cells for different fates in response to spatially organized signals lies at the heart of **pattern formation** – the generation of structure and form. Determination is a subtle concept and in most cases we have as yet little idea of what it entails at the cellular and molecular level. It is possible however, to define certain essential properties of a determined cell.

1. Once a cell has become determined it and its descendants never normally revert to a previous state or switch to an alternative state in another mutually incompatible pathway. Committed neural cells, for example, do not at some later stage produce epidermis, and mesodermal cells never produce neurones. A determined cell may, however, have further 'allowed' determined states superimposed later. For example, committed chick mesodermal cells may become muscle, cartilage,

bone, etc.; in terms of higher-level structure they may also become committed to being leg or wing (see later).

2. Determined cells pass on the memory of what they have become to all their descendants.

3. Maintenance of the determined state does not depend on exposure to a continuing signal. In this respect, determination contrasts sharply with the temporary changes in gene expression and metabolism that cells display in response to many stimuli or changes in external conditions.

The thread of **cell memory** spun out during development is essential to embryogenesis and multicellular life itself. It represents an intrinsic capability of eukaryotic cells, not shared to any extent by prokaryotes, without which multicellular organisms of any complexity could not have arisen. The signals that guide development (see below, Pattern formation and developmental signals) act at short range and for limited periods. Once determined in response to such a signal, cells must remember they are endoderm or ectoderm, muscle or neurone as they proliferate, migrate, and come under new influences. The basic body plan and the assignment of cells to the four main tissues (epidermis, neural tissue, endoderm and mesoderm) are established even for the largest and most complex animal when the embryo is only a few millimetres long, and are retained and elaborated on as the embryo grows. In regulative embryos this is achieved as additional states of determination are superimposed, gradually building up a final specification of a cell's fate. Without this remembrance of past history chaos would soon ensue.

As well as becoming determined for a particular type of tissue, cells can also become committed to contributing to a certain region or structure. If undifferentiated mesenchyme (mesodermal) tissue from a chick wing bud is transplanted to the leg bud when the limb buds are still tiny and apparently identical it will still develop as wing. This regional specification precedes determination for cartilage or fibroblast, which arises later in response to signals generated in the developing limb (see later, The developing chick limb). The precursors of the limb muscle cells, however, which migrate into the limb buds from the somites, are already determined as muscle before they reach the limb buds, where they too become further specified as wing or leg. This raises the intriguing possibility that the final molecular specifications of, say, a cartilage cell in the wing and an apparently identical one in the leg are in fact different and might eventually be detectable by molecular biology.

The point at which a cell, or in most cases a group of cells, becomes determined for a given fate is usually technically difficult to detect. The first detectable signs are generally the production of proteins characteristic of a particular specialization (e.g. muscle actin in the case of prospective muscle) or the beginnings of structural change. Commitment, however, has often (but not always) occurred much earlier.

In regulative embryos one way of finding whether a cell has become determined is by transplanting cells from one site to another and seeing what they develop into. In the amphibian embryo, which lends itself to such microsurgery, grafting experiments show that determination of the three primary germ layers begins in the blastula and by the beginning of gastrulation is complete. The distinction between epidermis and neural tissue is made later, at the beginning of gastrulation. A patch of ectoderm transplanted from the prospective neural area to replace a piece excised from a ventral site will settle in and develop as ventral ectoderm if the graft is made early in gastrulation. But if grafted later, the transplanted cells develop as misplaced neural tissue, showing that they have now become determined. This illustrates the subtle but crucial distinction between cells that can be designated as 'presumptive' or 'prospective' neural tube, mesoderm or whatever, because that is what in normal circumstances they will develop into, and those that have become irrevocably committed to that fate.

In some simple animals, the range of tissues and structures an early blastomere can give rise to is determined from the start. The fates of many of a nematode's cells, for example, are solely determined by their mitotic line of descent – their **lineage** – and appear not to be influenced at all by other cells or the local environment.

In animals with a largely regulative mode of development on the other hand, final choices can be left until very late indeed and are determined largely by the environment a cell finds itself in. Neural cells in the developing rat retina only become finally determined as rods, glia, or amacrine or bipolar neurones as late as the last cell divisions before the retina completes its development after birth. An undifferentiated retinal ventricular cell can even produce at its final division one daughter that differentiates into a rod photoreceptor cell and one that becomes a glial cell. Since the functional distinction between the non-excitable glia and excitable cells (sensory receptors and other neurones) is one of the most fundamental in the nervous system, it had been supposed that their lineages diverged at an earlier stage (whether they do in other parts of the nervous system is still an open question). Its initial specification as neural tissue apart, a mammalian retinal cell's line of descent appears to have little part to play in whether it will become photoreceptor, interneurone or glia. In this case the cells are thought to read their ultimate fate directly from signals provided by the local environment, possibly after the final cell division. In such a case the distinction between determination and overt and visible differentiation becomes blurred. How far a retinal cell's fate has become restricted to becoming one of these specifically retinal neural cell types by some earlier determinative event is not yet known.

The Genetic Basis of Determination and Cell Memory

Determination represents a limitation of a cell's developmental potential and is therefore presumed to reflect a self-perpetuating and heritable change in its ability to express selected portions of its genetic instructions. Some heritable record of the change must be made but we do not yet know if there is some universal molecular recording mechanism or whether, as is more likely, there are several different ways of registering determinative events. It is generally assumed, but not an absolute prerequisite, that some sort of chromosomal modification is made as it is otherwise difficult to envisage how the determined state could be heritable. Various threads of evidence suggest that irreversible changes in DNA structure of the nature of a mutation are, however, unlikely (see below, Differentiation).

There are various candidate mechanisms for cell memory, some of which are described in Chapter 3 (see Control of gene expression).

Most information on the genetic basis of determination comes from *Drosophila* and *Caenorhabditis* where mutations affecting determination have enabled many genes involved in the process to be identified. *Caenorhabditis* in particular with its entirely predictable, fully traced cell lineages, allows individual determinative events to be pinpointed with absolute precision and to be dissected in fine genetic detail (see Induction).

The current view of any given state of determination is that it may be described in terms of a particular set of active and potentially active genes. This gene set is selected in some way at the point of determination. At present, we know virtually nothing about what the esssential properties of, say, a mesodermal cell might be, let alone the state of 'legness' or 'wingness' in genetic terms. The phenomenon of **transdetermination** in *Drosophila*, however, suggests that commitment to and maintenance of a given determined state may sometimes be controlled by a relatively small number of 'master' genes.

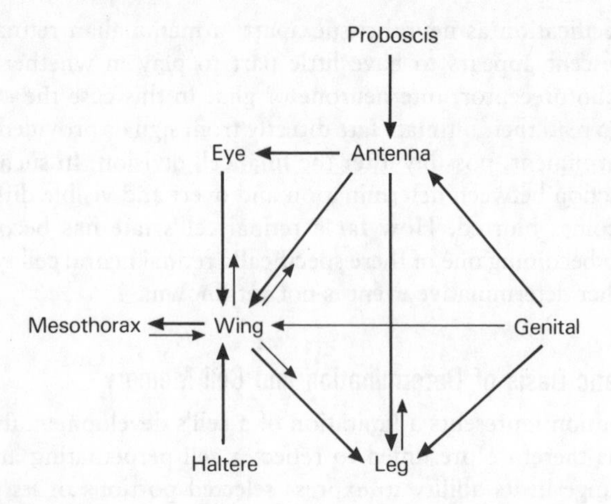

Fig. 10.13 Transdetermination in *Drosophila*. The heavy arrows indicate transformations that occur fairly frequently, the fine arrows those that occur rarely. (See Fig. 10.21 for names of parts of the insect body). (After S.A. Kauffman, *Science*, **181** (1973) 310–18.)

Transdetermination

Drosophila shows some rare exceptions to the rule that cells do not switch between alternative states of determination. In insects, adult epidermal structures (e.g. leg, wing, eye, antenna and genitalia) arise from groups of determined but undifferentiated ectodermal cells that are set aside in the larva as small sacs of epithelium – the **imaginal discs** – early in development (see Fig. 10.21). There is one disc for each structure. Their cells proliferate but remain undifferentiated through the various larval stages, until pupation and metamorphosis when the discs evert, expand and differentiate.

Under certain conditions of long-term culture small groups of cells or even a whole disc will very rarely **transdetermine** – that is, switch to an alternative state of determination, ultimately producing tissue characteristic of wing rather than leg, or leg rather than antenna, when they are allowed to differentiate. From the relative frequency with which different switches occur there appears to be a 'hierarchy' of determined states, with cells being able, for example, to switch between wing and leg or wing and eye but never directly between leg and eye. A model scheme has been proposed (Fig. 10.13), in which each state could be the result of different permutations of on/off states of a small number of genes (only four or five need be postulated). Each transdetermination would then represent the throwing of a molecular switch controlled by one of these genes. The basis of transdetermination is not yet known but some of these genes may have been identified by the amazing **homoeotic mutations** of *Drosophila* and other insects (see below, Homoeotic genes).

Differentiation

The final stage in a cell's developmental history is its overt **differentiation** as it takes on a specialized form and function. Differentiation in the general sense used in developmental biology describes the complete process of successive distinctions (whether morphologically distinguishable or not) that arise, for example, between endoderm, ectoderm and mesoderm. There is however, a final and quite distinct stage in this process. Embryonic cells eventually transform from rather 'primitive' cells without any unique structural features or particular physiological properties into cells with specialized internal structure and new and remarkable capabilities (Fig. 10.14).

After differentiation many cells (e.g. muscle and nerve) lose the ability to divide at all, and those that still do can only produce more cells like themselves, and their capacity for future division is also limited (in mammals at least). Unlike embryonic cells, which proliferate indefinitely in culture, differentiated human or mouse fibroblasts divide no more than 50–80 times, depending on the species and age of the individual they come from; and then the culture 'ages' and dies.

We take it for granted that our liver cells will give rise to more liver cells and fat cells to more fat cells, and this fact of multicellular life reflects the continuing workings of cell memory. The knowledge of what it has become has been programmed into a differentiated cell during its derivation from an unspecialized embryonic cell and does not simply reflect the influence of its present environment. When taken out of their natural environment and grown in culture, differentiated cells normally remain recognizably themselves and do not either revert to an embryonic state or change to another type of differentiated cell. The rare instances when they can be persuaded to change reveal factors that may be important in establishing and maintaining the differentiated state (see below, Transdifferentiation).

More than 200 cell types are recognized in the vertebrate body (see Table 10.1 for some of them). A differentiated cell is defined by its unique structural features and functional properties which are the result of the particular transport proteins, ion channels, receptors, structural proteins, etc. that it contains or is able to make on demand. Neurones, for example, generate and convey electrical impulses by

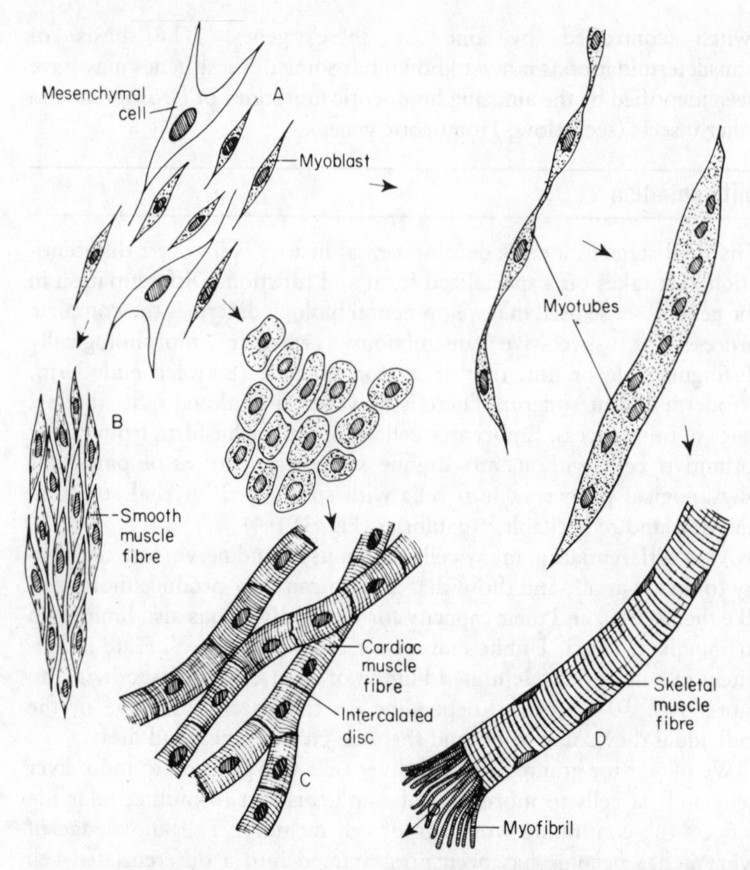

Fig. 10.14 Differentiation of mesenchyme into smooth, cardiac and skeletal (striated) muscle. Differentiated smooth muscle cells retain the power of division; skeletal muscle fibres are renewed from undifferentiated satellite cells (myoblasts) that persist amongst the muscle fibres. Cardiac muscle is not renewed. (Fig. 4.8 in J. McKenzie, *An Introduction to Developmental Biology*, Blackwell, Oxford, 1976.)

Table 10.1 A selection of specialized cells: the main types of cell in mammalian skin and eye

Skin

Keratinocyte (the keratin-containing cells of the epidermis, constituting the waterproof protective outer layer)
Basal stem cell of epidermis (from which keratinocytes are replaced)
Keratinocyte of fingernails and toenails
Basal stem cell of nailbed
Hair shaft cells (3 types)
Hair-root sheath cells (4 types)
Stem cell of hair matrix
Secretory cell of sebaceous gland in skin (secreting lipid-rich sebum)

Secretory cells of eccrine sweat glands (dark cells and clear cells secreting different components of sweat)
Secretory cells of apocrine sweat gland
Non-striated duct cells of sweat glands
Myoepithelial cells of sweat glands (contract around base of gland to expel contents)
Melanocyte (pigment cell)
Langerhans cells (macrophages)
Merkel cells (touch receptors associated with sensory nerve endings in epidermis)
Sensory neurones surrounding hair follicles
Smooth muscle associated with hair follicles
Sensory neurones responsive to temperature (heat-sensitive and cold-sensitive)
Sensory neurones that respond to painful stimuli
Fibroblasts (basal lamina and dermis) (secrete extracellular matrix and collagen fibres)
Endothelial cells forming walls of blood capillaries
Fat cells (dermis)

Eye
External corneal epithelial cell
Corneal endothelial cell
Corneal fibroblast
Hyalocyte (secreting vitreous humour)
Cells of ciliary epithelium
  Pigmented
  Non-pigmented
Epithelial cells of lacrimal gland (secreting tears)
Myoepithelial cell of iris
Lens epithelial cell
Crystallin-containing lens fiber
Endothelial cells of blood capillaries
Retina
  Retinal pigmented epithelium
Fibroblasts of ventricular membrane
Photoreceptors
  Rods
  Cones (3 types: blue-sensitive, green-sensitive, red-sensitive)
Neurones (connecting photoreceptors to optic nerve)
  Bipolar cells
  Amacrine cells
  Horizontal cells
  Ganglion cells
  Neurones of optic nerve
Glial cells (non-excitable supporting cells of nervous tissue)

virtue of certain combinations of ion channels in their plasma membrane. Each type of neurone also contains the enzymes for synthesizing a small number of neurotransmitters, and their synaptic membranes contain specific receptors for certain neurotransmitters, neuromodulators and hormones produced by other cells. In addition, neurones carrying out different functions (e.g. motor neurones, spinal interneurones and sensory neurones) have various adaptations of the 'basic' neurone morphology, fitting them for their particular tasks.

Different cell types may share some properties – for example, cells other than neurones produce many of the substances that are used as neurotransmitters, and virtually all cells produce the same set of basic 'housekeeping' enzymes and other proteins for essential processes such as respiration, transcription and translation. But the overall combination is unique to each cell type and there are many products that are made by only one type of cell. Only red blood cell precursors make haemoglobin; skeletal muscle produces one version of myosin (part of the contractile apparatus), smooth muscle synthesizes another; pancreatic ß-cells make insulin, pancreatic acinar cells synthesize digestive enzymes; mammary gland cells synthesize milk proteins, and so on. In the immune system, B cells make immunoglobulin antibodies and the virtually indistinguishable T cells make the related T-cell antigen receptors. Even some ubiquitous proteins such as non-muscle actin and various metabolic enzymes are produced from different genes in different tissues.

Each differentiated cell, therefore, appears able to express only a limited portion of the total genome and has lost the ability to make many of the proteins characteristic of other cell types. This raises the question of whether a differentiated cell has physically lost or irretrievably inactivated the genes it cannot express or whether they are still there but shut down for the duration. All the evidence shows that, with a few exceptions, each cell in the body of a multicellular organism still contains a full set of genes, and, again with some notable exceptions, that genes have not undergone any rearrangement or other fundamental structural alteration during differentiation or previous determination that would irreversibly activate or inactivate them.

The crucial experiments that decided this issue were carried out in the 1960s and set much of the agenda for future work on both differentiation and the earlier determinative steps in a cell's history. A nucleus taken from a fully differentiated intestinal cell of a *Xenopus* tadpole was transplanted into a fertilized *Xenopus* egg from which the nucleus had been removed. Even though the intestinal cell nucleus in its usual environment could express only a limited set of genes, the egg with its new nucleus could develop (admittedly rarely) into a normal fertile adult. Although these experiments have not been entirely repeatable with other animal eggs, they are taken as good evidence that differentiation does not involve the loss of any genes, and neither does it reflect some unalterable change in their structure or ability to be expressed.

All the evidence gained since on detailed gene structure by direct examination of DNA supports this view, as does the ability of some cells to transdifferentiate if subjected to certain treatments (see below). The changes seen as a cell differentiates are therefore the result of alterations in the pattern of gene expression and not in the genes themselves. The only exceptions to this rule that have been uncovered in mammals (there are isolated examples in other animals) are the extensive and irreversible DNA rearrangements that precede immunoglobulin gene expression in the B cells of the immune system and the similar changes that occur in the antigen receptor genes of T cells (see Chapter 12).

Having excluded gene loss, nuclear transplantation experiments raised many questions. How is the differentiated state maintained? How does it arise? What sort of molecular mechanism can switch off genes so that they are never expressed in a differentiated cell and its progeny, but are reactivated in the hospitable environment of the egg? What happens when the genome is reprogrammed in the egg? At the time many of these questions seemed technically unapproachable but with new techniques they are now being addressed with some success.

Work during the past few years has uncovered a host of gene-regulatory proteins in differentiated cells that control gene expression in a manner analogous to the better-understood bacterial repressor and activator proteins. Some of these regulatory proteins are only produced by specific cell types and are important in defining the properties of a differentiated cell. Although a differentiated mammalian cell expresses a large number of genes (on average 10 000 different mRNAs are produced) there are probably a relatively small number of 'controlling genes' that are crucial to specifying any given differentiated state. The topmost layers of control are now starting to be peeled away, and some of the regulatory mechanisms that are being exposed are discussed in more detail in Chapter 3 (see Control of eukaryotic gene expression). If a cell's past developmental history is recorded in a stable chromosomal record it may eventually be possible to reveal the successive layers of controls laid down as its precursors became determined in the early embryo.

Differentiation of Blood Cells

In contrast to some other aspects of development we probably know more about the differentiation of mammalian cells than of those of many other animals because of the obvious medical importance of the generation and differentiation of the various cells of the blood and immune system, muscle and nerve. Although most cells in a newborn animal are already fully differentiated or at least fully determined, mammalian blood and immune cells are continually being replaced throughout life from undifferentiated self-renewing **stem cells**. Erythrocytes (red blood cells), lymphocytes, and other leukocytes (white blood cells, which include several types such as granulocytes, macrophages and eosinophils) all derive from a single type of pluripotent self-renewing stem cell located (in mammals) in the bone marrow, and their differentiation has been studied intensively for many years. There are several points on these pathways where two different cells are derived from a common precursor, providing a late and reasonably accessible example of a type of determinative event (Fig. 10.15).

Differentiation of blood cells is a gradual process, occurring in successive stages over several cell generations. Some leukaemias, for example, represent a failure of the precursors of white blood cells to differentiate; instead the partly differentiated precursor cells go on dividing uncontrollably. Genes that may be involved in the generation of leukaemias (and of other cancers) have recently been uncovered (see Chapter 11: Oncogenes) and may throw much light on the mechanisms of differentiation.

Unfortunately, the initial pluripotent stem cell has not yet been unambiguously identified and so the first and crucial choice between becoming lymphocyte, red cell or white cell cannot yet be studied in detail. There are indications that it depends on cell-surface molecules on the cells of the bone-marrow stroma, the tissue in the bone marrow in which much of blood cell generation and differentiation takes place. Passage through subsequent stages depends largely on the actions of various protein growth and 'differentiation' factors produced by blood and immune system cells and other tissues, and which trigger changes in metabolism and alterations in gene expression within successive generations of differentiating cells. In some cases these signals specify choices between two alternative pathways of differentiation. Signals produced by one set of cells that influence the development of others are a universal feature of development. The embryonic production of specific signalling molecules – the **morphogens** and **inducers** of traditional developmental biology – analogous to these growth factors and hormones is suspected in many cases (see next section) although few have yet been isolated and identified.

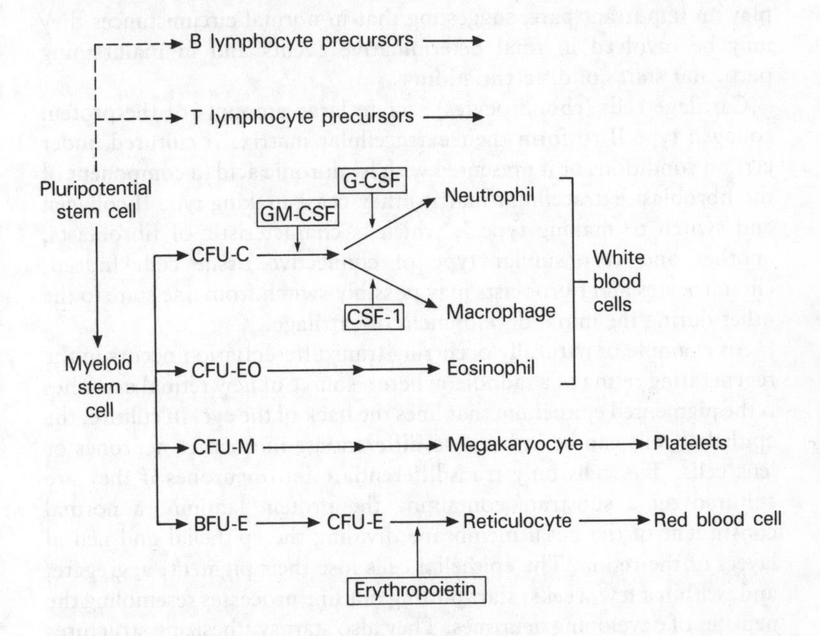

Fig. 10.15 Derivation of blood cells from a pluripotent stem cell in bone marrow. CFU-C, CFU-EO, etc. are 'colony-forming units' – cells that proliferate and differentiate into the particular blood cell shown. Proteins that induce differentiation of specific lineages are shown in boxes.

Transdifferentiation

Once differentiated, cells normally do not change their character. However, there are a few instances in which **transdifferentiation** or **metaplasia** takes place and cells take on another differentiated form. These rare cases are helpful in identifying external factors that may normally be involved in causing differentiation or in maintaining a particular differentiated state.

Transdifferentiation occurs naturally in a few rare cases, usually in regenerating tissue in response to damage, and it can also be induced in some cultured cells. In all cases found so far, the switch is usually to another product of the same primary germ layer. For example, no switch from an ectodermal to a mesodermal-derived cell has yet been encountered. In several instances proteins of the extracellular matrix play an important part, suggesting that in normal circumstances they may be involved in final determinative events and in maintaining particular states of differentiation.

Cartilage cells (chondrocytes) secrete large amounts of the protein collagen type II to form their extracellular matrix. If cultured under certain conditions or if presented with hyaluronic acid (a component of the fibroblast extracellular matrix) they cease making type II collagen and switch to making type I, which is characteristic of fibroblasts, another and very similar type of connective tissue cell. Indeed, chondrocytes and fibroblasts may possibly switch from one state to the other during the initial development of cartilage.

An example of naturally occurring transdifferentiation occurs in the regenerating retina of a tadpole, where a source of new retinal neurones is the pigmented epithelium that lines the back of the eye. In culture, the epithelial cells can be induced to differentiate into either neurones or lens cells. The cells only transdifferentiate into neurones if they are cultured on a substrate containing the protein laminin, a normal constituent of the basal membrane dividing the epithelial and neural layers of the retina. The epithelial cells lose their pigment, aggregate, and, within a few weeks, start putting out fine processes resembling the neurites of developing neurones. They also start synthesizing structures typical of neurones rather than epithelial cells.

How the cell senses the presence of laminin and how this signal is transmitted to the nucleus to switch off the 'epithelial' gene set and activate the 'neuronal' gene set is not yet known. The pigmented epithelium and neurones both derive from ectoderm, and their divergent development may normally be influenced by contact with the appropriate mesodermal tissues.

STRUCTURE AND FORM

Pattern formation and developmental signals

At the end of embryonic development the cells that have arisen from the zygote have arrived at their final differentiated state. How they end up as skin cell, neurone, muscle or bone in the correct relationship to each other is the question addressed by the study of **pattern formation.**

Any spatial organization – from the main body axes to the segmented structure of the insect body or the detailed disposition of bone, muscle and skin in a vertebrate limb – is arrived at by imposing differences in developmental potential on different parts of an initially homogeneous field. This entails specifying areas of the field as cells of different types in a particular spatial pattern. Differential growth, cell movements and even programmed cell death can then produce three-dimensional structure from this initial pattern, with further rounds of patterning being imposed as development proceeds.

In largely mosaic animals, much of the final pattern is already laid down in the cytoplasm of the egg (see previous sections). In highly regulative embryos, on the other hand, new patterns are continually being created by the developmental process itself.

Invariant Cell Divisions Create Structure

One way of creating structure is simply by the pattern of cell division. A group of cells dividing in predetermined orientations to each other will always generate the same structure, and in animals with invariant cell lineages such as the nematode the final body form is largely generated by a fixed sequence of cell divisions plus equally invariant migrations and programmed cell deaths.

Expression of a preset programme tells a cell when to divide, and possibly in what orientation, when to assign two daughter cells to different fates (i.e. patterns of cell division) and when finally to differentiate. If a blastomere from an ascidian or nematode is isolated and prevented from dividing, but its DNA is still allowed to replicate, it will, at around the time its progeny would normally differentiate, start to produce proteins characteristic of the range of differentiated cells it normally produces. Correct execution of the programme in normal development therefore probably depends both on further spatial parcelling out of persisting maternal determinants into successive generations of daughter cells and on the precisely timed production of new intracellular developmental determinants under the direction of the initial maternal programme. Cells can count time in a variety of ways but how they do so in this case is not yet known.

Positional Information

In regulative embryos, the developmental fate of a cell is set not only by its lineage – its memory – but also by additional **positional information** it receives from its environment. It acquires a **positional value**, which it then interprets according to its past history by entering a particular determined or differentiated state.

One way of presenting positional information is in the form of diffusing chemical signals emanating from a localized source. Diffusion away from the source sets up a chemical gradient and in some cases it is likely that cells read their position along the gradient from the level of chemical in their local environment (see below, The chick limb).

The prime example of a localized area of tissue influencing future development is the mesoderm of the dorsal lip of the blastopore – the primary organizer – (see above, Gastrulation). Despite an intensive search over the years the nature of the blastopore **morphogen** – the developmental signal the primary organizer is believed to produce – has not yet been uncovered. The search for the signals that specify the primary organizer mesoderm itself has, however, met with better luck (see below).

Cells can also read their fates from the extracellular matrix (see above, Transdifferentiation) and from the cell-surface glycoproteins displayed by their neighbours (see below). Internal 'clocks' that count time by the number of successive rounds of transcription necessary, say, to switch on a gene producing a particular regulatory protein, or by counting cell divisions may also exist.

A productive interaction between two tissues needs both a cell that is producing a signal, and a cell that is able to respond to it. A response to a diffusing chemical or cell-surface molecule usually requires in the first instance the presence of specific receptors. Gene expression must therefore be coordinated within the developmental programme to ensure that not only the signal but the appropriate receptor proteins and other cellular response machinery are produced at the right time and in the right cells.

The total field influenced by these signals is very small: in the case of diffusing signals it is of the order of a millimetre or less, in the case of direct cell–cell interactions, even smaller. The complete specification of position in a large vertebrate embryo must therefore be built up step by step by successive local signals. To acquire a positional value in a three-dimensional structure such as a developing limb, a cell also has to integrate information delivered to it along the different axes within the field – for example, base to tip and 'little finger to thumb' axes (see Fig. 10.19).

Induction

Throughout development there are many examples of a particular sort of one-way interaction in which one tissue or cell directs the development of another while itself remaining unchanged. **Induction** is generally presumed to be due to signal molecules (*inducers*) produced by the inducing tissue and to which the cells in the induced tissue become responsive at the correct time in development.

The classic example is that of the notochord in inducing the formation of the neural tube. Transplantation of mesoderm from below the prospective neural tube to the belly of an amphibian gastrula induces the ectoderm in that region to roll up into a misplaced neural tube. The signal emanating from the mesoderm is highly specific in positional terms as the character of the ectopic neural tube depends on exactly where the mesoderm comes from. If from the anterior end of the embryo it induces forebrain formation, if from the posterior end, a piece of spinal cord. It is commonly assumed that the notochordal tissue releases one or more inducing signals but their nature remains a total mystery. One complicating factor is that neurulation can be induced by an extremely wide range of apparently unrelated treatments.

The search for the signals that specify the notochordal mesoderm itself has been more successful. In the frog *Xenopus*, the animal half of the blastula becomes specified as ectoderm and the vegetal half as endoderm. Then mesoderm is induced from prospective ectoderm by the endoderm (see above, Specification of endoderm, ectoderm and mesoderm). Inductive signals are produced by different regions of the endoderm under the guidance of a subtle system of positional cues that is set up by the cytoplasmic movements that follow fertilization. Induction creates at least two regions of different developmental potential within the mesoderm.

The positional system (whose nature is unknown) is believed to create a 'dorsal signalling centre' in the endoderm, which then produces at least one signal that specifies the 'ectodermal' cells immediately adjacent as 'dorsal' mesoderm. This contains the primary organizing region (the region that initiates gastrulation, forms the notochord and also directs the correct development of the dorsal region of the embryo). The rest of the endoderm specifies an adjacent region as 'ventral' mesoderm, also by releasing a diffusible signal (Fig. 10.16). The organizing region is then believed to emit a signal that fine-tunes the dorsal–ventral pattern, possibly distinguishing various main regions in the mesoderm which will, for example (going from dorsal to ventral) become muscle, kidney and blood cells (see Fig. 10.5). These events occur probably around the 32/64 cell stage when development is still directed by maternal determinants.

The ventral signal is now believed to be a protein very similar to one of the mammalian fibroblast growth factors (basic FGF), and a protein related to mammalian transforming growth factor ß (TGFß) has tentatively been identified as at least part of the dorsal signal. At the same time, one of the maternal mRNAs, which becomes localized at the vegetal pole of the ovum (in at least broadly the right place to act as the

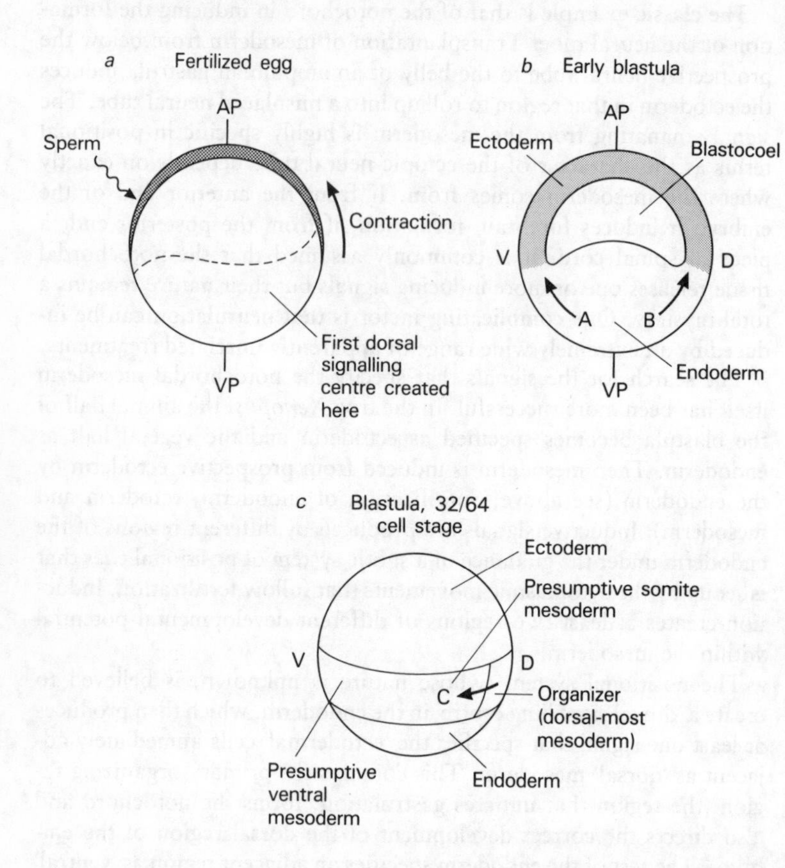

Fig. 10.16 Proposed scheme for generating the dorso-ventral axis of an amphibian embryo. *a*, Section of a fertilized egg. A contraction of the cytoplasm just under the plasma membrane (cortical cytoplasm) opposite the site of sperm entry generates (by means unknown) a 'dorsal signalling centre'. *b*, After cleavage has begun this emits a signal, B. A 'ventral' signal A emanates from another region of the endoderm. A and B induce mesoderm formation, B inducing the formation of the dorsal-most mesoderm as the organizer region. *c*, This in turn emits a signal (C) that specifies the fate of the mesoderm along the future dorso-ventral axis. (From H. Woodland and E. Jones, *Nature*, 332 (1988) 113–115.)

inducing signal), has been identified as the mRNA that encodes this TGFß-like protein. mRNA encoding the '*Xenopus* FGF' has also been identified in the oocyte. These represent two of the very few 'maternal determinants' that have yet been even provisionally identified outside *Drosophila*.

In mammals FGF and TGFß stimulate (and TGF also represses) the division of a wide range of cells (see Table 9.9). How they might act as 'morphogens' is not clear. Polypeptides related to growth factors are developmental signals in other animals. Another member of the TGFß family has indeed been identified as the product of the *dpp* gene which is involved in the specification of dorsal–ventral pattern in *Drosophila* embryos and the normal morphogenesis of imaginal discs at metamorphosis.

There are many cases of inductive interactions between mesodermal tissues and ectodermal tissues throughout development. The formation of feathers, scales or hairs by the epidermis is also induced by a mesodermal tissue – the underlying dermis. These structures are produced by the epidermal cells but their nature and arrangement are largely controlled by the dermis. If dermis from the back of a developing chick embryo is combined at an appropriate stage with epidermis from the leg (which normally produces scales) the leg epidermis is induced to form feathers in the exact arrangement seen on a chick's back. How such information is encoded in the signals produced by the dermis, or what those signals are, is not yet known but they have been highly conserved during evolution. If dermis from the snout of a mouse embryo is placed with chick epidermis, it induces the formation of feathers from the epidermis, but in an arrangement corresponding to the whisker pattern on a mouse's snout.

Induction in Caenorhabditis

Despite its largely determinate development, the nematode *Caenorhabditis* offers several examples of induction, one of which is being subjected to detailed genetic dissection.

The hermaphrodite worm possesses a single gonad that communicates with the exterior through an aperture – the vulva – in the body wall. Formation of the vulva is induced from mesodermal cells of the body wall (hypodermal cells) by a special gonadal cell – the anchor cell. The vulva consists of a simple structure of 22 cells that are derived in a fixed sequence of divisions from three precursor cells (Fig. 10.17). These cells are always the same in normal development, but they have been shown to belong to a set of six 'equivalent' potential vulval precursors each of which can if necessary go on to form part of the vulva. These six cells come to lie close to the anchor cell with the three normal vulval precursors nearest to it. The anchor cell then directs these three cells to embark on a programme of cell division that eventually forms the vulva. The determination of the vulval cells is particularly subtle. The cell nearest to the anchor cell receives the full force of the signal and follows one sequence of cell divisions. The outer two, which receive a weaker signal, follow an identical but different pattern. A single signal, acting at two levels, is thus specifying two different determined states.

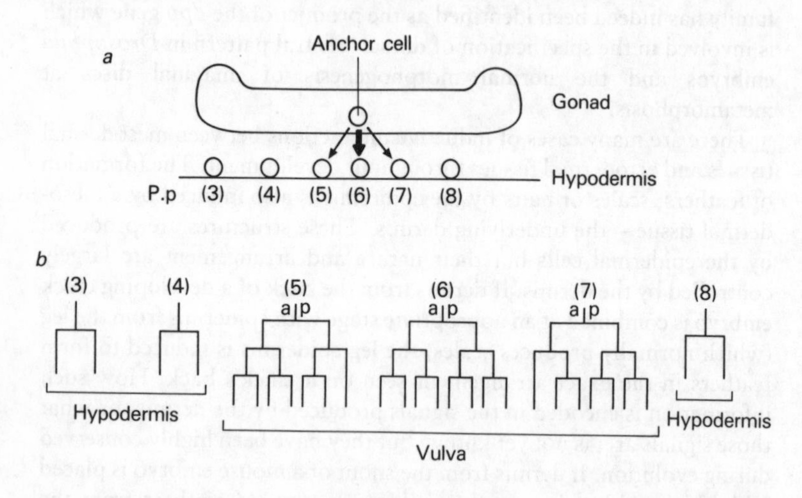

Fig. 10.17 Induction of vulval cell precursors and cell divisions leading to vulva formation in *Caenorhabditis*. *a*, Determination of vulval cell precursors by a signal emanating from the anchor cell. *b*, Subsequent patterns of cell division of determined cells. a and p denote anterior and posterior products of cell division respectively. See text for further explanation. (Adapted from P.W. Sternberg and H.R. Horvitz, *Annual Review of Genetics*, 18 (1984) 489–524.)

These cell divisions generate the simple vulval structure. The remaining three potential precursors do not come within the sphere of influence of the anchor-cell signal and after one further division join the layer of the body wall – the hypodermis – that underlies the outer cuticle. (In the absence of any anchor-cell signal, this is the fate of all six cells.)

Some genes involved in all aspects of this induction have now been identified by mutations that either abolish the vulva completely or result in multiple 'mini-vulvas', which arise when all six potential precursors are switched to a vulval pathway. Some genes are thought to encode products required for reading and interpreting the anchor cell signal, others to be concerned with the intracellular machinery that then commits the cell to a particular pathway. The detailed analysis now in progress should eventually identify many of the molecules and mechanisms involved.

Although the nematode may seem remote from the concerns of mammals, differential determination in response to a graded signal is exactly the mechanism proposed to explain many features of vertebrate development. The molecular mechanisms by which it occurs in *Caenorhabditis* should therefore be of universal interest.

Cell-surface Glycoproteins

Differentiating and differentiated cells carry a variety of stage-specific, cell-specific or tissue-specific membrane glycoproteins on their surfaces. Similar stage-specific and tissue-specific membrane components have been identified in a wide variety of developing tissues, and are believed to be important in tissue organization and morphogenesis. Some of these may represent receptors for morphogens and inducers, but others may have a more direct role.

Cells can recognize and adhere to (or repel) one another by their surface glycoproteins and this is one way of generating structure and preserving the integrity of separate tissues or structures. Combinations of surface glycoproteins unique to a particular tissue may direct cells bearing the appropriate receptor to follow a particular path of determination and/or differentiation. Successive layers of organization built up in this way could result in very regular multilayered structures such as the retina, where each layer of cells maps precisely onto the next.

One membrane glycoprotein involved in generating this type of regular structure is the product of the recently cloned *sevenless* gene in *Drosophila*. The protein is normally expressed in the developing adult compound eye as cells assume their final positions and differentiate. Mutant flies lacking the sev protein have a very specific defect. They fail to recruit a normal seventh photoreceptor cell into each facet of their compound eyes. Instead, the cell that would have become the seventh photoreceptor develops into a non-neural cell. The other photoreceptor cells develop normally. This suggests that the sev glycoprotein is an essential signal required by the seventh and only the seventh cell to position itself correctly and differentiate.

Compartments: building up a positional address

Insects have an apparently unique positional specification system that builds up a local 'address' for each cell in the embryo. This is then carried over to the imaginal discs (see Fig. 10.21) from which the insect's adult epidermal covering derives. Insects possess a particular type of developmental 'module' that has so far not been revealed in other animals. Each segment of the insect body can be shown to consist of several **compartments** – precisely delimited areas whose boundaries do not follow any particular morphological feature, and which each arise from the progeny of a small group of **founder cells**. The compartments in each segment are developmentally equivalent to each other although the morphological structures they produce are specific to each segment (Fig. 10.18).

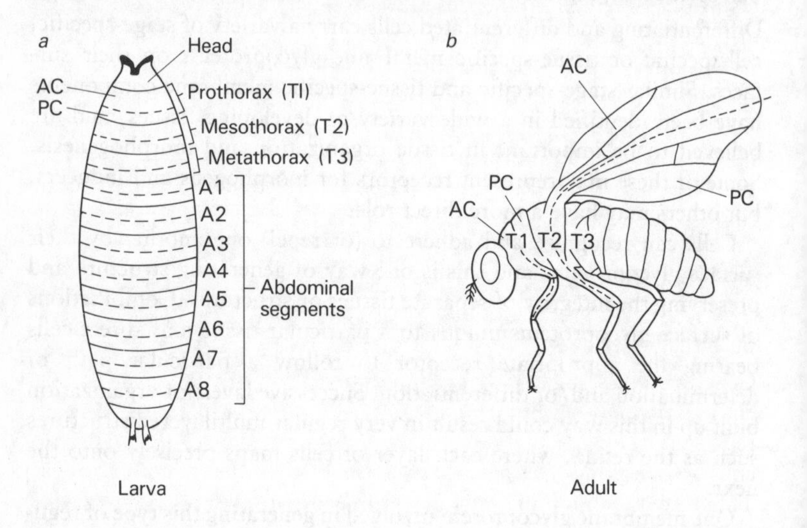

Fig. 10.18 Anterior (AC) and posterior (PC) compartments in *a*, larva, and *b*, adult fruitfly *Drosophila*.

In the fruitfly's wing, for example, at least six intersecting compartments have been distinguished. One compartment boundary runs in a straight line approximately along the long axis of the wing (not following the veining or any other morphological feature), dividing it into an anterior and a posterior compartment of roughly equal size. Another divides the wing into an upper and underside (dorsal and ventral compartments) and yet another divides most of the wing proper from the notum (the region at which it is attached to the thorax) to form a proximal (nearest to the body) and a distal (furthest away) compartment.

Compartment Boundaries

Each compartment is **polyclonal** – that is, composed of the progeny of several founder cells. Each clone arising from a single founder cell has an irregular outline except where it meets the compartment boundary. Compartment size seems absolutely invariant; fast-growing clones may fill a compartment to the exclusion of slower-growing companions but never enlarge it. The way in which compartments are physically delimited is not yet totally understood; the remarkable regularity of the boundaries may perhaps be the result of differences in cell-surface features between cells of neighbouring compartments that prevent them mixing.

The anterior and posterior compartments of each segment are one of the earliest distinctions to arise in the insect embryo (see final section on *Drosophila* for more detail). The specification is carried over into the imaginal discs. Compartments seem to be restricted to ectodermal tissue as little evidence of their existence has been found in internal organs (of mesodermal origin).

Selector Genes

Genetic studies suggest that compartments intersect, so that in both embryonic and adult structures a cell may acquire an 'address' representing a particular combination of on/off states of several sets of control (**selector**) genes, each set determining a different compartment. This address is built up gradually, with compartment boundaries being specified in a set order, anterior/posterior first, dorsal/ventral and proximal/distal later. The address is interpreted by segment-specific genes that specify the morphological structure at that point.

Genes affecting the choice between alternative pathways of compartment determination can be revealed by mutations that transform one compartment into another. The homoeotic mutation *engrailed*, for example, converts the posterior wing compartment into a mirror-image anterior wing compartment. The function of the normal *engrailed* gene is to act only in prospective 'posterior' cells of each segment, differentiating them from the basic anterior-type state: loss of function, therefore, has no effect on the anterior compartment. The compartment specification machinery is the same in each body segment, even though the morphological outcome is quite different; *engrailed*, for example, converts posterior to anterior compartment in both wing and leg.

A developmental feature such as compartments provides an elegant and economical way of building up precise positional specifications for individual cells from the actions of a relatively small number of genes. There is, however, no good evidence for the existence of compartments in other animals and they may be a unique feature of insect development.

THE CHICK LIMB

One experimental system that has yielded valuable insight into pattern formation and positional information in vertebrates is the **developing chick limb**. It illustrates many of the concepts introduced in previous sections.

Birds are unique in possessing quite different front and hind limbs (wings and legs) which although the same in general plan, differ markedly in shape and detailed structure at a very early stage of development. So they are ideal material in which to study how two rather similar structures, originating from apparently identical embryonic **limb buds** (which are evident after 3–4 days incubation of a fertilized egg) develop in different ways.

The way the limb bud develops after small pieces of tissue have been grafted from wing to leg and vice versa provides strong evidence in favour of positional specification of cells. When a piece of tissue is taken from the base of the leg bud (this would normally develop into thigh), and grafted beneath the ectoderm at the tip of the wing bud it develops not into a wing tip, nor into misplaced thigh but into toes. This shows that limb bud cells are already specified with regard to which limb they will make, but are not yet decided about which part of the limb they will make. That choice depends on their position in the limb just preceding the time they start to differentiate into skeletal and muscular tissues. Also notable in this experiment is the fact that the part of the wing bud that would normally make the wing tip does not, but because of its displacement relative to the ectoderm contributes to the forearm instead. The results also imply that the positional signals in both limbs are effectively the same, otherwise the cells of the leg bud would not be able to interpret the signals received in the wing.

The thumb–little finger (anterior–posterior) axis

Development along the different limb axes is apparently controlled by different parts of the limb rudiment. The 'thumb–little finger' (anterior–posterior) axis is generated by a small group of cells known as the **zone of polarizing activity** (ZPA). If this tiny piece of tissue is dissected out and grafted into a different position in another limb bud, that limb will develop abnormally, and from the type of abnormality produced various inferences can be drawn about the normal role of the polarizing region (Fig. 10.19).

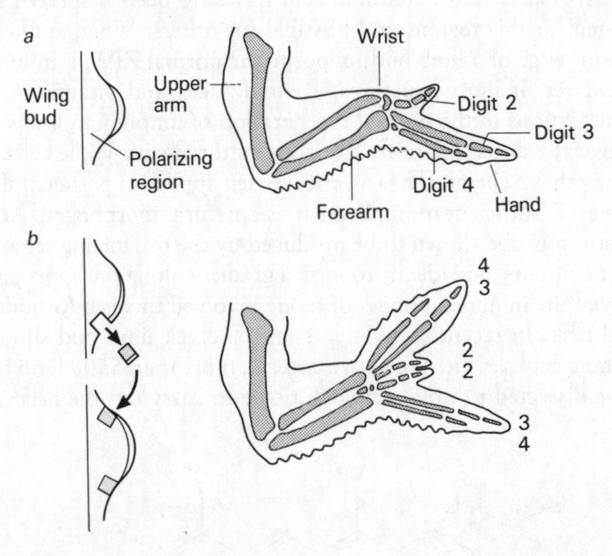

Fig. 10.19 Duplication of digits along the anterior–posterior axis of chick wing after transplantation of the zone of polarizing activity (ZPA). *a*, Normal; *b*, after transplantation of ZPA.

A wing bud with two polarizing regions produces a mirror image duplicated wing. This duplication is most clearly seen in the digits (the chick wing has three digits corresponding to the middle three digits of a five-membered hand). The precise degree of duplication depends on the position of the new polarizing region. The polarizing region itself does not contribute cells to the new wing and the simplest explanation is that it produces a chemical signal which diffuses to form a gradient. Cells become determined to produce the various structures by their position on this gradient.

Polarizing regions from mammalian (including human) and reptilian embryonic limb buds are also effective in chick limb buds showing that this developmental signal has been highly conserved throughout vertebrate evolution. This is welcome evidence that a study of one system does provide answers of more general applicability.

The ZPA morphogen

There is now extremely strong evidence that the morphogen produced by the zone of polarizing activity (ZPA) is **retinoic acid** (a derivative of vitamin A) (Fig. 10.20). Retinoic acid has long been suspected as a morphogen for this region, and possibly for others. When applied to the anterior edge of a limb bud, opposite the normal ZPA, it induces a duplicated set of limb structures. Retinoic acid and vitamin A can induce alterations in the normal regeneration of amphibian limbs and also trigger the differentiation of various cultured embryonic cells.

Showing that a chemical has an effect when applied experimentally is not, however, sufficient to implicate it as a natural morphogen. At the very least it must be shown to be produced by the organizing region in sufficient amounts, and ideally to form a gradient along the appropriate axis. Given the minute amounts of tissue involved this is a formidable technical task. In recent experiments on the chick limb bud showing that retinoic acid satisfies the above criteria, more than 5500 limb buds had to be dissected to isolate enough tissue to carry out the necessary analyses.

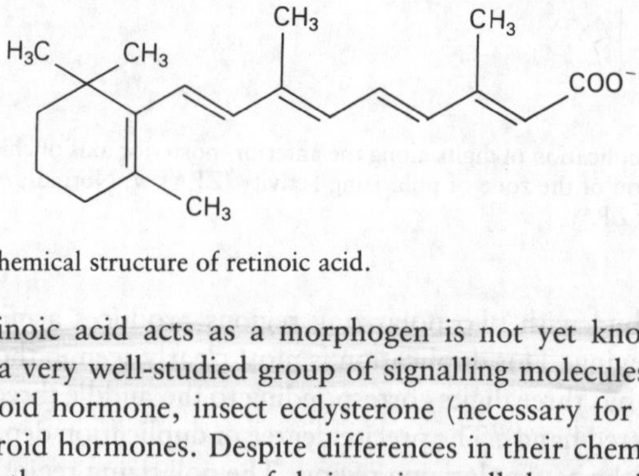

H_3C CH_3 CH_3 CH_3 COO^- CH_3

Fig. 10.20 Chemical structure of retinoic acid.

How retinoic acid acts as a morphogen is not yet known but it belongs to a very well-studied group of signalling molecules. This includes thyroid hormone, insect ecdysterone (necessary for pupation) and the steroid hormones. Despite differences in their chemical structure and in the consequences of their actions, all these molecules are believed to share a common mechanism of action. The steroid hormones and ecdysterone are known to act as transcriptional activators and, in the case of ecdysterone, probably as a repressor as well. The steroid hormones bind to intracellular receptors that then activate specific DNA 'response elements' to switch on particular genes. The thyroid hormone thyroxine also binds to an intracellular DNA-binding receptor, although its activity as a transcriptional regulator has not yet been proved.

Clues to the developmental actions of retinoic acid therefore probably lie in its receptor. By the application of lateral thinking and some ingenious molecular biology, the gene for the retinoic acid receptor protein, which bears some resemblance to that for thyroid hormone, has now been isolated and cloned. As well as providing large amounts of the receptor itself for biochemical study, it is not beyond the power of modern molecular genetics, given a certain amount of luck, to use this gene as a probe to pick out the genes on which the retinoic acid/receptor complex might be acting in chick and mammalian development (retinoic acid seems to have this role in vertebrate limb development generally). Ways of generating mutations to order in mammals are now also being developed, and could eventually make it possible to test the effects on development of an altered retinoic acid receptor gene.

The base to tip (proximal–distal) axis

An embryonic vertebrate limb grows and differentiates its internal structures successively from base to tip. Development of a complete limb requires the continuing presence of the thickened ridge of ectoderm – the apical ridge – visible on the tip of the limb bud. If this is removed at different stages as the limb bud elongates, an incomplete limb develops. If the apical ridge is removed from an early wing bud only the upper arm develops; if removed from a late bud, upper and forearm, but no hand develops.

The apical ridge determines the order of development from the base to the tip of the limb but it does not itself instruct the underlying tissue (mesenchyme) which structures to differentiate. It maintains a 'progress zone' of undifferentiated cells immediately beneath it that produce the parts of the limb in the correct order. The character of each limb segment seems to depend on the time spent in the progress zone by a particular group of cells, time possibly counted by the cells in terms of number of cell divisions.

DROSOPHILA

At the beginning of the century the American geneticist T.H. Morgan chose the fruitfly *Drosophila melanogaster* as a convenient subject for the new science of mendelian genetics. Had his choice fallen on some other organism the face of developmental genetics today would be very different. *Drosophila* is still the only organism of any morphological complexity whose genetics and embryology are both sufficiently well known to attempt to explain one in terms of the other. An unrivalled collection of mutant strains with specific developmental defects exists, all well-characterized both in genetic terms (and increasingly in molecular genetic terms) and in their morphological effects.

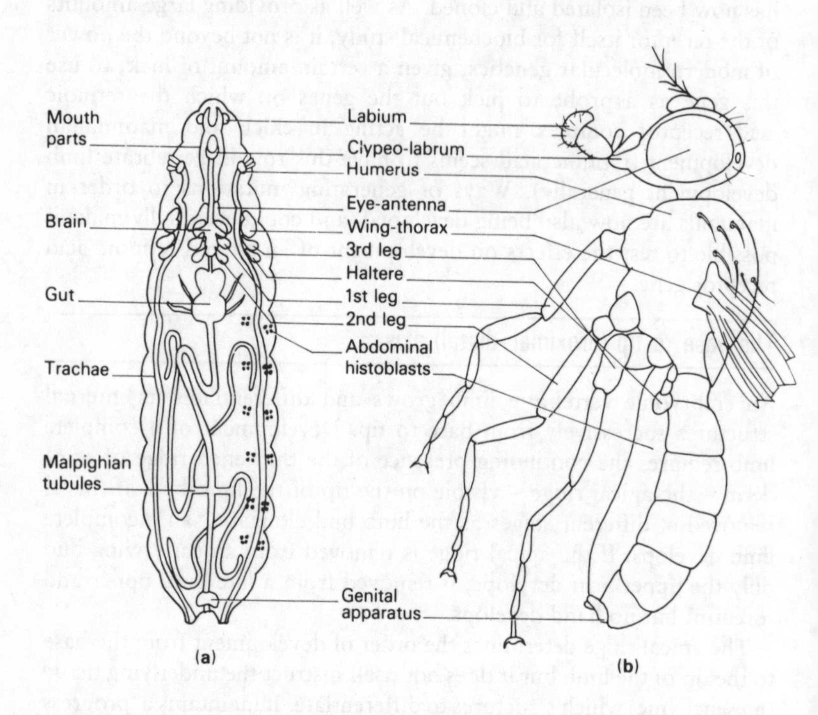

Fig. 10.21 The larval imaginal discs and imaginal precursor cells *a*, and the adult structures they form during the pupal metamorphosis *b*. (After H.R. Wildermuth, *Science Progress*, 58, (1970) 329–358. Reproduced in J.H. Sang, *Genetics and development*, Longman, Harlow, 1984.)

With much genetic sleight of hand fruitflies of virtually any desired genotype can be bred, and the species is particularly amenable to deliberate genetic manipulation. A recent technical advance, which has followed from the isolation and cloning of early acting developmental genes, is the labelling of embryos with gene-specific radioactive DNA probes. These specifically hybridize with the gene-specific mRNA to reveal the exact timing and spatial patterns of transcription from these genes (see Chapter 4: Identification, for background to DNA hybridization). In a similar way, once the protein products of genes are available, exquisitely specific monoclonal antibodies against them can be produced and used to detect their site of action (which is not necessarily always the same as the site of their synthesis). This has provided much information to back up conventional genetic and embryological analysis of early pattern formation.

Insects, like all arthropods, have a segmented body structure. This is based on repetition of a 'prototype' segment with various modifications. The *Drosophila* larva comprises a head, three thoracic segments (prothorax, mesothorax and metathorax) and eight abdominal segments. The head consists of several segments that become fused early in development. The segmented structure and the specification of each segment's identity is established in the embryo and is carried through the larval stages into the adult where it is most obvious in the epidermal tissues and the structures derived from them (see Fig. 10.21).

Drosophila takes only 9 days to develop from egg to adult. It has an embryonic stage in the egg, hatches out as a larva that goes through several moults, then pupates and undergoes morphogenesis into an adult fly. Specifically adult structures (leg, wing, eye, antenna and genitalia) arise from **imaginal discs**, sacs of ectodermal epithelium formed in the larva and which differentiate at metamorphosis to produce the epidermis clothing the adult body. The larval discs are already committed (determined) to becoming a particular segment and producing the structures appropriate to it. The epidermal covering of the adult abdominal segments is formed from similarly committed cells grouped in **histoblast nests** (Fig. 10.21).

Setting up the basic body plan

Insect eggs do not undergo immediate cleavage as do those of many other animals. Nuclear division proceeds rapidly after fertilization without cytoplasmic division to form a **syncytium** containing thousands of nuclei in a continuous cytoplasm (Fig. 10.22). The nuclei migrate to the periphery where, after the thirteenth nuclear division, a cell is formed around each nucleus to give the **blastoderm**. The blastoderm after this stage corresponds roughly to the blastula stage (see above, The blastula). It then gastrulates, the nervous system is formed, the larval organs develop and after a day the larva hatches.

There are two aspects to generating the basic segmented body plan of a fly. The anterior–posterior and dorsal–ventral axes are first established by maternal determinants supplied in the egg cytoplasm. Building on this grid of positional information the embryo's own genes (the

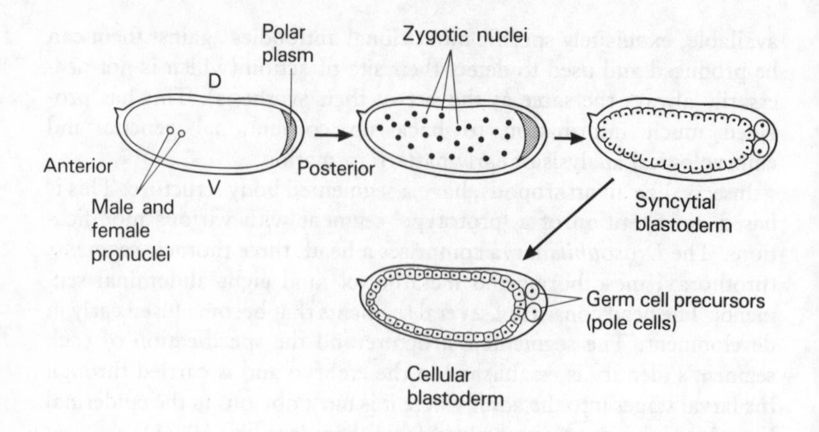

Fig. 10.22 Development of an insect egg to the cellular blastoderm stage. D, dorsal; V, Ventral.

zygotic genes) are activated to specify the number of segments, the position of their boundaries and finally, the identity of each segment and its potential course of development.

By painstaking analysis of as many mutations as can be obtained (and new genes are still being discovered) a picture of the sequence in which these genes normally act in development is being established. A hierarchy of states of determination is now emerging for body axis polarity, segmentation pattern, segment specification and compartmentation (see above, Compartments).

The Dorsal–Ventral Axis

The role of cytoplasmic determinants in specifying the initial body axes in *Drosophila* can be studied in great detail through the effects of a class of mutations called **maternal effect mutations**. These identify genes that must act correctly in the mother for embryonic development to proceed normally. Mothers with defects in these genes often produce embryos that do not develop properly and die at an early stage, even if the embryo itself contains a good copy of the gene inherited from its father. In these cases, lack of maternal gene product cannot be remedied by subsequent transcription from the embryo's own gene. Several maternal-effect genes are now known to code for the cytoplasmic developmental determinants whose actions are completed before transcription from the embryonic genome begins.

Maternal-effect genes involved in the initial formation of the body axes have been identified. One group of around a dozen (the 'dorsal' genes) encodes a set of cytoplasmic determinants that establishes the dorsal–ventral axis. Interactions between these gene products, which are localized to the prospective ventral portion of the zygote, produces a signal that distinguishes dorsal from ventral. In embryos lacking any of these determinants, the (largely mesodermal) ventral cells and structures are entirely or partly replaced by misplaced dorsal (ectodermal) tissue. The mutations rejoice in such names as *snake*, *pipe* and *tube*, which graphically describe the hollow tube of ectoderm the embryo develops into.

The dorsal state apparently represents the 'ground state' that must be over-ridden by the ventral signal. This is indeed consistent with the finding in other animals that mesoderm has to be induced from ectoderm and is not automatically specified by localized determinants.

The exact nature of the signal itself is not yet known. Several of the 'dorsal' genes have now been cloned and their protein products characterized, but the primary signal has not yet been identified. Dorsal genes which from genetic evidence are believed to be very closely involved in generating the signal have recently been cloned, and their DNA sequences may provide some new light. The product of another 'dorsal' gene – *snake* – was recently identified as a possible serine protease (by inference from the nucleotide sequence of the gene). Serine proteases are proteolytic enzymes of the sort that in mammals convert inactive enzyme precursors to active enzymes by cutting the polypeptide chain at a specific point. They are themselves made as inactive precursors which have to be activated by another protease. Enzyme cascades of serine proteases are a feature of the blood clotting system and complement system in higher vertebrates and a role for enzymes of this sort in activating the 'dorsal signal' can be envisaged.

The dorsal signal may be a chemical gradient of the sort set up in the chick limb by the ZPA (see p. 328), but there are several alternatives. An electrical signal could be propagated or some polarized cytoskeletal structure physically spanning the region could be set up. At this stage the blastoderm has not yet cellularized so the cytoplasm within the embryo is still continuous.

The Anterior–Posterior Axis

The anterior–posterior polarity is generated by a separate set of maternal gene products. Classical embryological manipulations suggest that the axis is generated from localized centres at each pole, but how it is established has not yet been finally worked out.

Segmentation

The intersection of the anterior–posterior axis and the dorsal–ventral axis throughout the fly's body is thought to set up a three-dimensional system of molecular coordinates that could in principle uniquely specify any position in each half of the embryo. Within this framework of positional information the embryo's own genes are activated to start to specify the segmented structure of the insect body. Segmentation is presumed to be guided chiefly by positional cues along the anterior–posterior axis.

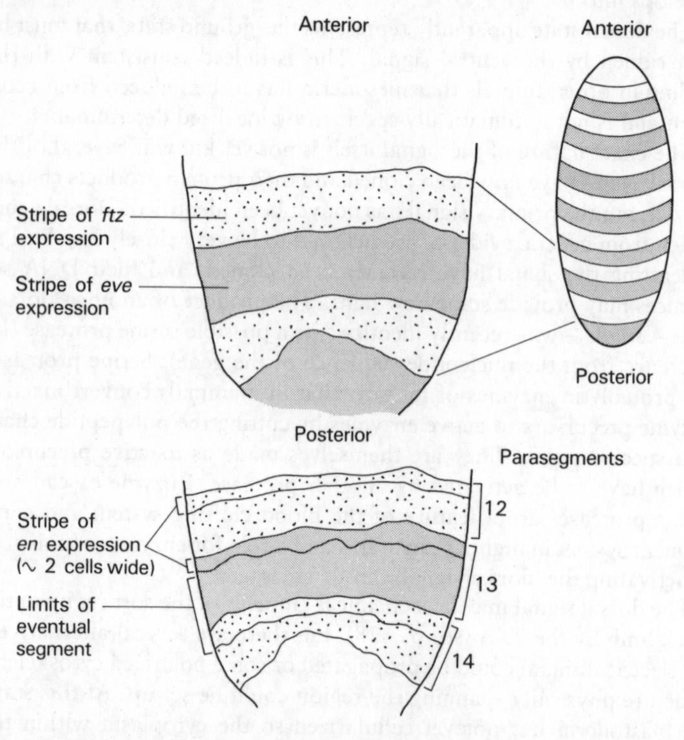

Fig. 10.23 Delimitation of parasegments. See text for explanation. (See P.A. Lawrence *et al.*, *Nature*, 328 (1987) 440–442.)

One set of genes first delimits broad regions and thus determines the eventual number of segments. If mutant they result in a shortened embryo physically missing chunks of contiguous segments, and are often termed 'gap' genes. Those that have been isolated encode gene-regulatory proteins and are presumed to both interact with each other and to activate the next sequence of zygotic genes.

Within this broad regional specification a second group of around 15 genes defines segment boundaries. Each is expressed in regularly repeating transverse stripes (of various numbers and widths) along the body. Their action marks out a unit of segmentation – the **parasegment** – which is the same width but out of register with the eventual segment unit.

The parasegment, rather than the segment, appears to be the basic module on which the insect body is constructed. This illustrates the important developmental point that domains of developmental action do not necessarily correspond to eventual obvious morphological divisions. Within each parasegment (which consists of the posterior compartment of one segment and the anterior compartment of the next) stripes of gene activity delimit the point at which the epidermis will later invaginate to form the visible constrictions between segments.

Figure 10.23 shows schematically the activity of two of the so-called 'pair-rule' genes that initially delimit the parasegments. The spatial relationship of their localized action with that of a gene specifying the posterior compartment of each segment shows the relationship of the parasegment to the visible segment. The pair-rule genes isolated so far also encode gene-regulatory proteins and it is extremely likely that in the case diagrammed the posterior compartment selector gene (*engrailed*) is activated by the products of the pair-rule genes. This is a means of maintaining the initial spatial relationship of the boundaries with absolute precision, cell by cell, as further levels of pattern are imposed, and is a motif repeated again and again in the segmentation and segment-specification process.

Specification of segment identity: the homoeotic genes

At this stage, the *Drosophila* embryo consists of a sequence of duplicated, essentially identical segments. The genes that give each segment its unique identity are now activated. They also act in overlapping zones of activity. Within the cells of the overlap the expression of one gene is suppressed (or in some cases possibly activated) by the product of the other and its effects are thus restricted to a region corresponding to a single segment.

The interaction between the *Antennapedia* gene, which specifies the second thoracic segment and the region anterior to it, and the *Ubx* gene which is involved in specifying the third thoracic and abdominal segments is such a case. Initially, *Antennapedia* is expressed throughout the anterior region plus part of the posterior region. Its activity is restricted to the mesothorax and parts anterior to it by the expression of the *Ubx* gene in all regions posterior to the mesothorax. *Ubx* has now been proved, as suspected, to produce a gene-regulatory protein that specifically switches off *Antennapedia*. Subsequent expression of other competing genes along the abdominal region similarly specifies the metathorax and abdominal segments.

Homoeotic Mutations

Mutations in segment-specifying genes such as *Antennapedia* and *Ubx* lead to incorrect segment identification and have bizarre effects. The dominant mutation *Antennapedia* produces an adult fly with tiny legs growing out of its head instead of antennae; another mutation, *ophthalmoptera*, results in a fly with minute wings where its eyes should be. The transformation of one part of the body into another is called **homoeosis**, the mutations that cause such transformations are **homoeotic mutations** and the genes they identify are **homoeotic genes**. Mutations that transform one compartment into another are also classed as homoeotic mutations. Homoeotic mutations have been known for many years and the detailed genetic analysis of their effects correctly predicted much of the segment specification mechanism that is now being demonstrated more directly from patterns of gene transcription and *in vitro* tests of the regulation of one homoeotic gene by another. All homoeotic gene products known so far are gene-regulatory proteins (see below).

As well as regulating each other's activity to set up a particular specification for each segment, the homoeotic genes are also presumed to activate, at the appropriate time, the genes that actually direct the epidermal cells of the imaginal discs to make structures (e.g. wings, legs and bristles) appropriate to the particular segment.

The homoeotic genes are of particular interest in the wider developmental context as they are the best-known examples so far of 'master genes' that establish and maintain a given determined state. They appear to act as genetic switches, specifying a choice between two different states, and are often also called **selector genes**. The effects of homoeotic mutations mimic the phenomenon of transdetermination (see above).

It is not clear how far we can assume that similar genetic switches will be acting in setting alternative states of determination in other animals.

Homoeoboxes

The homoeotic genes share a stretch of nucleotide sequence known as the **homoeobox** towards the 3′ end of the protein-coding sequence. This specifies a stretch of around 130 amino acids which binds specifically to DNA to regulate the transcription of other genes. Rather surprisingly, similar homoeobox sequences have been found in mammalian and human DNA. The genes in which they are located have been isolated but their role is so far unknown. The transcription of the homoeobox genes can however be followed in mammalian embryos by the techniques of *in situ* nucleic acid hybridization, using radioactively-labelled DNA probes corresponding to homoeobox sequences. Some homoeobox genes are expressed in particular regions of the mouse embryo at the gastrulation stage, before there is any evidence of tissue differentiation. During later development the same genes are also expressed selectively in particular organs and their expression declines rapidly after birth. Similar selective patterns of transcription have also been traced in human foetuses from around 5 weeks of gestation. However, the patterns seen provide no real clues yet to what role, if any, homoeobox genes might have in mammalian development. No homoeotic mutations have been found in mammals.

Progress in uncovering the genetic basis of *Drosophila*'s developmental programme is now so rapid that drosophila geneticists in their more optimistic moments predict that within the next 5 years they will have essentially solved the problem of the genetic regulation of early pattern formation in at least one multicellular animal of some morphological complexity.

The early genetic programme of *Drosophila*, geared as it is to producing a series of duplicate segments, is probably not directly applicable to vertebrate development (although the repetition of ribs and vertebrae suggest a segmental pattern somewhere in their development). However, the types of molecular mechanisms deployed, such as the cascades of intracellular transcriptional regulation, are probably of universal significance and may provide clues to which sorts of genes to look for in vertebrates.

FURTHER READING

See general reading list (Browder; Berril and Karp; Alberts et al.; and Watson et al. Vol II).

E.H. DAVIDSON, *Gene activity in early development*, 3rd edn, Academic Press, New York, 1986.

J.H. SANG, *Genetics and development*, Longman, Harlow, 1984. Includes useful background explanation of the classical developmental genetic work on which the recent explosion in *Drosophila* developmental genetics has built.

W.F. LOOMIS, *Developmental Biology*, Macmillan, New York, 1986.

Science **240** (1988) 1427–75. Articles explaining the usefulness of *Caenorhabditis*, *Drosophila* and *Xenopus* in developmental and other experimental biology.

R.A. RAFF and T.C. KAUFMAN, *Embryos, genes and evolution*, Macmillan, New York, 1983.

D.L. TURNER and C.L. CEPKO, A common progenitor for neurons and glia persists in rat retina late in development, *Nature*, 328 (1987) 131–6, and commentary by J. Scholes (p. 114, same issue).

S.C. CLARKE and R. KAMEN, The human hemopoietic colony-stimulating factors, *Science*, 236 (1987) 1229–37.

M. NOBLE et al., Platelet-derived growth factor promotes division and motility and inhibits premature differentiation of the oligodendrocyte/type-2 astrocyte progenitor cell, *Nature*, 333 (1988) 560–2; M.C. RAFF et al., Platelet-derived growth factor from astrocytes drives the clock that times oligodendrocyte development in culture, *Nature*, 333 (1988) 562–4. The role of a growth factor in the development and differentiation of glial cells.

J.R. PRIESS and J. NICOL THOMPSON, Cellular interactions in early C. elegans embryos. Induction of mesoderm in Caenorhabditis, *Cell*, 48 (1987) 241–50. An example of a very early induction event in the nematode.

H. WOODLAND and E. JONES, Growth factors in amphibian cell differentiation, *Nature*, 332 (1988) 113–15, and references therein. A commentary on the identification of amphibian growth factors and their role in specifying the mesoderm in the *Xenopus* blastula.

K. BASLER and E. HAFEN, Sevenless and *Drosophila* eye development: a tyrosine kinase determines cell fate, *Trends in Genetics*, 3 (1988) 74–9. A brief review of recent work on the sevenless gene.

G. EDELMAN, Cell adhesion and the molecular process of morphogenesis, *Annual Review of Biochemistry*, 54 (1985) 135–69.

J.M.W. SLACK, Morphogenetic gradients: past and present, *Trends in Biochemistry*, 12 (1987) 200.

M. ROBERTSON, Retinoic acid receptor: towards a biochemistry of morphogenesis, *Nature*, 330 (1987) 420–21, and references therein. A short review of developments following the identification of retinoic acid as the ZPA morphogen and the identification of its receptor.

V. ANDERSON, L. BOKLA, and C. NÜSSLEIN-VOLHARD, Establishment of dorsal–ventral polarity in *Drosophila*, *Nature*, **287** (1980) 795–801.

C. NÜSSLEIN-VOLHARD, H.G. FROHNHOFER and R. LEHMAN, Determination of anteroposterior polarity in *Drosophila*, *Science*, **238** (1987) 1675–81.

M. LEVINE, Molecular analysis of dorsal–ventral polarity in *Drosophila*, *Cell*, **52** (1988) 785–6, and references therein. Recent developments in identifying the mechanism for specifying the dorsoventral axis.

W.G. GEHRING, Homeoboxes in the study of development, *Science*, **236** (1987) 1245–51.

CHAPTER 11

UNLIMITED GROWTH: CANCER CELLS

The rapid growth typical of embryonic life slows down or even stops in many animals after development is complete, but cell division never ceases entirely. In the adult human body millions of cells are dividing every day, to repair accidental damage and compensate for natural wear and tear, and to replenish the short-lived cells of the blood and epithelial linings. Cell division is under strict control to match precisely the requirements for replacement or normal growth, and cells that escape from these controls and continue to proliferate can produce malignant tumours.

Ever since the early biochemists discovered metabolic differences between normal and cancer cells, much basic research on cell physiology and genetics has been devoted to searching for an underlying cause of the malign proliferation of cells that cancer represents. As difference after difference in metabolism, cell surfaces and internal structure of tumour cells were discovered (and often hailed as 'the' cause of cancer) the picture became increasingly murky. The general acceptance that most cancers arise from a single cell that has undergone one or more mutations provided a unifying framework, but until the advent of recombinant DNA techniques it was not possible to prove this directly or to find out which genes were affected.

Research into the cell and molecular biology of cancer has been one of the main beneficiaries of the new techniques for isolating and analysing animal genes at the DNA level. During the past decade cancer research has been set on a new and productive path by the discovery that certain genes are consistently found to be altered or malfunctioning in cancer cells. The identification of these **oncogenes** raises the hope of better diagnostic tests, and eventually more specific treatments that will seek out only those cells with damaged genes, in contrast to present chemotherapies and radiotherapy which depend on killing any rapidly dividing cell. Although such treatments are often successful they damage normal cells as well, and can have extremely unpleasant and debilitating side-effects, which limit their use and effectiveness.

Because oncogenes represent mutant versions of genes that are presumed to be normally involved in the regulation of cell proliferation and differentiation, they also provide cell biologists with valuable insight into the workings of these pathways, and may be able to identify some of the controls that usually keep cells in their place. Cancer research has also to concern itself with the wider question of the normal restraints to growth, of which very little is yet known. Without a clear picture of how the territorial integrity of tissues is maintained, it is difficult to understand how the changes found in cancer cells enable them to breach those defences.

This chapter outlines some of the ways in which cell biology and molecular genetics contribute to cancer research. The first section presents some general background on the normal regulation of cell division in the mammalian body. The remainder of the chapter introduces some of the current issues in cancer research. An introductory section covers present thinking on the origins of cancer. This is followed by sections on the nature of the cancer cell and the contribution of the tumour viruses to understanding the changes that occur in tumour cells. This is followed by sections dealing with the two main classes of 'cancer genes' that have been uncovered in human cancer.

LIMITS TO GROWTH: THE CONTROL OF MAMMALIAN CELL DIVISION

Undifferentiated embryonic cells can go on dividing indefinitely, but as they differentiate and take on their final form their capacity for cell division becomes limited. Cells such as skeletal muscle and neurones that have a highly organized internal structure cannot divide at all once mature, and neither can the terminally differentiated cells of the blood and immune system – such as red blood cells, mature leukocytes, antibody-secreting plasma cells and effector T lymphocytes.

Our complement of neurones, cardiac muscle and eye lens cells, once established in infancy, has to last a lifetime as these cells are never replaced. Most other tissues, however, are renewable, either through replacement from undifferentiated self-renewable **stem cells**, or by division of the differentiated cells themselves. Smooth muscle, hepatocytes (liver cells), the ubiquitous fibroblasts of the intracellular matrix, the endothelial cells that form the lining of blood vessels and the non-excitable glial cells of the nervous system are all differentiated cells that retain the ability to make more cells like themselves.

Skeletal muscle, skin, blood and bone, on the other hand, are renewed from stem cells that are already committed to becoming muscle or skin, etc. but remain in the immature undifferentiated state until needed. Skeletal muscle is replaced by the differentiation of **satellite cells** that persist in small numbers in muscle tissue throughout life. The cells of the blood and immune system are generated from stem cells in bone marrow. Cells continually being sloughed off the skin and the epithelial linings of the airways and digestive tract are renewed from stem cells in deeper layers. Stem cells produce one daughter stem cell that retains the ability to divide and one daughter that goes on to differentiate (Fig. 11.1).

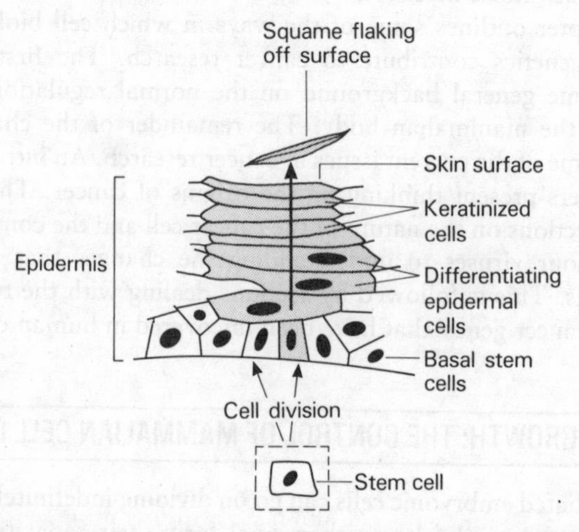

Fig. 11.1 Diagrammatic transverse section through the epidermal layer of the skin The basal layer contains dividing stem cells. One daughter cell continues to proliferate and differentiate, finally being shed from the surface. The other remains as a stem cell in the basal layer.

Red and white blood cells derive from a single population of stem cells in bone marrow. The differentiation of bone marrow cells into red blood cells (erythrocytes) on the one hand and the various white blood cells (lymphocytes, granulocytes, neutrophils, macrophages, etc.) on the other has been studied for many years as an accessible example of a complex differentiation pathway (see Chapter 10: Fig. 10.15). Protein 'factors' produced in response to physiological stimuli or infection act on bone marrow cells to induce them to become either red or white cells as the body requires. Blood cells turn over fairly rapidly compared to other tissues in adults, with different cells being renewed at different rates. A red blood cell has an average lifetime of 4 months in the bloodstream before it is replaced, whereas some leukocytes live only a few days. The memory cells of the immune system – the lymphocytes responsible for long-term immunity – can, however, persist for years.

We do not yet fully understand what regulates the division of many tissues *in vivo*, but work on cells in culture has identified some of the possible constraints, and also some of the signals that stimulate quiescent cells to divide. Chemical signals from the neuroendocrine system, endocrine glands and locally produced signals from other cells both stimulate and inhibit cell division, either as part of the body's normal physiology or in response to injury or infection. Cell division is also affected by other environmental factors such as physical crowding, available nutrients, and the molecular composition and physical properties of the surface on which they are growing. In the body this is represented by the material of the extracellular matrix and the basement membranes that divide one type of tissue from another.

The cell cycle

From the point of its formation at one cell division a eukaryotic cell must progress through an immutable sequence of events until it divides into two new daughter cells. This is the **cell cycle** during which it doubles its mass, replicates its internal organelles, duplicates its chromosomes, and then enters mitosis and finally divides. For cells steadily proliferating in culture, a full cycle normally takes around 18–24 hours but can be less. Progression through the cycle is under strict genetic control, each stage being dependent on completion of the previous one. External signals stimulate or inhibit cell proliferation by interacting with the intracellular controls that regulate a cell's progression through the cycle.

The eukaryotic cell cycle is traditionally divided into two main phases – the **M phase** of **mitosis** (when the duplicated chromosomes separate and are segregated into two daughter nuclei) and **cytokinesis** (cell division) on the one hand, and the growth phase or **interphase** on the other. Interphase in mammalian cells, and many other eukaryotic cells, is subdivided into conventional stages G_1, S and G_2 (Fig. 11.2).

In actively dividing cells, interphase is a period of intense biosynthetic activity. The cell mass doubles in preparation for the next cell division, new organelles are synthesized, and, during the **S phase**, the DNA is replicated to provide the two copies of each chromosome that will separate at mitosis. During G_2 the cell prepares for mitosis, which is heralded by condensation of the chromatin into chromosomes visible under the light microscope.

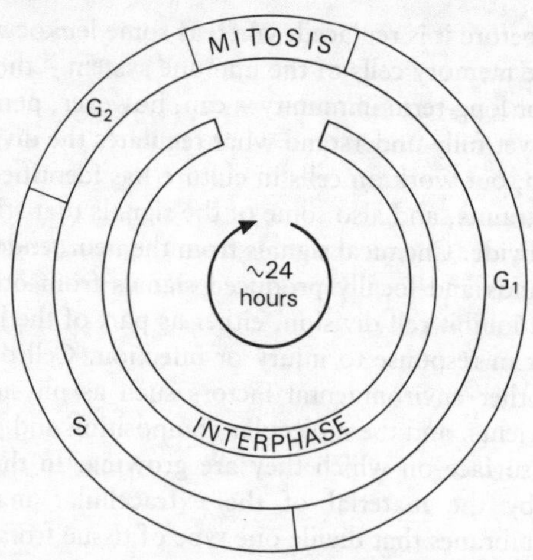

Fig. 11.2 The eukaryotic cell cycle.

Most adult tissues turn over their cell population at a characteristic rate, to cope with normal wear and tear, and their cells have characteristic **generation times** – the time each takes on average to complete a full cycle. Generation times range from as little as 8 hours for rapidly dividing cells in skin and gut linings that are subject to heavy wear and are being continually renewed, to several months for liver and pancreatic cells.

These differences are chiefly differences in the time a cell spends in the G_1 phase of its cycle. *In vivo*, cells not actively proliferating are arrested in a resting phase – or G_0 – at a particular point in G_1. They can stay quiescent in G_0 for months, carrying out their specialized metabolic tasks, and progress out of it only when they receive an appropriate stimulatory signal and the restraints on cell proliferation that operate *in vivo* are lifted.

The point at which cells are arrested occurs roughly in the middle of G_1 and is often known as the **restriction point** (**R point** or R) in mammalian cells. (In yeast cells, which have provided much information on the cell cycle, it is usually known as the **start**.)

The restriction point is crucial in the cell cycle, because it is at this point that the cell decides whether to continue through to cell division. If it has not reached an appropriate size, because of starvation for example, or does not receive an external signal to continue, it will pause, or enter the resting phase.

Once a mammalian somatic cell has passed the restriction point it is committed to continue automatically through DNA synthesis (S phase),

G_2, mitosis and cell division. Once a particular phase of the cycle has been embarked on it is always completed – cells do not suddenly stop in the middle of DNA synthesis for example, even if suddenly deprived of nutrients, but continue through to G_1, at which point they enter G_0. Once committed, most cells complete the S, G_2 and M phases fairly rapidly (in a matter of hours) and in much the same length of time, irrespective of their characteristic total generation times. The internal biochemical control mechanisms that determine whether a mammalian cell will pass the restriction point of the cell cycle and begin a new cycle are still largely unknown, but are now beginning to yield to a combination of genetic and biochemical approaches.

One widely held hypothesis is that passage of a cell through the R point occurs only after a rather unstable 'trigger' protein has accumulated to a certain level. Because this (so far hypothetical) protein is rapidly degraded, its accumulation to high-enough levels depends on a continuing high level of general protein synthesis, which may be stimulated by external signals (such as growth factors) and obviously also depends on a sufficient supply of nutrients. This type of mechanism would help to ensure that the cell does not embark on DNA replication and ultimately cell division until it has grown to an appropriate size and replicated its organelles. It also provides the cell with the means of indirectly taking stock of the nutrient supply to ensure that there will be sufficient to support the two new cells that will be produced.

Nuclear fusion experiments, in which a G_1 nucleus fused with one in the S phase also starts to synthesize DNA, show the presence of some positive factor or factors that initiate DNA synthesis. As soon as the DNA is replicated, however, it becomes unresponsive to these signals, possibly by becoming associated with protein, thus ensuring that the chromosomes are replicated only once per cell cycle.

Work with yeast cells has identified several genes (cell division cycle or *cdc* genes) and their products that are thought to be involved in regulating the passage from one stage of the cycle to the next. Genes and gene products corresponding to *cdc* genes are also beginning to be identified in animal cells. There are also several other genes, such as the histone genes, whose expression is regulated to coincide with particular points in the cycle when their products are needed.

Some identified **oncogenes** (genes involved in cells becoming cancerous, see next section) may also represent genes concerned with regulation of the cell cycle or the control of DNA synthesis. Mutated forms of these genes have been selectively maintained in the RNA tumour viruses, where their activity in infected cells leads to unregulated cell proliferation and thus to proliferation of the viral passenger.

External controls on cell proliferation

The full range of external controls on cell proliferation is unlikely to have been identified yet. As well as circulating and locally-acting signals such as hormones and growth factors, which are so far the best-understood mediators of cell division, the nature of the underlying substrate and the cell's contacts with it (see Chapter 8: Cell motility) are also important, as are its contacts with neighbouring cells. The invasiveness of malignant tumours, which infiltrate underlying tissues and slough off cells that settle elsewhere to make new tumours, is their most damaging and invincible weapon, and we know little about the normal restraints that keep cells in their place and how cancer cells are able to evade them. An important role, other than a purely physical one, is probably played by the basal laminas or basement membranes that divide, for example, the epidermal and dermal layers of the skin.

The general regulation of growth is under the overall control of the neuroendocrine system (Chapter 13: The neuroendocrine system; see also Tables 9.2 and 9.3) that produces hormones such as ACTH, thyroid-stimulating hormone, etc. which then stimulate the proliferation of the appropriate glandular cells and their synthesis of circulating hormones (e.g. the steroid sex hormones and thyroid hormone). In their turn, these hormones (amongst other effects) stimulate the proliferation of certain target cells. Their actions are sometimes indirect, stimulating cell division and other physiological changes through intermediaries. Growth hormone (somatotropin), for example, exerts part of its effect through inducing the liver to produce somatomedins (e.g. insulin-like growth factor I), which are the immediate activators of cell division in skeletal growth.

Regulated cell division in tissues such as skin and other epithelia is primarily under the control of the underlying mesodermal tissues – the dermis in the case of the skin. Local signals from the dermis (or the basement membrane separating dermis and epidermis) stimulate the epidermal stem cells to multiply, and there are also signals that, for example, control the correct spacing of hair follicles and pigment cells (melanocytes). At the same time, the epidermal cells are restrained from invading the dermis, a restraint that is evaded by the common skin cancer – basal cell carcinoma ('rodent ulcer').

Mitogens

Anything that directly stimulates mitosis and cell division is known as a mitogen, and the list of potential mitogens for mammalian cells is long.

Some are relatively non-specific in their effects, others affect only a small number of cell types. As well as some hormones, there are the multifarious polypeptide 'growth factors' that include the lymphokines of the immune system, and polypeptides not primarily thought of as mitogens (such as insulin). The vitamin A derivative retinoic acid (a suspected embryonic morphogen) and bacterial lipopolysaccharides are also powerful mitogens as are the relatively undiscriminating lectins (see below). B lymphocytes and T lymphocytes can be stimulated to divide by binding of antigen to the receptors they bear on their surface.

Few of the mitogens mentioned above can pass the barrier of the cell membrane (the steroid hormones, thyroid hormone and retinoic acid apart) and most mitogens appear to act through receptor proteins at the cell surface (see Chapter 9: Cell-surface receptors), activating various intracellular signalling pathways which eventually converge. Some proteins that bind to the cell surface in a relatively undiscriminating manner are powerful mitogens. These include the **lectins** – glycoproteins that were initially isolated from seeds. They bind to the sugar side chains found on many cell-surface proteins and are widely used experimentally to induce cell division.

The complete pathway from mitogenic signal to even the start of DNA synthesis is not yet known for any mitogen. There are gaps both in the understanding of the cell's intrinsic regulation of the passage from the resting to the synthesis phase, and of the immediate biochemical pathways stimulated by various mitogens. One common factor in their action is often the activation of protein kinase C, a phosphorylating enzyme that is encountered again and again in responses to external signals (see Chapter 9: Protein phosphorylation and protein kinases) and whose physiological targets are largely unknown, although it can phosphorylate a wide range of proteins. Another common factor in the intracellular response to some mitogens is the expression of two genes, *fos* and *myc*, which are both potential oncogenes. *fos* is activated transiently within minutes of exposure to platelet-derived growth factor and other growth factors. The protein it encodes acts in the nucleus but its role is not known. *myc* is activated later, within 1 or 2 hours of treatment with growth factors and other mitogens. It also encodes a protein of unknown function that acts in the nucleus.

Growth factors and other mitogenic polypeptides

Of particular interest at present in the regulation of cell division is a mixed group of protein 'factors' known generically as **growth factors**. They include the named growth factors such as epidermal growth factor (EGF), fibroblast growth factor (FGF), transforming growth factors (TGF-α and TGF-ß), and platelet-derived growth factor (PDGF) (see Table 9.9). There are also many other proteins that stimulate division and/or differentiation in various tissues – those involved in the differentiation of blood cells have been particularly well studied. The lymphokines of the immune system and some other biologically active polypeptides such as bradykinin, insulin, and α-thrombin that can induce cell division are often also referred to as growth factors.

Several growth factors were originally discovered as components of blood serum essential for the survival and proliferation of cells in tissue culture. Individual growth factors were originally isolated with great difficulty as they are typically present in relatively small amounts. 'Factorology' was for some time a rather controversial area of cell physiology, but with improved techniques for isolating and identifying proteins, the existence and identity of particular factors is now being clarified. At present there is intense research interest in growth factors and their mode of action, not least because derangements in growth factor signalling mechanisms may be one cause of cells becoming cancerous (see below, Oncogenes). Related polypeptides have also been discovered playing crucial roles as embryonic morphogens (see Chapter 10: Pattern formation and developmental signals).

It has become clear over the past few years that most growth factors do more than just stimulate or inhibit cell division, and are part of a complex regulatory network of polypeptide chemical signals. Many can either stimulate or inhibit cell division, depending on the type of cell and other growth factors present. They have been likened to the letters of an alphabet of control of cell function – but so far we know hardly any words.

Surfaces That Support Cell Division

Quite apart from any chemical signals it may be receiving the ability of a cell to divide is greatly influenced by its underlying substrate. Cells such as fibroblasts and epithelial cells will grow normally and divide in culture only if plated on a solid surface. This is apparently due to their need to be able to spread out flat on a surface in order to divide – **anchorage-dependent growth**. In suspension or in soft agar, fibroblasts draw in their cytoplasmic processes, becoming more rounded than usual, and cannot divide. Related to this is their inability to divide if they become hemmed in by other cells (the so-called phenomenon of **contact inhibition**). Fibroblasts in culture continue to proliferate until they have formed a confluent layer on the bottom of the dish, after which they stop dividing. If a path is cleared through the layer,

Table 11.1 Protein factors involved in differentiation of blood cells

Name	*Structure*	*Biological properties*
Multi-CSF (IL-3)	Glycoprotein of 133 amino acids	Stimulates growth of stem cells, white blood cell precursors of all types and mast cells
GM-CSF (granulocyte–macrophage colony-stimulating factor)	Glycoprotein of 118 amino acids	Needed continuously for proliferation and differentiation of progenitors of granulocyte–macrophage line
G-CSF	Glycoprotein, M_r 25 000	Stimulates formation of granulocytes
CSF-1 (macrophage colony-stimulating factor)	Glycoprotein, M_r 70 000	Stimulates production of macrophages from bipotential macrophage–granulocyte precursors
Erythropoietin	Glycoprotein of 166 amino acids	Produced chiefly by kidney in response to level of circulating red blood cells (erythrocytes). Stimulates production and differentiation of erythrocytes from already committed bone marrow precursors. Human recombinant erythropoietin is now available and may prove useful in treating some types of anaemia, especially where there is kidney damage that suppresses normal erythropoietin production (as in patients undergoing kidney dialysis)

however, the cells at its edge spread out into it and start dividing again until a continuous layer is reformed.

The fact that they stop dividing is now thought to be due not to specific signals they receive on contact with the other cells, but to their need to spread themselves out. The physical proximity of neighbours apparently forces the cell to draw in the projections that anchor it to the surface and by which it 'pulls' itself along and to become much more rounded (this is often termed **contact inhibition of movement**). How the change in cell shape, mediated presumably by changes in the internal cytoskeleton, affects its ability to divide is so far unknown.

One important distinguishing feature of tumour cells is that their growth is not constrained by contact inhibition. Cultured tumour cells continue to proliferate after they have formed a confluent layer, piling up on each other in a disorganized mass. They are also able to divide on surfaces such as soft agar that prevent normal cell division. One possible explanation is that for some reason so far unknown they do not need to spread out in order to divide. Tumour cells in culture do often have a permanently more rounded appearance than their normal counterparts and are far less mobile.

Limits on cell division

Normal differentiated cells, even when they can still divide, are not immortal. They appear to have a limited and predetermined lifespan in terms of the number of times they can divide. Cells from an adult animal will go through a set number of cell divisions in culture, ranging from about 25 to 100 depending on the animal and type of cell, after which they die. In general, the number of divisions corresponds roughly to the number the cells would normally make in the rest of the animal's lifetime. Cells taken from young animals will go through more divisions in culture than those taken from older animals, suggesting that the older animals have already used up some of their allotted divisions. As the culture 'ages' the rate of cell division also slows down. The reasons for this limited lifespan are still a total mystery.

Tumour cells, on the other hand, divide in culture indefinitely, and are for all practical purposes immortal. Embryonic cells also, if they are prevented from differentiating, will divide indefinitely in culture. Normal adult cells, if grown in culture for long periods, occasionally acquire at least some properties akin to tumour cells, and become capable of unlimited division – they are said to have become **immortalized**. Cells infected with tumour viruses or into which active oncogenes have been introduced become **transformed**; they show many of the properties of tumour cells, including unlimited growth (see below, The transformed cell). Immortalized cells, embryonic cells and cells derived from tumours form the permanent **cell lines** much used in biological research (Table 11.2).

Table 11.2 Some commonly used cell lines

Cell line	*Cell type and origin*
3T3	Mouse fibroblast
BHK 21	Syrian hamster fibroblast
HeLa	Human cervical epithelium (tumour)
L	Mouse fibroblast
L 6	Rat myoblast (immature muscle cell)
MPC	Mouse bone marrow myeloma cell
KB	Human nasopharyngeal epithelium (tumour)

THE ORIGINS OF CANCER

Cell division in the human (or animal) body is under strict control. After birth, cells only divide to renew worn out or damaged tissue or, in growing children, to fulfil a predetermined pattern of development. Occasionally, however, a cell becomes altered in such a way that it is able to escape from the normal restraints and proliferate to form an abnormal mass of cells – a **tumour** or **neoplasm.**

Some of these tumours are non-invasive or benign. Common warts (benign papillomas) stop growing once they reach a certain size, and other benign neoplasms, although they may continue to grow and require surgical removal, do not invade other tissues and once removed do not grow again. But in some cases the tumour is **malignant** or **cancerous**. It continues to grow uncontrollably, invades underlying tissues and usually eventually **metastasizes**, spreading to other parts of the body by way of cells that slough off the tumour and migrate, disregarding the normal territorial defences that prevent the invasion of one tissue by cells of another. Death is caused when these metastases affect the function of some vital organ.

The most common spontaneously occurring forms of cancer in adults (accounting for more than 90 per cent) are solid tumours of the epithelial tissues (skin, linings of gut and airways and the linings of glands); these are called **carcinomas**. Epithelial tissues are often undergoing continual renewal through cell division, and this, together with the fact that they are the point of contact with the environment probably accounts for the predominance of tumours affecting them. Skin cancers alone account for around 50 per cent. Tumours of the connective tissues (muscle, fibrous tissue, and bone) which are called **sarcomas**; the **lymphomas**, cancers of immune system cells; and **leukaemias**, a diverse group of proliferative disorders of white blood cells and their precursors, together account for around 8 per cent of human cancer in adults.

Pathologists further subdivide these basic types in terms of the sites at which they occur and the cell type involved. Adenocarcinomas, for example, are carcinomas involving glandular epithelium, gliomas are tumours of the glial cells of the nervous system and myelomas are tumours of bone marrow cells. Cells such as neurones which do not divide once mature do not form tumours except, rarely, in infants, whose nervous system is not yet fully mature, and in the rare 'embryonal' cancers which represent a persistence of embryonic tissue after birth. Apart from a few rare types, cancer is predominantly a disease of older adults, the incidence of the commonest cancers rising sharply between the ages of 50 and 80.

Cancer is generally held to originate in mutations in a single cell that alter it in some way to allow it to override normal controls on cell division and become **tumorigenic**. This genetic change is then inherited by all the cell's progeny, which proliferate to form the tumour. Evidence for the mutational origin of cancer comes from many sources. Many agents, whether chemical, biological (tumour viruses) or physical (e.g. radiation) can cause cancer. Chemical **carcinogens** comprise a diverse group of compounds but all are also **mutagens**, which damage or alter DNA, as do physical carcinogens such as X-rays and ultraviolet light. Tumour viruses also appear to exert their effects either by inducing mutations within the infected cell's genes or by possessing their own abnormal versions of cellular genes that become incorporated into the cell's DNA.

There is now convincing evidence that a malignant cell can be produced as a result of mutations in one or more of a set of genes that are believed to be vital in regulating normal cell division and differentiation. (A disruption of normal differentiation, rather than a direct disruption of the regulation of cell proliferation, is thought by some researchers to be the underlying defect in cancer cells. Cancer cells show many characteristics that can be interpreted as a reversion to a rather undifferentiated state, including continuing proliferation and their ability to establish themselves in other tissues.)

The link between mutation and cancer is not a straightforward one. This is clear from cases where an unusually high incidence of cancer can be unequivocally linked to previous exposure to a carcinogenic agent –

a chemical or radiation. In all instances, there is a considerable delay, sometimes of decades, between the exposure, which may have been for a very short time, and the appearance of cancer. The dramatic increase in the incidence of lung cancer seen this century in the Western world lagged some 20 years behind the widespread adoption of the habit of cigarette smoking, and appeared first in men, to be followed by a rise in its incidence in women that paralleled a later increase in smoking amongst women.

From these case histories, many studies of experimental carcinogenesis in animals, and the incidence and distribution of spontaneously occurring cancer in the human population, it is now believed that all cancers arise by a complex multistage process. The initial mutation probably occurs long before overt cancer appears and has been termed the initiation stage. This alteration, although necessary for carcinogenesis, is not usually sufficient. By itself, such a mutation might, for example, lead to a benign tumour or to a 'precancerous' state characterized by abnormal cell growth and organization in tissue. Further changes in the cell, which may take years to happen, are needed before a fully malignant state is reached. What these changes may be, and why they take so long is still far from clear but they probably involve further mutations.

The many causes of cancer

Most human cancer is now believed to be 'environmentally' caused, by exposure to (as yet largely unidentified) carcinogenic and precipitating factors as a result of what might be loosely called lifestyle – a particular type of diet, reproductive history, environmental conditions, exposure to disease and cultural practice. By far the commonest cancer in the developed world today is lung cancer, which has been unambiguously linked to heavy cigarette smoking. Other common cancers (cancer of the large intestine and breast cancer) also seem to have predisposing environmental components – diet and, for breast cancer, reproductive history.

Different cancers predominate in various parts of the world (Table 11.3) and there is strong evidence supporting the idea that this reflects differential exposure to environmental carcinogens as well as possible genetic differences between populations. Even where particular 'risk factors' have been identified, however, very little is known about how they may predispose to cancer.

How several different factors may contribute is well illustrated by the role of the Epstein–Barr virus in Burkitt's lymphoma and nasopharyn-

Table 11.3 The geographical incidence of some common cancers

Type of cancer	*Region of highest incidence*	*Risk up to age 75 (%)*	*Range of variation**	*Region of lowest incidence*
Men				
Skin	Queensland (Australia)	>20	>200	Bombay
Oesophagus	North-east Iran	20	300	Nigeria
Lung	Great Britain	11	35	Nigeria
Stomach	Japan	11	25	Uganda
Liver	Mozambique	8	70	Norway
Prostate	USA (blacks)	7	30	Japan
Colon	Connecticut (USA)	3	10	Nigeria
Mouth	India	>2	>25	Denmark
Rectum	Denmark	2	20	Nigeria
Bladder	Connecticut	2	4	Japan
Women				
Cervix	Colombia	10	15	Israel (Jews)
Breast	Connecticut	7	15	Uganda

Source R. Doll, *Nature*, 265 (1977) 589–596.
*The highest incidence observed in any country divided by the lowest observed incidence.

geal cancer. Burkitt's lymphoma, a cancer of lymphoid cells, is found throughout the world but is fairly rare except in children in certain parts of Africa. Burkitt's lymphoma seems to be associated with previous infection with a herpesvirus – the Epstein–Barr virus (EBV) – earlier in childhood. But this cannot be the sole cause as the virus is widespread throughout the rest of the world where it usually results in asymptomatic infection in infancy. However, if primary infection occurs later in life, it causes infectious mononucleosis (glandular fever). As well as early infection with EBV the environmental factor in Africa that predisposes to or precipitates lymphoma seems to be malaria, as the area of high Burkitt's lymphoma corresponds to that in which malaria is endemic. Many of the symptoms of glandular fever result from the determined efforts of the immune system to attack the proliferating virus-infected lymphoid cells and concurrent infection with malaria may suppress this reaction, allowing the virus to persist and eventually turn the cells into cancer cells. In a different genetic and environmental background (in southern China) persistent EBV infection of cells of the pharynx appears to be the primary cause of a different cancer – nasopharyngeal cancer – the commonest type in this region.

The role of viral infection varies widely from cancer to cancer in humans. In animals, many **RNA tumour viruses** (see below) have been identified as the direct and sole cause of tumours. In humans only a very few similar RNA tumour viruses are known. One is HTLV-1, the cause of adult T cell lymphoma, a cancer commonest in Japan and parts of the Caribbean. There are, however, several common cancers, such as liver cancer, where virus infection (in this case by the DNA virus hepatitis B) is the major predisposing cause in areas where the virus is endemic.

The cancer cell

Cells isolated from human tumours can sometimes be established in culture and their peculiar properties studied more closely. Cultured tumour cells differ from normal cells in many ways, most characteristically in their ability to divide indefinitely and to override normal restraints on cell proliferation in culture such as contact inhibition. Normal fibroblasts and epithelial cells require a solid surface on which to grow and multiply, whereas tumour cells from similar tissues have lost this **anchorage-dependence** and also proliferate in suspension. Tumour cells often produce novel cell-surface proteins, some of which, such as carcinoembryonic antigen, resemble proteins normally found on embryonic but not mature cells. They can also lose cell-surface proteins such as the major histocompatibility antigens, which may make them 'invisible' to the immune system (see Chapter 12). They show changes in metabolism, shape (tumour cells are more rounded or more spindly than normal cells in culture and their internal cytoskeleton is disorganized), and sometimes start to secrete hormones, growth factors or other proteins such as tumour angiogenesis factor (this is involved in attracting capillaries to infiltrate a tumour providing it with a supply of nutrients).

Many normal cells are responsive in a regulated way to growth factors, hormones or other biologically active molecules they secrete themselves, and such *autocrine* mechanisms of growth control are probably important physiological regulatory mechanisms. When such responses become deregulated, as when a cell starts continuously secreting a growth stimulatory hormone or growth factor for which it carries receptors, it enters a cycle of self-stimulated proliferation – so-called autocrine transformation. Tumours can also become inappropriately responsive to circulating endocrine hormones or substances produced locally by other cells, if they produce the necessary receptor.

Tumour cells often show chromosomal aberrations – loss of all or part of a chromosome, translocations of part of one chromosome to another, rearrangement within chromosomes and repeated duplication of selected portions of a chromosome. The karyotype – the number and type of chromosomes present – is often difficult to discern. Cancers of blood cells have provided most evidence on chromosomal abnormalities. In some cases, a certain abnormality is linked consistently with a particular type of cancer, suggesting that it might be one of the underlying causes of the cell becoming cancerous. It has proved difficult, however, with this as with all the other alterations displayed by tumour cells, to sort out which changes are at the root of the malignant nature of the cell, and which are secondary to its altered pattern of growth.

The Transformed Cell: A Model Cancer Cell

When normal mammalian cells growing in culture are infected with tumour viruses or treated with carcinogens they may become **transformed**, showing new features that resemble those of cancer cells – most typically, the ability to override contact inhibition, to proliferate indefinitely and to grow on soft agar. The link between transformation (also called **neoplastic** or **malignant transformation**) and cancer was established when it was found that some transformed cells transplanted into experimental animals could produce tumours. Within certain limitations, transformed cells provide a model cancer cell, which can be readily manipulated and studied in culture. The ability to transform cells is an invaluable preliminary test for potential cancer genes, carcinogens and tumour viruses.

An important distinction with regard to mechanisms of tumorigenesis is between cells that have simply become immortalized (i.e. they can grow indefinitely in culture but show no other characteristics of a transformed cell and do not produce tumours when transplanted) and those that have become fully transformed (see below, DNA tumour viruses, for more on this point).

Tumour viruses

Much of our understanding of basic mechanisms of tumorigenesis and the nature of the changes that take place in cancer cells comes from studies of cells infected with tumour viruses. A virus is defined as a tumour virus if it can be shown, by the usual criteria for determining whether an infectious agent is a cause of a particular disease, to be the sole or partial cause of a cancer. This does not imply any particular mechanism of action and different types of tumour virus cause cancer in very different ways.

DNA Tumour Viruses

There are two main groups of DNA tumour viruses, one of which comprises some common DNA viruses that are in certain circumstances major contributory causes of human cancer; the other, although of no practical importance in naturally occurring cancer, has provided much of our basic understanding of tumorigenesis.

SV40, polyoma and adenovirus

This latter group consists of two related viruses, **SV40** (simian virus 40) and **polyoma**, and the unrelated **adenoviruses** (some of which cause a mild respiratory infection in humans). All normally produce a mild infection in susceptible hosts but can transform certain cultured cells and cause tumours in highly susceptible newborn animals. They have been studied intensively as laboratory models to uncover basic events in tumorigenesis. Because they have been so well characterized, they are also much used as convenient models of general eukaryotic gene expression, and have provided several 'firsts' in eukaryotic genetics, most notably the discovery of split genes and RNA splicing, and of enhancers – a type of eukaryotic gene control element.

In their normal 'permissive' hosts these viruses go through a conventional infective cycle, transcribing and replicating their DNA, producing new virus particles and destroying the host cell in the process. In some cells of other species, however, which do not permit their replication, and are therefore said to be 'nonpermissive', they (rarely) integrate their genomic DNA into the host chromosomes, where it continues to direct the production of some viral products (Fig. 11.3) – the so-called **transforming antigens** (e.g. the **T antigens** of SV40 and polyoma). These proteins are normally involved in regulating viral transcription and replication, but when produced in quantity interfere with the host cell's normal growth regulatory mechanisms and cause it to become transformed. Their actions are reversible – if the T antigens cease to be produced the cell reverts to normal.

The polyoma T antigens ('large', 'middle' and 'small' T) are all derived from differential transcription and RNA splicing of the 'early' region of the virus – this is the region usually transcribed early in infection before viral replication commences. They have proved particularly useful in separating two aspects of tumorigenesis. They show that simply becoming immortal (i.e. capable of indefinite proliferation) is not in itself sufficient to confer tumorigenicity on a cell. On its own, polyoma large T antigen (which is a DNA-binding protein acting in the nucleus) immortalizes cells but cannot completely transform them – they cannot, for example, produce tumours on transplantation. The additional function of the middle T antigen (which is located on the inner surface of the plasma membrane) is needed to make the cells tumorigenic. Conversely, middle T on its own cannot transform normal, freshly isolated cells; it requires the prior action of large T to make them sensitive to its action. Middle T is believed to transform cells by interacting in some way with the membrane tyrosine kinase encoded by *src*, a proven potential oncogene (see below, RNA tumour viruses).

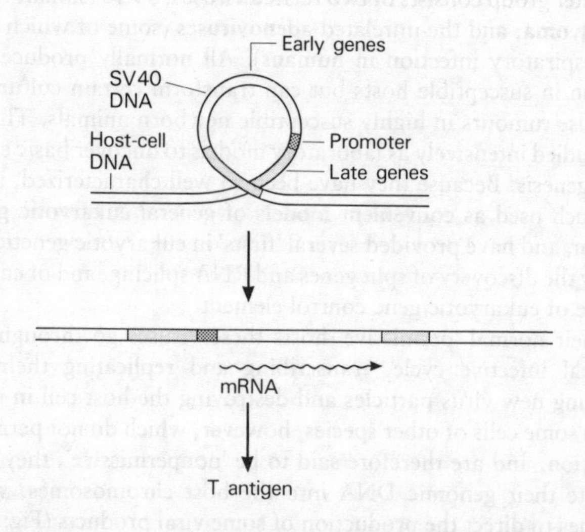

Fig. 11.3 The integration of the circular genome of simian virus 40 (SV40) into the host cell's DNA. When recombination takes place within the 'late' genes they are disrupted in the integrated DNA and can no longer be expressed. New virus particles cannot therefore be produced. The early genes, which encode the large T antigen, can still be expressed.

This is a direct experimental demonstration of the need for several steps in tumorigenesis, and ties in with the multifunctional and multistage nature of cancer and the observation that there is often a 'benign' precancerous stage of abnormal proliferation, which sometimes, but not always, progresses to a fully fledged cancer.

The SV40 large T antigen, on the other hand, carries out both functions and can fully transform cells on its own. It is found mainly in the nucleus where it interacts with a host-cell protein (p53) that is implicated in control of the cell cycle. Although T antigen can stimulate host cell DNA synthesis on its own, this is not in itself sufficient for transformation, and the persisting high levels of p53 that result from large T activity are needed. In large amounts, p53 is itself an oncoprotein, capable of transforming cells. The discovery of p53's involvement showed that a cell's own genes can be implicated in tumorigenesis if they become deregulated or inappropriately expressed, an important principle that is now firmly established.

Other DNA tumour viruses

The other 'DNA tumour viruses' comprise a disparate group of common and widespread viruses that usually cause some more-or-less severe disease (e.g. hepatitis B and glandular fever) but in some circumstances go on to become a contributory cause of cancer. The link between the herpesvirus **Epstein–Barr virus** and Burkitt's lymphoma and nasopharyngeal cancer has already been mentioned above (p. 353) and is also dealt with later in the chapter.

For many years virus infection has also been suspected as a cause, or at least a predisposing factor, of human cervical cancer, one of the commonest cancers in women worldwide. Strong candidates are two of the **papillomaviruses** commonly found in the genital tract. Papillomaviruses in general cause benign neoplasms (including common warts). In the case of cervical cancer the factors that lead to the transformation of a benign precancerous state, which often regresses naturally, to the full-fledged carcinoma are not yet clear.

Hepatitis B infection is one of the major causes of liver cancer worldwide. Liver cancer (hepatoma, or hepatocellular carcinoma) is common in areas where hepatitis B is endemic, such as parts of Africa, China and South-east Asia. Hepatitis B virus, like Epstein–Barr virus, can establish a persistent carrier state, with its genome integrated into host DNA. Liver cancer, which occurs in a proportion of persistent carriers about 25–30 years after infection, is believed to result from a combination of the chronic liver damage that occurs in some carrier states and the persistent attempts of the liver to repair itself, the possible action of the integrated hepatitis DNA as a mutagen or oncogene activator, and 'environmental' factors such as diet. Factors other than chronic hepatitis must be required as the geographical distribution of high incidence of hepatitis does not exactly parallel that of high rates of liver cancer. Effective vaccines are available against hepatitis B, but are still expensive; getting them to those who need them most is now a political and economic question.

RNA Tumour Viruses

RNA tumour viruses contain RNA as their genetic material and cause malignant tumours in a wide range of mammals and other vertebrates. Only a few human RNA tumour viruses have been unambiguously identified, amongst them HLTV-1 (human lymphotropic virus 1) which causes a disease of the immune system, adult T-cell lymphoma. Despite this apparent lack of relevance to human cancer, RNA tumour viruses have been studied for many years because of the light they shed on some of the basic mechanisms of tumorigenesis. In particular, they have guided cancer biologists towards genes in human cells that are implicated in the development of cancer (see next section, Oncogenes), a discovery widely hailed as one of the most important developments in cancer studies for many years.

RNA tumour viruses have a distinctive and unusual means of propagating themselves and persisting in infected cells. They belong to the **retroviruses**, a family of RNA viruses that have the unique ability to copy their RNA genome into DNA, which then enters the host cell's chromosomes and persists there as a **provirus** (Fig. 11.4) being inherited by the cell's progeny like a normal gene. They may even, if they infect the reproductive cells (the germline), be passed on from generation to generation (**vertical transmission**). They can also be transmitted vertically by infection passing across the placenta from mother to embryo. Retroviruses carry a gene for a unique DNA polymerase **reverse transcriptase**, which unlike other DNA polymerases, uses RNA as a template to make DNA. Many integrated retroviruses also direct the formation of new virus particles that are released from the cell, often without destroying it, to infect other cells or other animals (**horizontal transmission**).

Not all retroviruses cause cancer. Many seem to exist as apparently harmless parasites of the genomes of vertebrates and there are some of medical importance other than the tumour viruses – for example, the human immunodeficiency virus (HIV), responsible for AIDS, which infects and destroys a subset of T lymphocytes in the human immune system.

RNA tumour viruses induce tumorigenesis in various ways. One is represented by the widespread naturally occurring tumour viruses that cause (sometimes epidemic) disease in pets and domestic animals. This group is exemplified by feline leukaemia virus (FeLV), a major cause of illness in pet cats, and avian leukosis virus (ALV), which causes B-cell leukaemias in chickens. These viruses are fully capable of replication and are transmitted horizontally as well as vertically. Where the mechanism of tumorigenesis has been elucidated, they appear to act by inserting a strong viral promoter or enhancer near certain host-cell genes. (Promoters and enhancers are control sites in DNA at which genes are switched on – see Chapter 3 for full definitions.) Under the influence of the viral element the cellular gene is overexpressed and transformation of the infected cell results. In domestic fowl the avian leukosis virus that causes B-cell leukaemias is found integrated within the cellular gene *myc*, which from other evidence is a proven oncogene (see next section). The mouse mammary tumour virus (MMTV), which always causes mammary tumours, is found inserted near either or both of two separate genes *int-1* and *int-2* (which encode proteins whose role is not yet known).

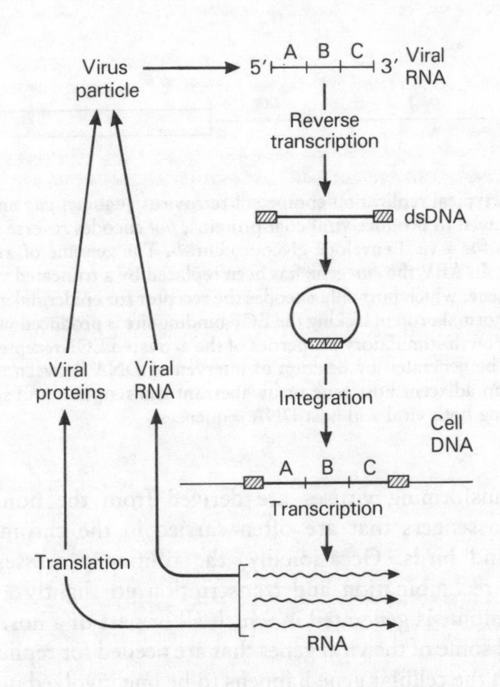

Fig. 11.4 Multiplication of a retrovirus via a DNA intermediate integrated into the host cell DNA. When the viral RNA is reverse transcribed into DNA, repeated sequences (LTRs) are generated at either end of the DNA. These contain promoter sites.

These viruses all show the lag time between infection and tumour appearance characteristic of naturally occurring cancer, indicating that a multifactorial or multistage process is operating. The switch on of these cellular genes is probably only the first step in the pathway leading to a tumour.

In complete contrast are the second group of RNA tumour viruses, which first pointed the way to the existence of potential oncogenes in human and animal cells. These **acute-transforming RNA tumour viruses** are found mainly in laboratory animals. They are highly tumorigenic; a single virus particle can cause a cell to become transformed and animals often develop tumours within a few days of infection.

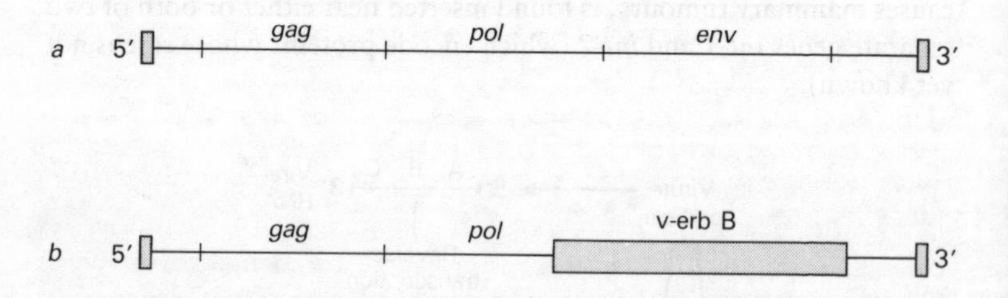

Fig. 11.5 *a*, A typical replication-competent retrovirus genome. *gag* encodes a protein that is cleaved to produce viral coat proteins, *pol* encodes reverse transcriptase and *env* codes for a viral envelope glycoprotein. *b*, The genome of avian erythroblastosis virus. In AEV the *env* gene has been replaced by a truncated version of the cellular *erbB* gene, which normally encodes the receptor for epidermal growth factor (EGF). An abnormal protein lacking the EGF binding site is produced which appears to retain the growth stimulatory properties of the activated EGF receptor. New viral genomes may be generated by deletion of intervening DNA between an integrated provirus and an adjacent host gene or by aberrant transcription and splicing of an RNA containing both viral and host DNA sequences.

Acute transforming viruses are derived from the non-oncogenic retroviral passengers that are often carried in the chromosomes of mammals and birds. Occasionally, the routine processes of DNA replication, recombination and transcription go slightly awry and a new viral genome is generated in which all or part of a host cell's gene has replaced some of the viral genes that are needed for replication (Fig. 11.5). When the cellular gene happens to be one involved in regulating cell growth and proliferation, its subsequent inappropriate expression as part of the provirus overrides normal regulatory controls and causes the cell to become transformed. In the course of becoming incorporated into the virus the cellular gene may also become truncated or otherwise mutated, giving rise to a novel protein that is unresponsive to the factors that normally regulate its action. The new viral **oncogenes** can be shown to be solely responsible for the cell becoming cancerous. In this lies their great interest: they represent the only known instances (the T antigen gene of SV40 apart) where the action of a single gene can be unambiguously pinpointed as a cause of cancer *in vivo*.

Apart from Rous sarcoma virus (RSV), the first of this type to be identified more than 70 years ago, the acute transforming retroviruses cannot replicate. (RSV has simply acquired an additional gene without losing any of its own.) Although they are incapable of replicating themselves, new virus particles containing the defective genome can be produced if a normal 'helper' endogenous virus is present that packages the defective viral genome in its own coat proteins.

Oncogenes and human cancer

The excitement of the past decade in cancer research has stemmed largely from the identification of genes in humans and animals that are consistently involved in causing cancer. These are normal genes, which if lost, mutated or expressed in an unregulated way, are (at least in part) responsible for a cell becoming cancerous. Cancer researchers believe they have at last identified a set of genes that are the targets for mutations leading eventually to a fully-fledged cancer cell.

The first group are those commonly called **cellular oncogenes** or **transforming oncogenes**. If their structure or normal expression is altered by mutation they become active **oncogenes**, with the power of transforming cells *in vitro* and (at least partially) causing cancer *in vivo*. More than 30 potential oncogenes have now been identified in mammals and birds although only a few have yet been linked to actual cases of human cancer (Table 11.4). Altered or 'activated' forms of these genes have been found in around 15 per cent of human tumours of various sorts (in some reports as many as 40 per cent of tumours harboured an altered oncogene). The presence of an active oncogene in a tumour cell can be detected by introducing its DNA by transfection into a susceptible mouse cell line, which then becomes transformed. The oncogene responsible can then be isolated by using smaller and smaller fragments of the transformed cell's DNA to transform further cells and so on.

The first clues to the existence and possible nature of these cellular oncogenes came from RNA tumour viruses. Some carry a single gene –

Table 11.4 Some potential oncogenes in the human genome. Those that have been implicated in human cancer are marked with an asterisk

Gene	*Normal product*
	Growth factor
sis	B chain of platelet derived growth factor (PDGF)
	Growth factor and hormone receptors
fms	Membrane receptor for the growth factor, monocyte colony stimulating factor (CSF-1)
erb-A	Receptor for thyroid hormone (acts in nucleus, see p. 279)
*erb-B**	Receptor for epidermal growth factor (EGF); has tyrosine kinase activity (see Table 9.9.)
	Other proteins with tyrosine kinase activity
*abl**	Encodes a protein with potential tyrosine kinase activity, although not detectable in its normal form. Found rearranged in the Philadelphia chromosome of chronic myelogenous leukaemia (see p. 369)
src	Tyrosine kinase located on the inner surface of the plasma membrane. Expressed in a wide range of cells, including non-dividing neurones. Role unknown
	Guanine-nucleotide binding proteins
*H-ras** *K-ras** *N-ras**	Encode the p21 family of proteins located on inner surface of the plasma membrane (see p. 277). Normal biological role not yet known, but they are probably a link between membrane receptors for some growth factors and their intracellular response machinery. They have been found mutated in a wide variety of human tumours (see text)
	Nuclear proteins
*myc**	Short-lived nuclear protein of unknown function expressed in many cell types. It is expressed transiently within an hour or two of stimulating quiescent cells with serum or growth factors, and at varying levels throughout differentiation. Implicated in control of normal cell division, possibly by effects on DNA replication. It becomes overexpressed in Burkitt's lymphoma due to translocation (p. 370). A related gene *N-myc* is found amplified and overexpressed in a high proportion of neuroblastomas

Table 11.4 contd.

Gene	*Normal product*
*myb**	Nuclear protein involved in control of cell proliferation, exact role unknown. Rearranged *myb* genes have been found in human colon and bone marrow tumours
fos	One of the first genes to be (transiently) expressed when resting cells are stimulated to divide by serum or growth factor. Its transcription is dependent on activation of a specific enhancer. Deregulated *fos* in transgenic mice can cause bone tumours
*trk**	Human oncogene (no viral counterpart known) produced by fusion of the gene for non-muscle tropomyosin and a truncated tyrosine kinase gene. (Originally called *onc-D*)

an **oncogene** – which is directly and solely responsible for its ability to cause tumours. The oncogene differs from one type of tumour virus to another. These genes turned out to be incomplete or otherwise mutated versions of genes that are part of the normal genetic complement of many eukaryotes.

Although cellular oncogenes encode a wide variety of proteins, they are all linked by a proven (or suspected) involvement in regulating cell division and/or differentiation. The control of cell proliferation is still one of the great white areas remaining on the map of cell physiology, and oncogenes provide a short cut to some of the pathways involved.

The oncogenes that can be isolated by their ability to transform cultured cells represent only one set of genes involved in human cancer. From a quite separate line of research involving the genetics of rare, inherited predispositions to cancer has come the identification of a second and different type of 'cancer gene' (see below, The familial cancers: clues to 'cancer suppressor' genes) and possible alternative routes to a cancerous cell.

Identifying and isolating some of the genes involved in cancer are now relatively easy with recombinant DNA techniques. The task of unravelling their effects on cell growth, division and differentiation and how they cause cancer cells to behave in the uncontrolled way that they do, is much more difficult.

Cellular or Transforming Oncogenes (Table 11.4)

The standard test defining an oncogene of this sort is in the first instance whether, when introduced into a susceptible strain of cultured cells (the mouse cell line NIH 3T3), it causes them to become transformed. Even the unaltered, normal form of the gene can show up in this test if it becomes overexpressed (as a result of the artificiality of the experimental conditions) relative to its usual state in a cell of this particular type and developmental stage.

Genes that pass this test or closely resemble proven oncogenes are all loosely termed oncogenes, whether they come from human, animal or even yeast or green plants. Some have mutated counterparts in known viral oncogenes (see RNA tumour viruses) some have no known viral counterpart. Viral oncogenes are termed v-*onc* (*onc* may be any of a wide range of genes), their normal cellular counterparts c-*onc*, or simply *onc* and are often termed **proto-oncogenes**. This terminology reflects the idea that they are the 'parental' genes from which viral oncogenes have been derived. It is important to emphasize that cellular oncogenes are part of a cell's normal genetic complement and only become oncogenic as a result of mutation or altered expression.

The viral oncogenes derived from these genes are usually mutated, partial or rearranged copies. In the case of those encoding growth factor receptors, for example, the viral oncogenes have lost the DNA encoding the growth-factor binding site or other regulatory portions of the receptor and, therefore, the receptor may be permanently active in stimulating cell division even in the absence of any stimulatory signal. Similar sorts of changes are found in 'activated' cellular oncogenes. The unusual ability of some viral oncogenes to cause cell transformation and tumorigenesis on their own may also be because the viral oncogene is usually being grossly overexpressed under the influence of the strong viral promoter.

Altered oncogenes in human cancer cells

An important step forward in linking oncogenes with human cancer was the finding that suspected oncogenes were active, and often mutated, in tumour cells from cancer patients but not in their unaffected cells. Only a few of the many potential oncogenes identified from RNA tumour viruses have been linked with human cancer in this way, most notably the *ras* gene family.

The human genome carries three *ras* genes – *H-ras, K-ras* and *N-ras* – (and two inactive *ras* pseudogenes). They encode three related proteins located on the inner surface of the plasma membrane. The ras proteins (p21) are guanine-nucleotide binding proteins (see Chapter 9:

G proteins). Their biological role is not yet known but they are thought to be involved in the intracellular pathways that receive and interpret signals from cell-surface receptors, in this case from receptors for growth factors.

Active altered *ras* genes have been detected in around 10–35 per cent of samples of various human tumours. *ras* genes become oncogenic after single point mutations, chiefly in codons 12, 13 and 61, and genes altered in these positions have been isolated from a wide variety of human tumours, including carcinomas of the colon, liver, lungs and bladder. Altered *N-ras* has also been found in some leukaemias.

The way in which the mutant *ras* proteins contribute to tumorigenesis is not yet clear. The mutation at codon 12 abolishes the GTPase activity of the protein. By analogy with other G proteins this might make *ras* permanently active, in the absence of any signal from its (presumed) linked receptor. Many viral oncogenes represent aberrant forms of proteins involved in such signalling pathways.

Oncogenes and chemical carcinogenesis

Altered oncogenes in human tumours might, however, be simply an effect rather than the cause of a cell becoming cancerous. In order to establish that oncogene activation can be at the root of the tumorigenic process, workers have turned to laboratory animals in which tumours can be induced in controlled conditions using chemical carcinogens, and cause and effect is much easier to distinguish.

The best direct evidence so far that activated oncogenes can cause human cancer, therefore, comes from work on the *ras* genes in laboratory rats and mice. The altered *ras* genes found in various human tumours often carry the same mutations, one of the most common being a change of a G to A in codon 12 (which changes the twelfth amino acid in the ras protein from glycine to glutamic acid). It was necessary to demonstrate experimentally that this mutation could produce an active *ras* oncogene. This type of G to A mutation can be experimentally induced by the chemical carcinogen NMU (*N*-methyl-*N*-nitrosourea). Using a special strain of laboratory rats sensitive to MNU, and in which the resulting tumours also carry an altered *ras* gene, it has been possible to show that the mutation in codon 12 does activate the *ras* oncogene, and that it is involved at a very early stage of carcinogenesis – that is, it is almost certainly a cause, rather than an effect of carcinogenesis. Mutations at various other sites also make the *ras* gene oncogenic.

Because much human cancer is believed to be linked to the action of carcinogens (the link between smoking and lung cancer is the best known, but by no means the only example), the demonstration that suspected oncogenes are the genes affected in chemical carcinogenesis greatly strengthens the case for their universal importance, whether the immediate carcinogen is a chemical or radiation or a 'spontaneous' mutation in the cell.

Additional evidence that an alteration in *ras* genes is an early event in some human cancers comes from studies of 'preleukaemic' patients. These patients suffer from myelodysplastic syndrome, with low blood counts, abnormal bone marrow cells and a higher than normal risk of progression to acute leukaemia. In several patients who progressed to acute leukaemia, the mutation activating the *ras* oncogene was already present in their bone marrow cells (but not in other body cells) in the preleukaemic stage.

In the real world, a mutation in *ras* by itself, or any other single oncogene, is almost certainly not sufficient to lead directly to cancer. Mutated *ras* genes on their own are unable to transform freshly isolated normal cells (as opposed to the already 'immortalized' mouse cell line used in the standard oncogene test). But if they are transfected together with another potential oncogene (e.g. *myc*, or *myb*, see Table 11.4) transformation results. The oncogenes that complement *ras* are those able to immortalize cells – that is, to make them capable of continued division, but not necessarily to transform them. All the evidence from both spontaneous and some forms of viral carcinogenesis (see DNA tumour viruses) suggests that immortalization is an initial, necessary but not sufficient step in tumorigenesis.

Chromosomal Aberrations That Activate Oncogenes

Many cancer cells have a visibly abnormal set of chromosomes. Several types of aberration have been reported, including loss of chromosomes, deletions, inversions and the translocation of part of one chromosome to another. Chromosomal abnormalities have been best characterized in cancers of blood cells – leukaemias and lymphomas.

The idea that some at least of these chromosomal abnormalities are the underlying cause of the cell becoming tumorigenic, by disrupting a potential oncogene perhaps, or unmasking a recessive 'cancer' gene, is attractive but up to now has been impossible to prove or disprove. Cancer cells show many structural and metabolic abnormalities that are most probably a consequence of their cancerous state rather than the cause, and chromosomal abnormalities could easily arise after the cell has become cancerous.

The Philadelphia chromosome

Some cancers, however, are characterized by a consistent and specific abnormality, amongst which the association between the Philadelphia chromosome translocation and chronic myelogenous leukaemia (CML) is the most consistent yet discovered. At least 90 per cent of patients with CML carry an abnormal chromosome – the Philadelphia chromosome – in their leukaemic cells, but not in the rest of their body. The Philadelphia chromosome is generated by a **reciprocal translocation,** which exchanges a large piece of the long arm of chromosome 22 for a small bit at the end of the long arm of chromosome 9, the shortened chromosome 22 being the Philadelphia chromosome (Fig. 11.6).

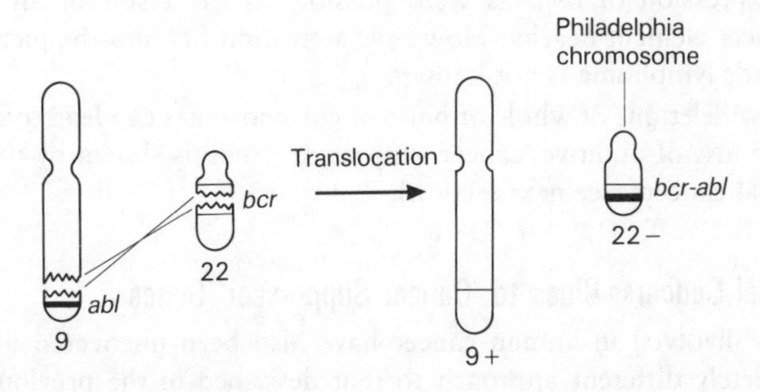

Fig. 11.6 The reciprocal translocation that results in the Philadelphia chromosome typical of chronic myelogenous leukaemia cells (see text for explanation).

Now that it has become possible to look directly at the DNA sequences involved in the Philadelphia translocation it has been found that a new protein with altered activity is indeed produced as a result of the translocation. As the altered protein is derived from the known oncogene *abl* it seems extremely likely that the translocation is indeed at least partly responsible for originating a leukaemic cell.

The breakpoints on both chromosomes vary in every case, but that on chromosome 22 always occurs within a region of around 6 kilobases known as the *bcr* – breakpoint cluster region. This is part of a protein-coding gene (the *bcr* gene, of unknown function). The breakpoint on chromosome 9 always lies just to the centromere side of the potential oncogene c-*abl* or possibly just within it. When this portion of chromosome 9 is translocated to chromosome 22 it forms a hybrid transcription unit involving part of *bcr* and *abl*. This is translated into a protein that contains a large portion of bcr protein fused to an abl protein lacking its first 25 amino acids. The new protein, like the viral oncoprotein encoded by the gene v-*abl*, now exhibits unregulated tyrosine kinase activity, but how this relates to the development of leukaemia is not known in either case.

Burkitt's lymphoma

Another well-studied chromosomal translocation involving activation of a known oncogene occurs in Burkitt's lymphoma. A reciprocal translocation between chromosomes 8 and 14 brings the *myc* gene on chromosome 8 into the sphere of influence of the immunoglobulin heavy chain genes on chromosome 14. The immunoglobulin genes are being actively expressed in the lymphoma cells and this leads to overexpression of *myc* as well, possibly as the result of an active enhancer element nearby. How *myc* activation fits into the picture of Burkitt's lymphoma is not known.

How deletions of whole or parts of chromosomes can lead to cancer by the loss of putative 'cancer suppressor genes' is shown by the rare familial cancers (see next section).

Familial Cancers: Clues to 'Cancer Suppressor' Genes

Genes involved in human cancer have also been uncovered using a completely different approach to that described in the previous sections. Several rare types of cancer are inherited (they are known as **familial cancers** to distinguish them from the much more common sporadic or spontaneous cancers). It is more accurate to say that what is inherited is a very strong predisposition to the early onset of a certain type of cancer. Overt cancer is often preceded by 'benign' tumours at the site – for example, polyps of the colon in inherited colon cancer – that have a very high risk of becoming malignant.

The predisposition to many of these cancers is inherited as a dominant, single-gene Mendelian trait. This means that someone only has to inherit a defective gene from one of their parents to be at high risk of developing cancer unusually early in life. Some 10 years ago the suggestion was put forward that the onset of the tumour was caused by loss or mutation of the 'good' copy of the gene inherited from the other parent, thus unmasking the recessive 'cancer' allele (Fig. 11.7). (Different forms of the same gene are known as alleles.) The early onset was explained by the fact that in contrast to most cancer, only one mutation was needed in this case, as the patient had inherited one already. During the past few years this theory has been proven correct and its general application extended beyond the rare familial cancers.

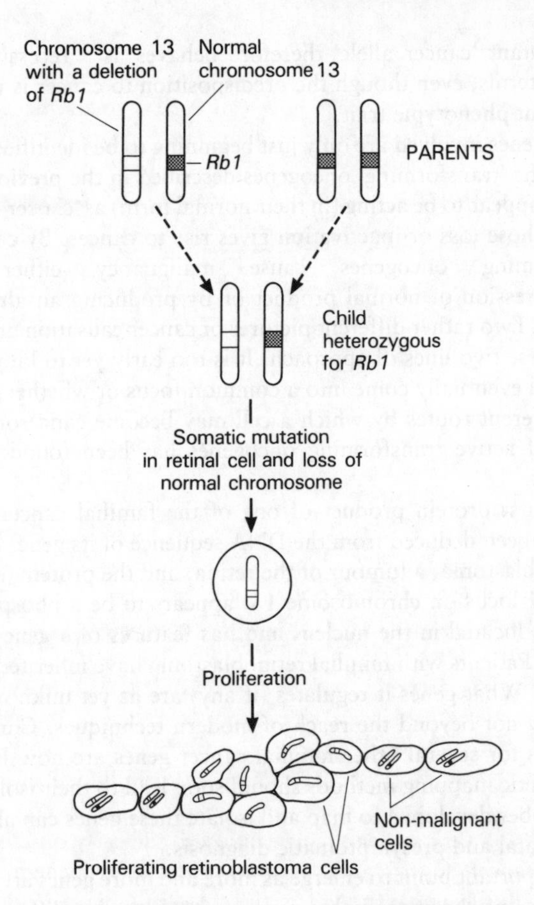

Fig. 11.7 The inheritance of the familial cancer, retinoblastoma, and the events that lead to appearance of the tumour. *Rb1* is a gene on chromosome 13 that when lost or inactive on both homologous chromosomes leads to retinoblastoma, a rare tumour of the retinal tissue of the eye.

By direct analysis of DNA the normal body cells of a familial cancer patient can be shown to be heterozygous, containing one defective 'cancer' allele and one corresponding normal allele. But their tumour cells appear also to have 'lost' the normal allele. This may happen either as a result of a complete loss of the good chromosome (the loss of all or part of a chromosome is rare but not unknown in somatic cells) or by mutation at the level of the gene. This unmasks the mutant allele, which may itself represent a physical loss of the gene, or a mutant form.

The mutant 'cancer allele' therefore behaves as a recessive allele in genetic terms, even though the predisposition to cancer is inherited as dominant phenotypic trait.

The genes involved are only just beginning to be identified. They are unlike the 'transforming' oncogenes described in the previous sections as they appear to be acting (in their normal form) as 'cancer suppressor' genes whose loss or inactivation gives rise to cancer. By contrast, the transforming oncogenes cause malignancy either through overexpression of normal product or by producing an altered active protein. Two rather different pictures of cancer causation are emerging from these two lines of approach. It is too early yet to know whether they will eventually come into a common focus or whether they represent different routes by which a cell may become cancerous. No evidence of active transforming oncogenes has been found in familial tumours.

The first protein product of one of the familial cancer genes has recently been deduced from the DNA sequence of its gene. The cancer is retinoblastoma, a tumour of the retina, and the protein (encoded by the *Rb-1* locus on chromosome 13) appears to be a phosphoprotein, which is located in the nucleus and has features of a gene-regulatory protein. Patients with familial retinoblastoma have inherited a deletion of *Rb-1*. What genes it regulates (if any) are as yet unknown but are probably not beyond the reach of modern techniques. Chromosomal locations for several other familial cancer genes are now known and new genetic mapping methods should soon lead to their isolation. The gene probes developed to map and isolate these genes can also be used for prenatal and presymptomatic diagnosis.

An important point to emerge as more and more genes are mapped is that the same chromosomal regions are implicated in different types of cancer – for example, the p21 region of chromosome 3 in renal cell carcinoma, melanoma and small-cell lung cancer.

The next crucial question was whether these genes might be involved in the far-commoner sporadic cancers. There are a few cancers (e.g. colorectal cancer and retinoblastoma) for which there is both a rare familial form and a much commoner sporadic form. On the assumption that the same genes might be involved in both cases, gene probes developed to analyse the familial case have been used to analyse DNA from sporadic tumours of the same type. There is now evidence that the relevant alleles have indeed also been lost in some sporadic forms of the cancers. Loss of the p21 region of chromosome 3, for example, has now been specifically implicated in many types of sporadic lung cancer, and similar results have been obtained for other tumours.

The familial cancers provide support for a long-held theory that, in genetic terms, the cancerous state generally is recessive. This idea was first put forward 20 years ago after it had been found that cultured tumour cells artificially fused with normal fibroblasts lost their typical malignant character. The best explanation for this was that the genetic changes in the tumour cell were recessive, that is, they could be compensated for by the presence of a good copy of the corresponding genes provided by the other partner in the hybrid cell. Recessive mutations are often 'loss of function' mutations rather than ones leading to an altered and active form of gene product.

Tumour promoters

Early attempts to induce skin cancers in experimental animals identified two quite separate classes of agents that could act together to produce a tumour. A carcinogen such as benz-pyrene painted on the skin of a laboratory rat will eventually produce a tumour but only after heavy and prolonged application. If, however, an initial treatment with benz-pyrene is followed by application of, in the original instance, the blistering agent croton oil, a tumour develops rapidly with no need for further applications of carcinogen. In itself, croton oil is non-carcinogenic and it, and other substances that have this effect, are called **tumour promoters.** It was subsequently found that treatments in general that provoked inflammation or cell proliferation (such as excision of part of the liver, which induces rapid division of liver cells to replace it) had a tumour-promoting action.

The discovery of tumour promoters was one of the first clues to the multi-stage nature of carcinogenesis. In this particular model of carcinogenesis, the belief is that once a carcinogen has produced an 'initiating' mutation, the increased division of the initiated cell induced by tumour promoters provides conditions in which the events necessary for the cell to become fully cancerous are more likely to happen. These could include mutations in the homologous copy of the already mutated gene or in other genes by somatic recombination or loss of parts of chromosomes, all of which can happen during DNA replication and mitosis in somatic cells.

Although this version of events seems to hold only for the particular case of chemical carcinogenesis and skin tumours, the intracellular actions of one class of tumour promoters, the **phorbol esters** (usually exemplified by **TPA** – 12-*O*-tetradecanoyl phorbol acetate) continue to be studied for the light they may throw on general mechanisms controlling cell proliferation.

The phorbol esters are of especial interest as they appear to be able to enhance the expression of certain genes by interacting with a specific transcriptional activator protein – AP-1. Whether this is related to their tumour-promoting activity is not yet clear. However, a protein (JUN, the product of an avian RNA tumour virus oncogene) has been found recently that mimics AP-1 and has oncogenic potential, which suggests that AP-1 is involved in switching on genes involved in the control of cell division. Another intracellular target of the phorbol esters is the ubiquitous enzyme **protein kinase** C, which is increasingly being implicated in pathways that stimulate cell division.

FURTHER READING

The literature on oncogenes and their role in cancer and in normal cellular processes is now enormous. A good place to start is Watson et al., Vol. II (see general reading list) and Bishop (see below). See Alberts et al. (general reading list) for the cell cycle and control of cell proliferation, and the role of the proto-oncogenes in normal cell proliferation and differentiation.

R.W. RUDDON, *Cancer*, Oxford University Press, Oxford/New York, 1987. An up-to-date overview of cancer research, including oncogenes.

J. CAIRNS, *Cancer, Science and Society*, W.H. Freeman, San Francisco, 1976. Although written before the discovery of cellular oncogenes, this book for a general readership provides an excellent introduction to many aspects of the epidemiology and genesis of cancer.

M.G. LEE and P. NURSE, Cell cycle genes in the fission yeast, *Science Progress (Oxford)*, 71 (1987) 1–14.

J.L. MARX, The *fos* gene as a master switch, *Science*, 237 (1987) 854, and references therein.

M.B. SPORN and A.B. ROBERTS, Peptide growth factors are multifunctional, *Nature*, 332 (1988) 217–19.

R. WEISS, N. TEICH, H. VARMUS and J. COFFIN, *RNA tumour viruses*, 2nd edn, Cold Spring Harbor Laboratory, New York, 1985.

M.A. EPSTEIN and B.G. ANCHONG, The epidemiological evidence for causal relationships between Epstein–Barr virus and Burkitt's lymphoma: results of the Ugandan prospective study, *Nature*, 272 (1978) 756–61.

P. TIOLLAIS, C. POURCEL and A. DEJEAN, The hepatitis B virus, *Nature*, 317 (1985) 489–95.

J.M. BISHOP, The molecular genetics of cancer, *Science*, 235 (1987) 305–11. A review of transforming oncogenes involved in human cancer.

B. PONDER, Gene losses in human tumours, *Nature*, 335(1988) 400–2, and references therein. A commentary on recent developments in the search for tumour suppressor genes.

A. KNUDSON, Genetics of human cancer, *Annual Review of Genetics*, 20 (1986) 231–51. A review by the originator of the "loss of alleles" concept of familial cancer, dealing with retinoblastoma and other familial cancers.

G. KLEIN, The approaching era of the tumour suppressor genes, *Science*, 238 (1987) 1539.

See Chapter 3 for references to recent work on AP-1/JUN.

CHAPTER 12
IMMUNOLOGY

Ever since its 'discovery' towards the end of the nineteenth century, the **immune system** that forms the heart of the body's defence against infection has been a source of endless interest to biologists. This elaborate defensive network ramifies throughout the body, its millions of cells primed to detect and respond to invasion by the bacteria, viruses and other parasites that cause disease. A fully-fledged adaptive immune system is present only in vertebrates and, in contrast to the body's other defence mechanisms which are undiscriminating in their action, it mounts a precisely-targetted, highly selective attack on invading microorganisms If successful, this often results in a long-lasting **acquired immunity** to reinfection. The phenomenon of acquired immunity itself was recognized long before either the nature of infection or the immune system had been discovered, and led to the development of the first vaccinations (against smallpox) in the eighteenth century.

As the workings of the immune system were gradually uncovered, however, its remarkable properties proved of far wider biological interest than the prevention of infectious disease. It poses fundamental questions of biochemistry, genetics, cell biology and physiology, some of which have only been resolved in the past decade, while others still remain to be answered.

The definitive feature of all immune responses is their **specificity** – infection in childhood with mumps confers a lifelong immunity to the mumps virus but none at all to those of measles or chickenpox. The other properties of the immune system that have fascinated immunologists for nearly a century are its apparently inexhaustible repertoire of these different specific responses, and its ability to **remember** previous encounters with disease, as demonstrated by acquired immunity.

The most familiar type of immune response is the production of specific protective **antibodies** – proteins that are produced in response to invasion by disease-causing bacteria and viruses, and that bind specifically to the agent that has induced them (the **antigen**), aiding its clearance from the body. Remarkably, the body has the ability to produce a virtually unlimited variety of different antibodies, not only against the components of pathogenic microorganisms but against practically any protein, many polysaccharides and many other natural and synthetic compounds. How this diversity of specific responses is generated – the immunological problem of **generation of diversity** – has occupied immunologists for most of this century and has only been finally solved in the past 10 years with the aid of recombinant DNA technology.

The synthesis of specific antibodies is, however, only one of the responses the immune system makes. During the past 30 years the focus of immunological research has been gradually shifting towards **cell-mediated or cellular immune responses**, and, in the process immunology has acquired a whole new vocabulary. Cell-mediated immune responses are mounted by cytotoxic (killer) cells of the immune system against the cells of its own body that have become altered or 'foreign' in some way, usually as a result of virus or other intracellular parasite infection, or as a result of becoming cancerous. Tissues or organs transplanted from another individual are also rejected by cell-mediated responses. Cell-mediated responses are in their own way as specific and discriminating as antibody responses and result in a specific cell-mediated immunity to the agent that has provoked them.

These two types of immune response represent the two faces of the immune system. The triumph of modern immunology has been to start to unravel the intricate cooperative interactions between the cells of the immune system in both antibody-mediated and cell-mediated immune reactions and to reveal the underlying principles that unify them. The specificity of cell-mediated responses has been explained by the discovery of a second type of variable antigen-specific protein, which is similar but not identical to antibodies and is carried by a subset of the cells of the immune system (see below, T-cell receptors). Cell-mediated immune responses seem to predominate in many diseases and these resist prevention by conventional vaccines which often stimulate mainly antibody production. The greater understanding of cell-mediated responses built up over the past decade has brought hope that vaccines can be developed that are more effective in inducing cell-mediated immunity.

At the heart of the immune system's ability to fight infection lies its ability to discriminate between 'self' and 'non-self' and to recognize certain incoming antigens as 'foreign'. The ubiquitous **histocompatibility antigens**, which provide nearly every cell in the body with an identical immunological fingerprint, play a crucial part in this process.

Histocompatibility antigens are cell-surface proteins first discovered in the 1950s as the **transplantation antigens** responsible for the rejection of tissues transplanted between different individuals. The most important are the major histocompatibility complex antigens or **MHC antigens** (in humans called the HLA antigens). Although the MHC antigens were first discovered in the artificial context of transplant rejection, they have since been proved to play a vital part in all normal immune responses, and provide the key to explaining much of the workings of the immune system.

The body does not usually attack its own constituents, even though many of them are highly antigenic and will provoke immune responses if experimentally injected into another individual or animal species. How the body acquires its normal self-tolerance and what causes its breakdown leading to autoimmune disease is still one of the major unresolved problems in immunology, despite the enormous advances that have been made in understanding the immune system since the principle of self-tolerance was first recognized by the early immunologists at the end of the nineteenth century.

Behind the recent advances in immunology lie improvements in cell biological techniques that have made it possible to distinguish and isolate the different types of cell that take part in immune responses and to study their individual roles in detail in cell and tissue culture. Before the 1960s most of the cells involved in specific immune responses were simply described as 'lymphocytes' – of several rather ill-defined types – on the basis of their anatomical location, appearance under the light microscope and some indications of their immune function as a result of animal experiments. Today, some half-dozen different lymphocytes are distinguished, each with their distinctive role. In addition the parts played by the other white blood cells (e.g. neutrophils, eosinophils, basophils and monocytes and the related mast cells and macrophages in tissues) as 'accessory' cells in immune responses has also become much clearer.

There are still several questions to be answered, either completely or in part. How does the immune system avoid reacting against the body's own antigens? How do the different types of cells within the immune system interact with each other, both by direct contact and by means of chemical signals (lymphokines)? How is it affected by hormones and signals from the nervous system? What part does it play in preventing cancers arising?

The answers to these questions will have important implications for the treatment of the many diseases that are known to have or suspected of having a strong immunological component. As well as providing a basis for improvements in vaccine design, the growing understanding of the workings of the immune system, together with technical advances in other fields, has also led to a revived interest in other immunologically based therapies.

The remarkable ability of antibodies to distinguish their corresponding antigens has long been used as the basis for clinical diagnostic tests and assays. One of the most dramatic developments in recent years in this field is the large-scale production of pure antibodies of single specificities – **monoclonal antibodies**. Already, this has become a multimillion pound industry and has opened up a wide range of new applications in basic research, medicine and industry.

The first part of this chapter provides brief definitions of basic terms and concepts in immunology, including the solution of the problem of generation of diversity. This is followed by a brief outline of the architecture of the immune system. A typical immune response is then followed from first encounter with antigen to the final mopping up operations. Tolerance and self-tolerance are then briefly discussed. The section on the major histocompatibility antigens outlines their structure and role in transplant rejection and disease susceptibility; their role in immune responses is covered more fully in the preceding sections. The final section concentrates on monoclonal antibodies and their uses, and on vaccine development.

INNATE IMMUNITY

The exquisitely specific immune interactions outlined in the following sections are by no means the body's only defence against infection. All animals possess mechanisms of **innate immunity**. These are non-specific with respect to antigen, relatively undiscriminating in their effects and, even in vertebrates, form the first line of defence against infection. The skin, the mucous membranes lining internal surfaces, and the antibacterial substances in many body secretions (e.g. lysozyme in tears) prevent many would-be pathogens gaining access to internal tissues and ever encountering the immune system. The body's own resident microflora also plays a part in suppressing the growth of 'opportunistic' pathogens.

The phagocytes that scavenge dead cells and debris from damaged tissues and rapidly clear foreign matter from the bloodstream will also ingest microorganisms they find in their way. They are attracted in large numbers to sites of tissue damage caused by invading micro-organisms where they participate in the inflammatory reactions that attempt to wall off and contain the infection. Some bacterial pathogens directly activate the body's complement system, which induces a local inflammatory response and destroys them without the intervention of the immune system. Viral infection elicits interferons, polypeptides that amongst their many other effects, prevent virus multiplication. All these mechanisms keep potential pathogens at bay and help to contain those that get through while the second line of defence – an antigen-specific immune response – is developing.

Phagocytic cells also play a crucial part in triggering specific immune responses. As they take up and digest invading microorganisms or damaged cells they process and present the constituent foreign antigens to the immune system in a form it can recognize (see below, Antigen presentation). After an immune response is underway phagocytes dispose of antibody-coated bacteria, cells lysed by complement or destroyed by killer lymphocytes. Other white blood cells are activated to killer status by the presence of antibody (irrespective of its specificity). The physiological consequences of immune responses (e.g. inflammation), the destruction and elimination of microorganisms and infected cells, and hypersensitive reactions such as the familiar allergies are the result of highly complex interactions between antibodies or T cells and the body's non-specific defences.

THE BASIS OF IMMUNE SPECIFICITY

The exquisite specificity of a vertebrate immune response lies in the **antibodies** and **T-cell receptors** produced respectively by the **B** and **T lymphocytes** of the immune system. These individually recognize and interact with surface features of incoming 'foreign' molecules (**antigens**), which in natural circumstances are components of bacteria, viruses and other pathogens.

Antigens

Any substance that is capable of provoking a specific immune response is termed an antigen. Not all substances are antigenic: the best antigens are large molecules with a varied surface structure, such as virtually all proteins and many large polysaccharides. Small molecules such as sugars, nucleotides, etc. are not antigenic on their own but often become so if associated with a carrier protein or other macromolecule. Small molecules that are only antigenic in this form (the dinitrophenyl group is the classic example) are termed **haptens**.

Substances that can provoke an immune response on their own are also often called **immunogens**. All immunogens are by definition antigens but a given antigen may not necessarily be **immunogenic** in all circumstances. Whether it actually provokes an immune response depends on how 'foreign' it is to the organism in question. The body will not normally react to its own proteins, polysaccharides, etc. even though they may be highly antigenic if introduced into another species, or even, for some molecules, another individual of the same species. Immunogenicity also depends on the dose of antigen encountered, whether its route of entry brings it into contact with the appropriate part of the immune system, and whether it is in a form that is efficiently presented to the immune system.

Antigenic Determinants (Epitopes)

Antigen is an operational definition that can denote anything from a single molecule or fragment of a molecule to a large multimolecular complex, such as a virus or a bacterial cell surface. The features of an antigen that are recognized and responded to by individual cells of the immune system are the **antigenic determinants** or **epitopes**. Many antigens – especially proteins with their varied surface topography – carry a number of different antigenic determinants, each of which will provoke a distinct response from the immune system (Fig. 12.1)

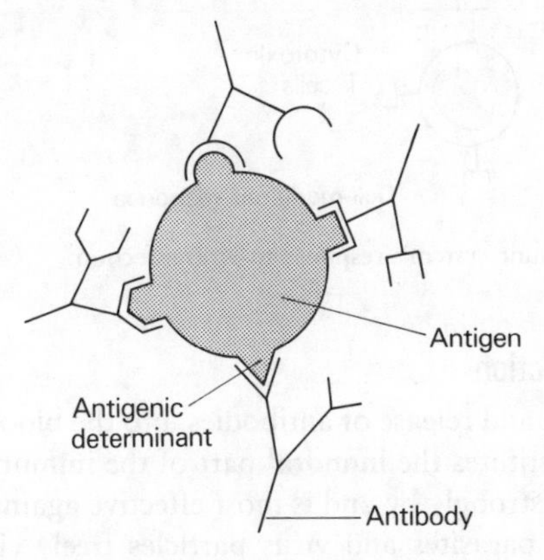

Fig. 12.1 An antigen carrying several antigenic determinants will provoke the production of a number of different antibodies, each specific for one determinant or epitope.

Immune responses

The immune system mounts two types of specific response against an incoming antigen. Natural infections usually provoke both the production of specific **antibodies** and **cell-mediated** or **cellular** immune responses. Which one predominates and which is most effective at actually combatting the infection depends on the way the particular invader establishes itself in the body and presents to the immune system (Fig. 12.2).

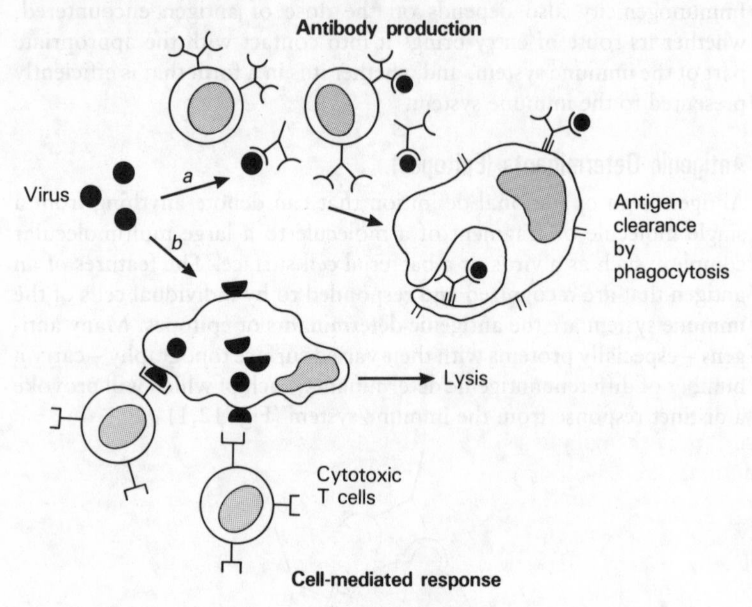

Fig. 12.2 The immune system's response to viral infection.

Antibody Production

The production and release of antibodies into the blood and lymphatic circulation constitutes the **humoral** part of the immune response. It is provoked most strongly by and is most effective against bacteria, their protein toxins, parasites and virus particles freely circulating in the bloodstream, in extracellular tissue fluids and on tissue surfaces, and is usually the main response to injected purified immunogens.

The body can produce millions of different antibody molecules, each of which 'fits' only one or a very few antigens (or more precisely antigenic determinants). Antibody molecules recognize their cognate (corresponding) antigen by way of an antigen-specific **antigen-binding site**, complementary in shape and physical properties, especially surface charge, to a corresponding antigenic determinant. Antibody-producing cells derive from the **B lymphocytes** (B cells) of the immune system, and antibody production usually also requires the cooperation of another class of lymphocyte, **helper T cells** (see later). Each B cell produces antibody of a single specificity.

Antibodies exert their protective effects by binding specifically to the antigen that induces their synthesis, forming an antigen–antibody complex that is more efficiently scavenged by the body's phagocytic cells than the antigen alone. Antibodies bound to target antigens on the surface of a bacterial cell also present sites at which other components of the body's non-specific defences, such as complement, can act to destroy the bacterium. As well as circulating antibody, locally produced antibody in saliva and other secretions also provides protection against many pathogens.

Cell-mediated (Cellular) Immune Responses

Cell-mediated or **cellular immune responses** are mounted by the **T lymphocytes** of the immune system, and are directed chiefly against the body's own cells that have become infected with bacteria or viruses, fungi, or intracellular protozoan parasites such as *Leishmania* and the malaria-causing *Plasmodium*, and that consequently display a 'foreign' surface to the immune system. Tumour cells displaying altered surface antigens are also attacked in cell-mediated responses. Infected cells are recognized and destroyed by direct contact with antigen-specific killer lymphocytes – **cytotoxic T cells** – often with the cooperation of helper T cells. Each T cell carries antigen receptors of a single specificity on its surface. Transplanted tissues and organs from incompatible donors are also rejected through a cell-mediated immune response (see below, The major histocompatibility antigens). Antigen-nonspecific, although not totally undiscriminating, killing of infected or cancerous cells by non-T lymphocytes and leukocytes (e.g. natural killer cells) also forms a part of cell-mediated defence mechanisms (see below, Immunology and cancer).

Antibodies

All antibody molecules have a similar overall structure but differ most in the detailed amino acid sequence of the parts of the molecule that make up the antigen-binding site (Fig. 12.3). Antibodies are **immunoglobulins**, one of several classes of **serum globulins** which are distinguished from each other electrophoretically. An immunoglobulin (Ig) molecule consists of four polypeptide chains of two different types – two identical **heavy (H) chains** and two identical **light (L) chains**. Each chain has a **constant** (C) and a **variable** (V) portion.

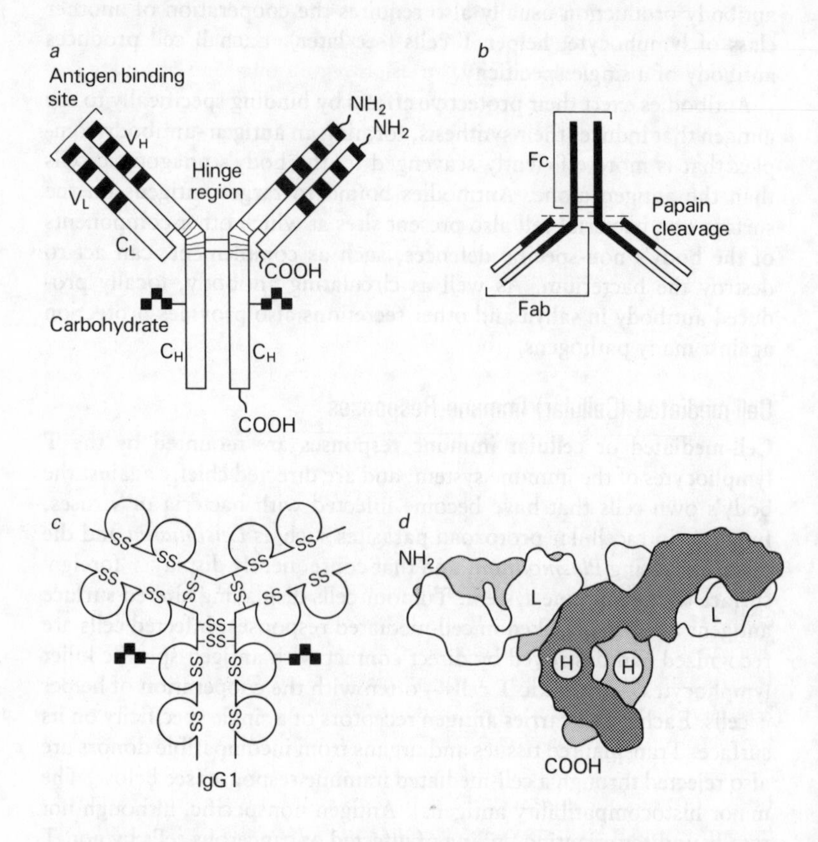

Fig. 12.3 *a*, Conventional representation of an antibody molecule showing heavy (H) and light (L) chains and their constant (C) and variable (V) regions. *b*, Fc and Fab regions. Antibodies are cleaved by papain digestion into several fragments. The Fc fragment comprises most of the constant regions of the heavy chains. The Fab fragments contain the antigen-binding sites. *c*, Arrangement of immunoglobulin domains. An immunoglobulin molecule is built up from a prototype repeating unit (often called the immunoglobulin unit domain), that also occurs in other related molecules (e.g. MHC antigens and T-cell receptors.) The antibody represented here is IgG. IgM and IgE have a third unit in the constant region. *d*, Overall shape of an immunoglobulin molecule reconstructed from X-ray crystallographic data.

The variable region differs in amino acid sequence from one type of antibody to another. A polypeptide chain folds into a three-imensional configuration that is determined entirely by its amino acid sequence. The folding of the variable regions of the light and heavy chains therefore forms a unique **antigen-binding site**, which differs from antibody to antibody, and which determines its specificity. A particular antigen-binding site can usually bind only one or a very few closely related antigenic determinants. The basic unit antibody molecule possesses two identical antigen-binding sites.

The variable region of heavy and light chains consists of a 'framework' of relatively constant amino acids into which are slotted three hypervariable regions. These regions differ greatly from antibody to antibody, and are the main determinants of the antigen-binding site.

There are only two types of light chain (as determined by the structure of the invariable region) – λ and κ. An antibody molecule contains either κ or λ chains, never both. There are, however, several different types of heavy chain (see below). All heavy chains are encoded by one cluster of genes (on chromosome 14 in humans), the λ light chain by a separate gene cluster (on chromosome 22), and the κ light chain on chromosome 2 (see below, Generation of diversity).

Antibody Class

The Fc portion of the heavy chain constant region determines what **class** an antibody belongs to, irrespective of its antigen specificity. There are five classes of antibodies in humans (IgG, IgM, IgD, IgA and IgE), each bearing a different type of heavy chain (respectively γ, μ, δ, α and ϵ), along with either κ or λ light chains. They appear at different times and in different circumstances during an immune reaction. The constant region determines the particular physiological role of the antibody through its interactions with non-specific components of the immune system such as complement and other cells of the body (see Table 12.1 and Hypersensitivity reactions for more detail).

Allotypes, isotypes and idiotypes

Immunoglobulins and T-cell receptors are excellent antigens and carry various antigenic determinants that can be revealed if, for example, molecules from one species are injected into another, or into a different inbred strain of the same species.

These determinants provide a convenient way of distinguishing the various classes of antibody and the different genetically determined subclasses of, for example, κ and λ light chains and γ heavy chains. Within each species, determinants that distinguish the constant regions of the different types of heavy chain, and the κ and λ chains, are known as **isotypes**. They are not antigenic in any member of the species but are antigenic in other species.

The "determinants" of an antiBody include
Isotypes – which differ between species. Allotypes
which differ within a species and Idiotypes
which are different for each antibody

Table 12.1 Classes of immunoglobulins

IgM	First class of immunoglobulin produced in an immune response. After a B cell has become genetically programmed to synthesize antibody of a particular specificity, IgM molecules of that specificity are produced and remain bound to the cell membrane. When antigen is encountered and an immune response initiated, the cell switches to producing a form of IgM that is released (secreted) by the cell as antibody and is found in large amounts in the circulation during the early stages of a primary humoral immune response. Secreted IgM is composed of a pentamer of antibody molecules held together by a J, or joining, polypeptide (not to be confused with the J (joining) regions in V genes (see Generation of diversity)). Because of its large size IgM does not cross the placenta. It mediates complement fixation and stimulates phagocytosis. 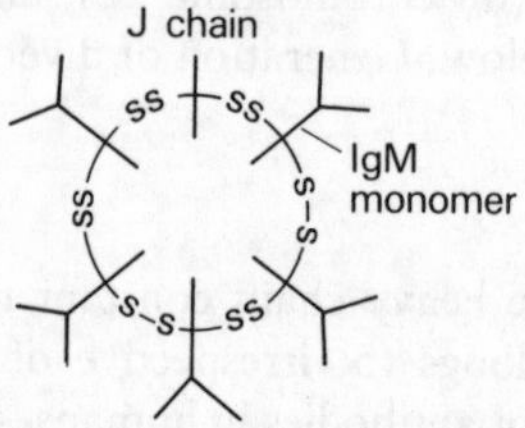
	J chain IgM monomer
IgG	Main type of antibody produced and released into the circulation during the later stages of a primary humoral immune response against large doses of antigen, and in secondary immune responses IgG antibodies can also cross the placenta to provide passive immunity for the developing foetus. Bacteria covered with IgG are readily destroyed by complement. Bacteria, viruses and other antigens complexed with IgG are more readily taken up by scavenging phagocytes.
IgD	Appears as surface-bound immunoglobulin on B cells subsequent to and in conjunction with IgM. Only very small amounts secreted. Function unknown.
IgA	Antibody produced by certain types of lymphoid tissue and secreted locally in gut and in saliva, tears, etc. It is the predominant antibody in milk and colostrum. It combines with microorganisms to prevent them attaching to epithelial surfaces, lining of gut, etc. and is important in preventing gastrointestinal infections in newborns. Occurs as a monomer and as a dimer held together by a J chain.
IgE	Antibody involved in local hypersensitivity (allergic) reactions and reactions against internal parasites. IgE was formerly known as reaginic antibody or homocytotropic antibody.

There are also a number of alternative genetically-determined forms of each type of chain within a species as a whole. Any individual will only express one of these variants of each chain. (They may inherit two different gene variants but because of the peculiarities of immunoglobulin gene expression they will only use one to make antibody.) The determinants that distinguish these are known as **allotypes** as they are antigenic in another individual of the same species who carries a different form of that particular gene. They are generally localized in the constant region. Examples of allotypic determinants in humans are the alternative forms of κ chain Inv(1), Inv(1,2) and Inv(3) and the Gm series of markers for the various γ chains.

The third set of antigenic determinants on antibodies are known as **idiotypes** and are determined by the variable region. They differ from antibody to antibody (see Anti-idiotypic networks).

T-cell receptors

The second type of antigen-specific recognition molecule within the immune system is the **T-cell antigen receptor**. These are borne in large numbers on the surfaces of T lymphocytes, which are the lymphocytes involved in the initial recognition of antigen entering the body and in the cell-mediated or cellular arm of the immune response. Each T cell carries receptors of a single antigen specificity.

Like antibodies, T-cell receptors are very variable molecules, each one specific for one or a small number of antigens. The molecular

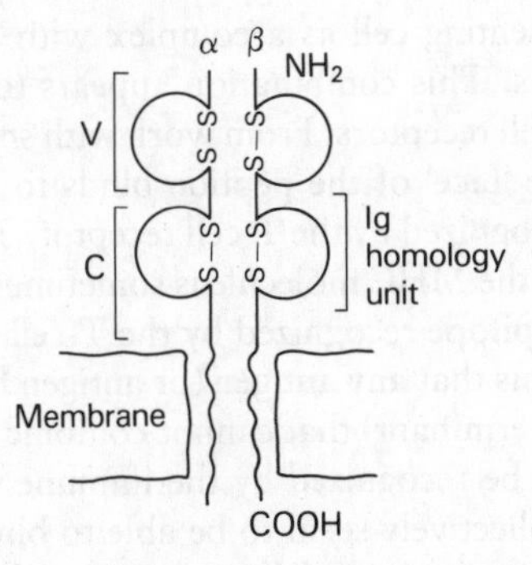

Fig. 12.4 The ß/α T-cell receptor. A second type of receptor consisting of polypeptides γ and δ is borne by the earliest distinguishable foetal thymocytes and, in the adult, by a population of T cells present in intestinal epithelium.

structure of the T-cell receptor was only determined in 1984 (that of antibodies has been known since the 1950s). It is a membrane glycoprotein. The antigen receptors on most T cells in adults consist of two polypeptide chains – α and ß – of similar size, each of which has a constant and variable portion. The variable portions of the molecule form a unique **antigen-recognition site** (Fig. 12.4). A crucial operational difference between antibodies and T-cell receptors is that the T-cell receptor only recognizes antigen when it is presented to it as a complex bound to certain of the major histocompatibility antigens that are carried on the surfaces of most cells (see below).

MHC proteins

The third molecular components of most immune responses are the **MHC glycoproteins** (the major histocompatibility antigens). These are ubiquitous cell-surface glycoproteins encoded by genes of the major histocompatibility complex. Every cell in the body carries a set of MHC proteins (MHC antigens) which together make up the 'tissue type' of the individual – in humans they are called the **HLA antigens** (human leukocyte associated antigens) although they occur on all body cells. They were initially discovered as the **transplantation antigens** that provoke rejection of transplanted tissue.

The vital role of MHC proteins in normal immune responses has only been fully revealed in the past decade. Most antigens entering the body are not recognized directly by lymphocytes but must be processed and presented to them by non-lymphocyte 'antigen-presenting cells'. During this processing antigen fragments are displayed on the surface of the antigen-presenting cell as a complex with one or other of the MHC glycoproteins. This combination appears to be recognized as a single entity by T-cell receptors. From work with small peptide antigens it appears that one 'face' of the peptide binds to the MHC molecule while another is recognized by the T-cell receptor. The peptide determinant recognized by the MHC molecule is sometimes called the **agretope** in contrast to the **epitope** recognized by the T-cell receptor.

It follows from this that any antigen (or antigen fragment, representing an antigenic determinant) that cannot combine with an MHC molecule is unlikely to be recognized by the immune system. In practice, MHC molecules collectively seem to be able to bind (in a non-specific fashion) an enormous array of different antigen fragments. There are however instances where genetically-determined unresponsiveness to a particular antigen has been traced to its inability to bind with the particular MHC antigens that animal carries. Conversely, the rapid rejection of tissues and organs transplanted from a person with another tissue type represents an all-too-efficient recognition of the different MHC antigens on their cells.

MHC proteins belong to the same protein 'superfamily' as immunoglobulins and T-cell receptors. They are not antigen-specific in the same sense, although some MHC molecules will bind certain antigens but not others. They represent a more primitive type of recognition system than the highly specific antibodies and T-cell receptors. Many lower animals, and even vascular plants, show rejection responses to 'incompatible' tissues of another strain or to multicellular parasites which are mediated by cell-surface glycoproteins (e.g. the rejection of incompatible pollen by the stigma). Details of the different types of MHC antigens, their structure, and their role in transplantation rejection and in some diseases are covered in a later section, and their role in antigen recognition is covered more fully on p. 405.

Clonal selection

The immune system's ability to produce a specific antibody or cell-mediated response within a day or two of being challenged by any of a virtually unlimited number of different antigens is a remarkable feat. How is it achieved? One theory current in the 1940s held that the incoming antigen acted as a physical template around which a matching antibody was moulded. A better understanding of protein synthesis soon showed this to be unlikely, if not impossible. It was replaced in the 1950s by the concept of **clonal selection** – a cornerstone of modern immunology. As first propounded in relation to antibody synthesis, it proposed that there already exists in the body a pool of lymphocytes, each primed to produce antibody of a particular specificity. When the immune system encounters an antigen, those lymphocytes of the corresponding antigen specificity are stimulated to multiply and start producing antibody.

Clonal selection in its modern guise applies equally to B and T lymphocytes. A lymphocyte acquires its antigen specificity at random, early in its development, quite independent of any encounter with antigen, and, once acquired, the particular specificity is passed on to all its progeny. Small **clones** of cells of identical antigen specificities are thus formed, made up of the progeny of a single cell. The mature lymphocyte population of the body contains millions of these clones, each primed to produce a different antibody (or T-cell antigen receptor). Antigen challenge activates only corresponding clones of antigen-specific lymphocytes which start to proliferate rapidly. At a certain point, after the initial clone has expanded considerably, the lymphocytes undergo terminal differentiation into **effector** cells – antibody-secreting **plasma cells** in the case of B lymphocytes, and active T cells of various types.

Clonal selection obviously requires that lymphocytes display their antigen specificity in some way recognizable by the challenging antigen. Once programmed, both T and B lymphocytes display antigen-specific cell surface receptors. In the case of B lymphocytes, it is a membrane-bound form of the antibody their progeny will eventually produce (see later sections).

The question of how a lymphocyte acquires its antigen specificity and how such an extensive repertoire of different specificities can be generated was not resolved until the late 1970s (see below, Generation of diversity).

Immunity and immunological memory

When the immune system encounters an antigen for the first time it takes several days to mount a **primary response**. It is this short delay that often allows infectious agents to multiply within the body and cause overt disease. The total repertoire of randomly generated antigen-binding sites in the lymphocyte pool contains many that differ only very slightly in their shape and binding properties, and which can bind the same antigenic determinant, albeit more or less strongly. The initial primary response activates many clones of greater or lesser antigen specificity, from which those with a better fit are selectively expanded as the immune response develops. The immune system can remember, however, and does not have to go through this long-winded process of selection next time it encounters the same antigen. A second exposure, even years later, often provokes an immediate and much stronger response, the **secondary immune response**, which prevents the infection taking hold. This is the basis of the long-lasting **immunity** that we develop to the common infectious diseases of childhood.

The immune system's memory consists of clones of highly specific B and T cells that have been selected during the primary response but are then set aside as **memory cells** without developing into antibody-producing cells or functional T cells. Memory cells, unlike effector T and B cells, are long-lived, persisting for months and even years without dividing. They are immediately activated by a subsequent challenge with the original antigen and rapidly proliferate and differentiate.

Generation of diversity

Antibody molecules are the most variable proteins known and the ability of a single individual to produce an apparently limitless variety proved for many years extremely difficult to explain in terms of 'conventional' genetics.

The number of different antigen-binding sites an individual can produce has been estimated as between 10^6 and 10^8. The antigen-binding sites on each molecule are formed by the combination of a light chain and a heavy chain. Allowing for the fact that any heavy chain can associate with any light chain, if each possible polypeptide chain were to be specified by a separate gene, very large numbers (running into many thousands) of heavy and light chain genes would be required to provide the estimated number of antigen-binding sites. Separate encoding of every possible polypeptide chain would also entail a somewhat extravagant repetition of the gene sequences for the constant regions, of which there are no more than a dozen or so variants.

Until the mid-1970s, when it became possible to analyse immunoglobulin gene structure directly, an unambiguous answer to the problem of antibody diversity had not been forthcoming. There were two possibilities. One was that all possible antibody specificities are already encoded in the basic genetic complement of an animal (its genome) and are passed from generation to generation through the germline (the reproductive cells). This became known as the **germline theory** of antibody diversity. The alternative view held that there are only a relatively small number of prototype antibody genes hard-wired into the genome, and that during lymphocyte development, these genes undergo random mutations in the portions encoding the variable regions of the antibody chains to generate the repertoire of antigen specificities. This was the **somatic mutation** theory of antibody diversity, which refers to the fact that the mutations are presumed to occur only in somatic cells and not in the reproductive cells of the organism. The advent of recombinant DNA techniques finally made it possible to look directly at what was happening at the DNA level, which turned out to be more complicated that anyone had envisaged. As it transpired, both theories were partly correct.

Immunoglobulin Gene Structure and Assembly

The immunoglobulins (and T-cell receptor proteins) turn out to be exceptional proteins in that each of their polypeptide chains is encoded by several quite separate DNA sequences (**gene segments**), each of which is present in a set of variant copies in the genome (Figs 12.5,

12.6). Before an immunoglobulin molecule or a T-cell receptor can be synthesized, extensive physical rearrangements in the DNA bring together one each of these separate gene segments to produce a complete functional gene for each polypeptide chain. (This is a different process from the splicing of RNA transcripts to assemble a functional mRNA from the typical eukaryotic 'split gene', and which indeed occurs later during immunoglobulin gene expression.)

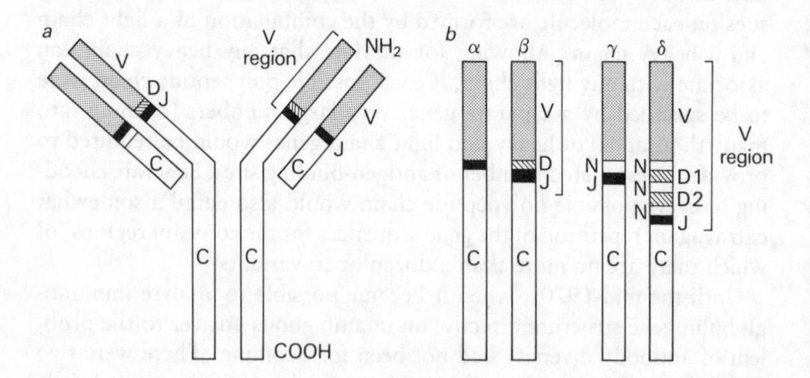

Fig. 12.5 Antibody and T-cell receptor showing the parts encoded by different gene segments. *a*, Antibody. For the heavy chain, for example, a V segment encodes the N-terminal signal sequence that is necessary for the molecule to be secreted from the cell, and which is enzymatically removed before release, as well as the 96 N-terminal amino acids of the heavy chain variable region. A D (diversity) gene segment encodes from 1 to 15 amino acids, comprising the most variable part of the heavy-chain variable region, and a J gene segment codes for the last 10 (or so) amino acids of the variable region. The whole constant region in both light and heavy chains is encoded by single C segments. *b*, T-cell receptors (see Fig. 12.6 and caption for explanation of N segments).

The immunoglobulin and T-cell receptor genes are the only vertebrate genes that have been found to undergo irreversible structural changes during cellular differentiation. In general, differentiation involves changes in gene expression – genes being switched on or off – rather than structural alterations. However, such controlled gene rearrangement is not unknown elsewhere. For example, parasitic protozoan trypanosomes produce a novel cell-surface antigen every generation (antigenic variation) by physical DNA rearrangement – and in this way, ironically, keep one step ahead of their host's immune system.

For simplicity the following outline of gene rearrangement deals primarily with the immunoglobulin genes. The evidence so far available for the T-cell receptor genes indicates a similar sequence of events in the assembly of the variable regions of α and ß genes. T-cell receptor gene rearrangement appears to take place in the thymus.

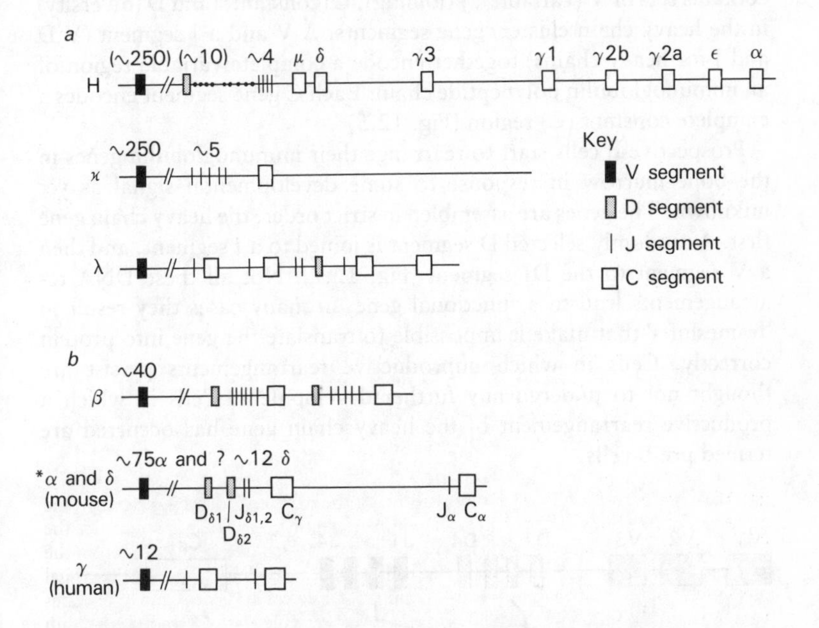

Fig. 12.6 Arrangement of germline genes for *a*, immunoglobulin and *b*, T-cell receptor polypeptide chains. In the mouse δ chain region there seem to be a relatively small number of V segments (no more than a dozen) and two each of D and J, but great potential for diversity is introduced by the use of both D segments and random insertions of up to six nucleotides between V and D1, D1 and D2 and D2 and J1 during rearrangement. These inserted nucleotides are termed N segments and are not encoded in the germline. Similar insertions occur in the γ chain between V and J, and contribute to diversity.

By combining gene segments in different permutations a vast number of different antigen-binding sites can be encoded by a relatively small number (several hundreds rather than thousands) of genetic elements. Additional variation is also introduced into the immunoglobulin genes, but not, as far as we know, the T-cell receptor genes, by further somatic mutations at a much later stage in lymphocyte development.

In the earliest lymphocyte precursor stem cells in bone-marrow, antibody genes are in the unrearranged germline configuration, as they are in all other body cells. In mammals there are three clusters of immunoglobulin genes, located on different chromosomes. One encodes the heavy chain, the other two encode two alternative forms of the light chain, known as kappa and lambda. Each gene cluster or **gene family** contains sets of V (variable), J (joining), C (constant) (and D (diversity) in the heavy chain cluster) gene segments. A V and a J segment (V,D and J for heavy chains) together encode a complete variable region of an immunoglobulin polypeptide chain. Each C gene segment encodes a complete constant (C) region (Fig. 12.5)

Prospective B cells start to rearrange their immunoglobulin genes in the bone marrow in response to some developmental signal as yet unknown. The genes are assembled in strict order, the heavy chain gene first. A randomly selected D segment is joined to a J segment, and then a V segment to the DJ segment (Fig. 12.7). Not all these DNA rearrangements lead to a functional gene; in many cases they result in 'frameshifts' that make it impossible to translate the gene into protein correctly. Cells in which unproductive rearrangements persist are thought not to undergo any further development. Cells in which a productive rearrangement of the heavy chain gene has occurred are termed **pre-B cells**.

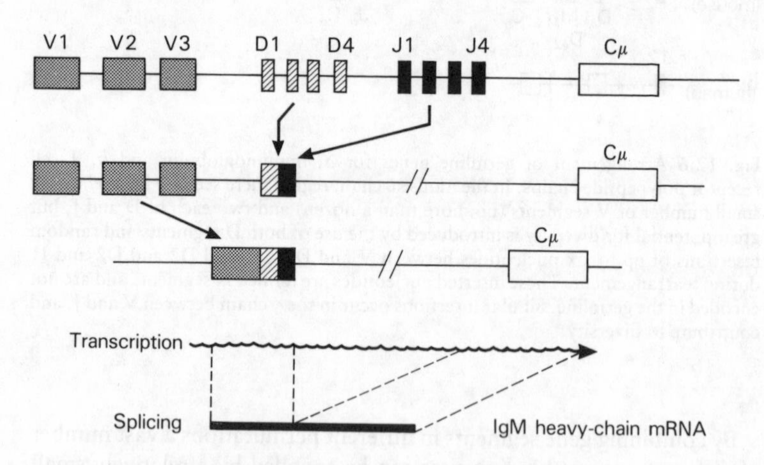

Fig. 12.7 Stages in the assembly of a functional immunoglobulin heavy-chain gene and production of mRNA.

Once a functional light chain gene has also been assembled, by a similar sequence of rearrangements to those described above, the antigen specificity of the cell is determined. In any individual B cell light chains of only one type (κ or λ) are produced.

The recombination enzymes that cut and rejoin DNA to stitch together V, D, and J segments recognize **recognition sequences** present in the DNA on either side of each segment. The process appears to be rather imprecise, with a leeway of several nucleotides either way. Far from being a drawback, this imprecise joining is a further source of variation for the antibody molecule.

Immunoglobulin Synthesis: Heavy Chain Class

The rearranged heavy chain variable region now lies adjacent, although separated from, the cluster of C gene segments. The first immunoglobulin synthesized is always IgM. The heavy chain is made from a transcript that includes the rearranged variable region and the μ constant region gene, which lies nearest to it.

A particular scheme of transcription and RNA splicing produces a class M heavy chain containing a stretch of carboxy-terminal amino acids that anchors the immunoglobulin molecule in the plasma membrane instead of allowing it to be secreted. The B cell is thus able to advertise its antigen specificity to other cells of the immune system and to incoming antigen. Later, after antigenic stimulation through these **IgM receptors**, the B cell will proliferate and its progeny differentiate into plasma cells. A change in the pattern of RNA splicing now generates heavy chains lacking the membrane anchor and which combine light chains to form secreted IgM antibody.

Allelic exclusion

B cells, like all somatic cells, contain two copies of each chromosome, one inherited from the father and one from the mother. For most genes, the copies on both chromosomes are expressed. In B cells however, only the immunoglobulin genes on one of the chromosomes of the relevant pairs are expressed – a phenomenon known as **allelic exclusion**. Allelic exclusion ensures that a cell only produces antibody of a single specificity, and is now explained by the fact that once a productive DNA rearrangement has taken place on one chromosome the process stops.

The heavy chain switch

Antibodies of different classes (IgG, IgA, IgE) are made by further irreversible DNA rearrangements, the **heavy chain switch**, which joins the heavy chain variable region to different constant regions. DNA rearrangements involving a 'switch' region preceding the γ, μ, and α C genes results in a new combination of the rearranged V region and a C gene. This occurs in the progeny of cells already activated by antigen. IgM, for example, is invariably replaced by IgG in the later stages of an immune response and is the class of antibody produced straight away by committed memory cells in a secondary response. Excessive IgE production often leads to a hypersensitive response (see later). The stimulus for heavy chain switching is not yet known.

Somatic Hypermutation: Fine Tuning the Immune Response

Immunoglobulin gene assembly occurs before cells encounter antigen. At a much later stage, probably some time between antigen stimulation and terminal differentiation, **somatic hypermutation** in the variable regions of the heavy chain and light chain genes leads to further changes in the amino acid sequence of the antibody-binding site. These changes may not materially alter antigen specificity but will affect properties such as strength of binding to antigen. The mechanisms underlying somatic hypermutation are not yet known, but rather than gross changes involving whole stretches of DNA (as with the DNA rearrangements outlined above) they seem to be point mutations, altering a single nucleotide at a time.

Cells producing antibodies with a high affinity for their target antigen, as a result of the changes introduced into the antigen-binding site by somatic hypermutation, will then be selected and their numbers increased in preference to those producing low-affinity antibodies. This allows fine tuning of the immune response to occur once antigen has been encountered. The IgG antibodies produced in a secondary immune response in a previously immunized animal are typically of much higher affinity than the IgM antibodies produced in the primary response. Somatic hypermutation does not appear to occur in T-cell receptor genes.

Potential for Diversity

The potential for antibody diversity in this system is immense. A simple calculation based on the variable region immunoglobulin genes known to be present in mice – 250 V genes, 10 D genes and 4 J genes in the heavy cluster, 250 V genes and 5 J genes in the κ cluster and 2 V and 3 J genes in the λ cluster, gives $250 \times 10 \times 4 = 10\ 000$ possible different heavy chain V regions and $250 \times 5 = 1250$ possible κ light chain V regions. (The λ chain genes can be disregarded as they do not contribute greatly to antibody diversity in this example.) Adding in a conservative estimate of additional variation as a result of the flexibility of joining at each junction increases these numbers to around 3000–4000 for light chains and 90 000 for heavy chains. Since any light chain can combine with any heavy chain this gives a total of 90 000 × 4000 = 360 000 000 (3.6×10^8) possible different antigen-binding sites that can be generated from fewer than 600 germline genes. Similar calculations for the δ/γ T-cell receptor give a staggering 10^{17} possible antigen-binding sites (largely due to variation introduced by the N regions, see Fig. 12.6).

Some of these hypothetical antigen-binding sites will, of course, not actually be able to bind anything, and many will be of the same broad antigen specificity, differing only in their affinity (strength of binding) for antigen. Even so, the random joining of different gene segments, random association of light and heavy chains and the further variability introduced by somatic mutation are thought to be able to account for the antibody repertoire, and the rearrangements in T-cell genes for the T-cell repertoire.

THE IMMUNE SYSTEM

The immune system ramifies throughout the body providing a network of millions of cells that can detect antigen and respond to it wherever it may gain entry. The only organ isolated from the immune system is the brain as the blood–brain barrier does not allow the passage of cells or antibodies. The cells involved in protecting the body against infection are the antigen-specific **T** and **B lymphocytes,** and the various types of **leukocytes** (neutrophils, eosinophils and basophils and monocytes), non-specific **killer cells, macrophages** (a type of phagocyte) and **mast cells,** which are non-specific in their actions (Fig. 12.8).

Leukocytes typically circulate in the blood but can migrate into tissues to take part in inflammatory reactions. Lymphocytes are the main constituent of the various **lymphoid tissues** (see Fig. 12.9) and also circulate in blood and lymph. Tissue macrophages, which along with polymorphonuclear leukocytes are the chief 'professional' phagocytes of the body, are part of the **reticuloendothelial system.** This is a network of fibrous tissue supporting macrophages and macrophage-like cells in both lymphoid and non-lymphoid organs (such as liver, lungs and kidneys) and is concerned with the uptake and clearance of foreign matter from the body. Macrophage-like cells are also plentiful in the dermis of the skin.

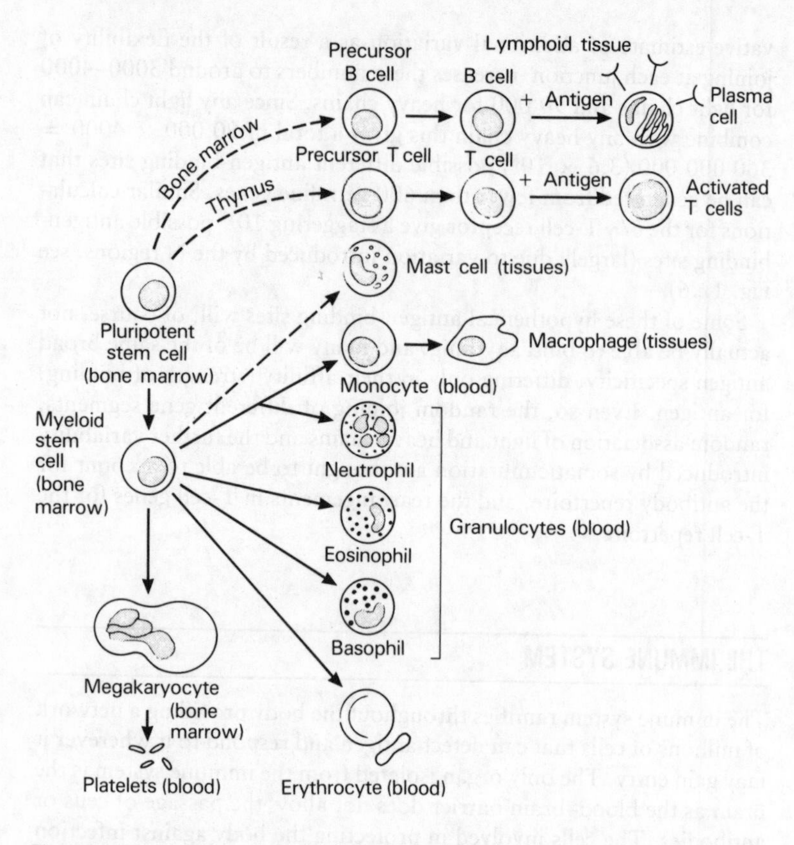

Fig. 12.8 The origin of cells of the immune system. Each solid line represents a separate differentiation pathway from a pluripotent stem cell in the bone marrow. Dashed lines indicate pathways that are not yet fully worked out. In all cases, several generations of differentiating cells intervene between the stem cell and the mature, fully differentiated cell.

The lymphoid system

The anatomical core of the immune system is the **lymphoid tissue** (Fig. 12.9). In mammals, the **primary lymphoid organs** are the **bone marrow** and the **thymus**, which produce B and T lymphocytes respectively, but which do not participate in immune reactions directly. Both B and T lymphocytes (along with the other blood cells) originate from stem cells that are present in the embryonic yolk sac and later in bone marrow.

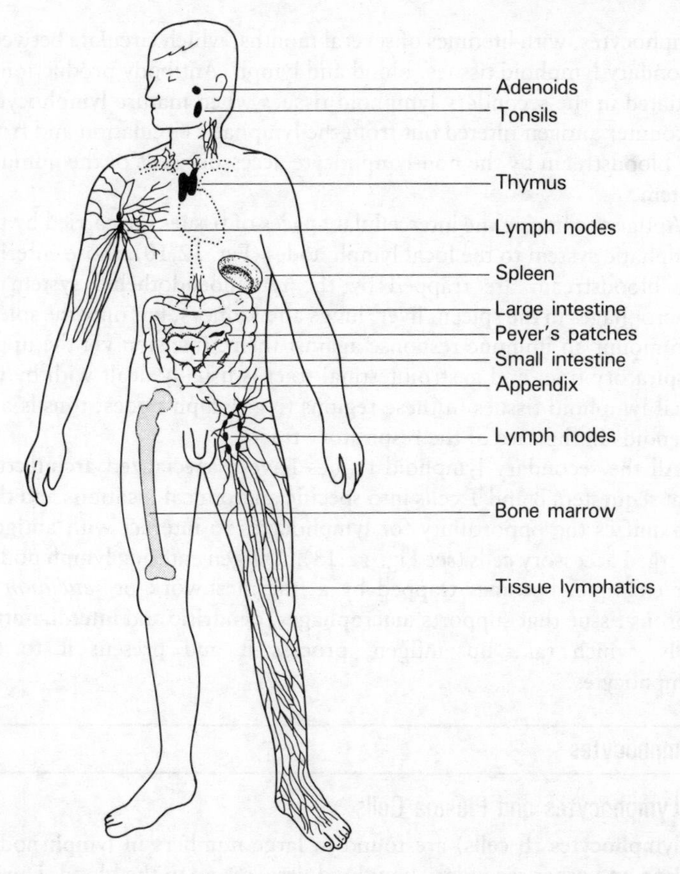

Fig. 12.9 The human lymphoid tissues. (From C.C. Blackwell and D.M. Weir, *Principles of infection and immunity in patient care*, Churchill Livingstone, Edinburgh, 1981).

The embryonic thymus is populated by these cells, and in later life is also replenished by fresh T cell precursors from bone marrow. Large numbers of lymphocytes are produced daily by bone marrow and thymus.

Immature short-lived 'virgin' lymphocytes (i.e. lymphocytes that have acquired antigen specificity but have not yet encountered antigen) migrate from bone marrow and thymus to the **secondary lymphoid tissues – lymph nodes, spleen, tonsils, Peyer's patches, adenoids** and **appendix**. Here they are thought to mature into long-lived lymphocytes, with lifetimes of several months, which circulate between secondary lymphoid tissues, blood and lymph. Antibody production is initiated in the secondary lymphoid tissues when mature lymphocytes encounter antigen filtered out from the lymphatic circulation and from the bloodstream by the non-lymphocyte accessory cells of the immune system.

Antigens entering the intercellular spaces of tissues are carried by the lymphatic system to the local lymph nodes (Fig. 12.10), those entering the bloodstream are trapped by the reticuloendothelial system of macrophages in the spleen, liver, lungs and kidneys, but only the spleen can mount an immune response against them. Invasion via the upper respiratory tract and gastrointestinal tract is usually dealt with by the local lymphoid tissues in these regions (e.g. lymph nodes, tonsils and adenoids in the case of the respiratory tract).

All the secondary lymphoid tissues have a specialized architecture that sequesters B and T cells into specific anatomical locations and that maximizes the opportunity for lymphocytes to interact with antigen-charged accessory cells (see Fig. 12.13). Antigen entering lymph nodes, for example, becomes trapped by a fine meshwork or *reticulum* of fibrous tissue that supports **macrophages, dendritic** and **interdigitating cells**, which take up antigen, process it and present it to the lymphocytes.

Lymphocytes

B Lymphocytes and Plasma Cells

B lymphocytes (B cells) are found in large numbers in lymph nodes, spleen and other secondary lymphoid tissues, and in the blood. During its differentiation in the bone marrow an individual B cell becomes programmed to produce antibody of a particular antigen specificity (see above, Generation of diversity).

After encounter with antigen in secondary lymphoid tissues, B cells proliferate and after several cell generations differentiate into antibody-producing **plasma cells**, whose cytoplasm is packed with rough endoplasmic reticulum at which immunoglobulins are being synthesized, assembled and started off on their journey out of the cell. Plasma cells secrete antibody at the rate of around 10 000 molecules per second. They do not divide further and have a lifetime of only a few days.

T Lymphocytes

T cells or **T lymphocytes**, unlike B cells, which are concerned solely

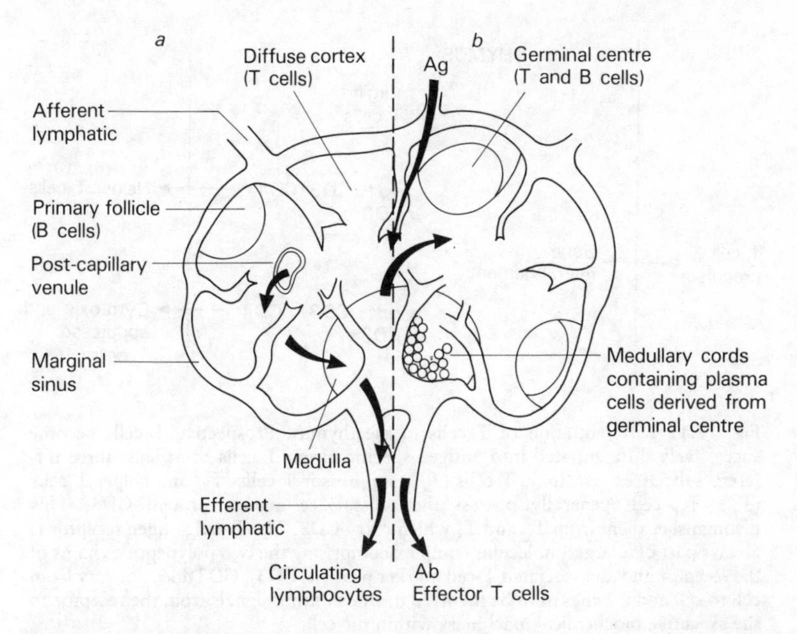

Fig. 12.10 Schematic diagram of lymph node: *a*, unstimulated; *b*, after antigen stimulation. (See text for explanation.)

with antibody production, come in several different varieties with different functions within the immune system. **Cytotoxic T cells** recognize and kill cells carrying a corresponding antigen by direct contact, **helper T cells** recognize antigen on antigen-presenting cells and then interact with B cells and other T cells to enable them to respond to antigen, and **suppressor T cells** are involved in suppressing immune responses and the general regulation of the immune system.

All T cells are small lymphocytes, superficially indistinguishable from each other and from B lymphocytes. The various types can be distinguished by their behaviour in experimental *in vitro* tests and by the typical 'marker' protein molecules they bear on the cell surface (often known as **T-cell differentiation antigens**).

The thymus largely acquires its lymphocyte population as it develops in the embryo and is also later replenished by cells that migrate from the bone marrow. Although it begins to atrophy from puberty onwards it apparently continues to produce lymphocytes throughout life. In the thymus, immature **thymocytes** proliferate, acquire antigen specificity and differentiate into the different types of T cell (Fig. 12.11). 'Virgin' T cells migrate from the thymus to the lymph nodes, spleen and other

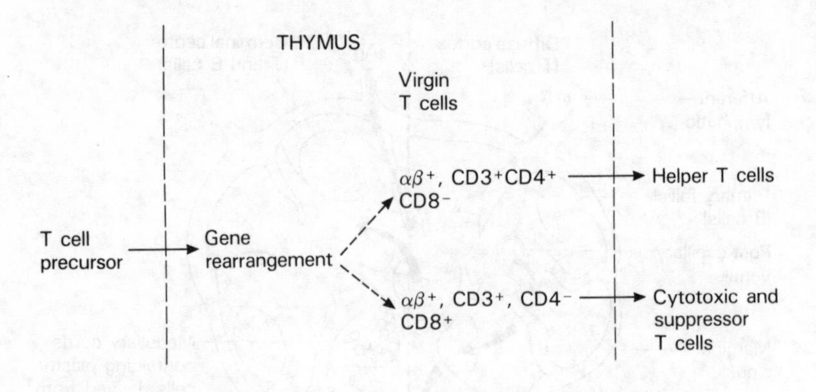

Fig. 12.11 Differentiation of T cells in the thymus. Prospective T cells become successively differentiated into antigen-specific virgin T cells of at least three different subclasses: cytotoxic T cells (T_C), suppressor T cells (T_S) and helper T cells (T_H). T_H cells generally possess the cell-surface marker protein CD4. This distinguishes them from T_C and T_S which carry CD8. The T-cell antigen receptor is always part of a larger molecular complex comprising the two polypeptide chains of the receptor and an associated T-cell marker protein (CD3). CD3 does not vary from cell to cell and is believed to be involved in transmitting signals from the receptor to the executive biochemical machinery within the cell.

lymphoid tissues where they mature and circulate in large numbers in the blood.

T cells as a separate class of lymphocytes were first distinguished in the 1960s, when the essential role of 'thymic' lymphocytes in immune responses was recognized. Large numbers of the lymphocytes that are continually being produced by the thymus never emerge (of some 10^8 cells produced daily only around 2×10^6 leave the thymus). The reason for this wastage is still not clear but may be linked to the rigorous selection process that T cells have to undergo in the thymus. Only those T cells whose receptors satisfy the criterion of being able to recognize antigen in conjunction with certain MHC antigens are allowed to develop further.

Helper T cells (T_H) are in many ways the front-line cells of the immune system. In general they seem to be the first antigen-specific cells to be activated by incoming antigen and in turn they stimulate the corresponding antigen-specific 'executive' cells of the immune system to carry out their functions – antibody production in the case of B cells, cell killing in the case of cytotoxic T cells, and suppression of the activity of T and B cells by suppressor T cells.

Cytotoxic T cells (T_C) are responsible for the antigen-specific recognition and killing of cells infected by viruses or other intracellular parasites and which are therefore displaying foreign antigens on their surface. They also attack tissues transplanted from an incompatible donor, and therefore displaying foreign histocompatibility antigens, and tumour cells displaying an abnormal cell surface. Cytotoxic T cells do not become cytotoxic until they have become activated by antigen. Antigen recognition stimulates T_C cells to proliferate and become responsive to signals from cooperating cells of the immune system that induce them to express their cytotoxic potential. Once activated, they range through the circulation seeking out and destroying cells bearing the appropriate antigen.

A single killer T cell can kill many cells without being itself destroyed. How they kill their target cells is still somewhat of a mystery. Cells rapidly disintegrate (lyse) on attack by killer T cells, which appear to make direct physical contact with them. One idea is that they produce molecules that in effect punch holes in the cell membrane and destroy its integrity. Alternatively, they may in some way induce various metabolic aberrations within the target cell that cause it to self-destruct.

THE IMMUNE RESPONSE

The immune response to any antigen represents a complex network of interactions between immune system cells, both antigen specific and non-specific. In most natural infections and especially those mediated by viruses and other intracellular parasites, the production of antibodies against circulating virus particles and antibody- and cell-mediated responses to parasite-infected cells develop in parallel, although one may be more effective than the other in combatting the infection, depending on the life style of the microorganism involved. This section follows the progress of a 'typical' immune response in terms of the cooperative cellular interactions that occur at each stage, from the initial encounter with antigen to the final mopping-up operations.

Antigen recognition

Processing and Presentation

Before most incoming antigens can trigger an immune response, they have to be 'processed' and presented to the immune system in a form it can recognize. The suggestion that antigens are presented to lymphocytes by macrophages and other non-lymphoid cells was first made in the 1930s and is now a well-established preliminary to both

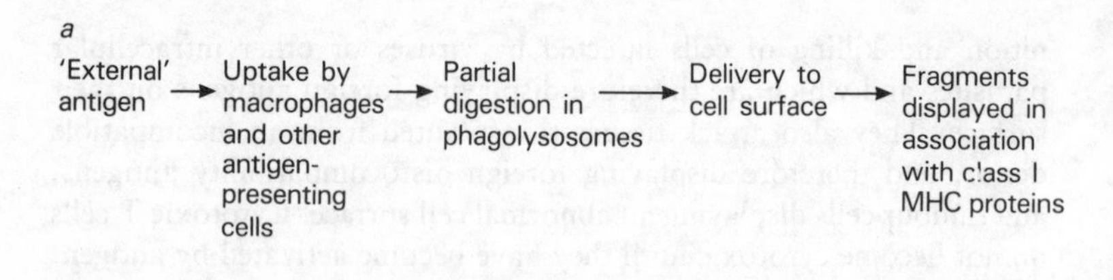

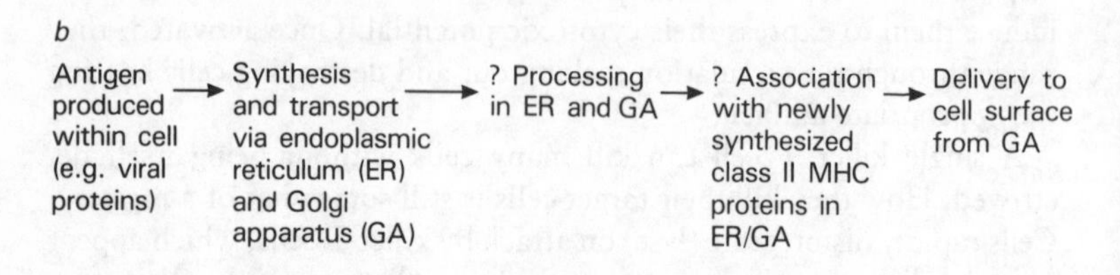

Fig. 12.12 Antigen processing. Possible routes by which antigen may be processed and reappear on the cell surface in association with MHC proteins.

antibody-producing and cell-mediated immune responses.

Antigens entering the body are trapped, not directly by lymphocytes, but by the phagocytic cells of the blood (**monocytes** and **polymorphonuclear leukocytes**) and the **macrophages** and macrophage-like cells of the **reticuloendothelial system** in lymphoid tissue, lungs, liver and kidneys which are derived from blood monocytes. Foreign particles (such as carbon) entering the bloodstream can be shown to be rapidly sequestered into these organs. This represents a preliminary scavenging by the body's non-specific defence system. Antigen that ends up in macrophages in lymphoid tissues then encounters the immune system.

Although not antigen specific in their actions, macrophages are not entirely undiscriminating. They carry surface glycoproteins that recognize in a relatively non-specific fashion the carbohydrate on some bacterial cell-surface components and are activated by some bacterial lipopolysaccharides. They also carry receptors specific for complement and for the Fc portions of antibodies, through which their later involvement in immune responses is mediated. Activated macrophages produce a variety of signal molecules (lymphokines) that have stimulatory effects on other immune system cells.

Macrophages taking up an antigen such as a virus particle or a bacterial cell by phagocytosis, digest and destroy most of it. However, some components escape complete digestion and are recycled, reappearing on the surface of the macrophage (Fig. 12.12). In this form they are recognized by the helper T cells of the immune system, which can only see a foreign antigen when it is complexed with the MHC molecules that the macrophage (like all cells) carries on its surface. Macrophages and other cells infected with viruses or intracellular parasites may also display processed antigen on their surface and thus act in a similar way as antigen-presenting cells (APCs) to the cellular arm of the immune system. Other antigen-presenting cells are the dendritic cells of lymph nodes and the macrophage-like Langerhans cells in the skin, and in some cases possibly B cells. In all cases, the antigen-presenting cell is thought to display processed fragments of the original antigen molecules. Processing is presumably needed to break down a large antigen into small pieces that can combine with an MHC molecule. A single type of MHC protein appears able to form at least a temporary complex with a wide range of small peptides.

Lymphocyte cooperation and role of the MHC

Associated Recognition and MHC Restriction

The interactions between T cells, B cells and accessory cells outlined below guide the immune response down the path most likely to lead to elimination of the challenging antigen.

Antigen specificity is obviously one key to restricting lymphocyte interactions to their appropriate target. The other constraint on lymphocyte cooperation, which ensures that a particular type of lymphocyte reacts only to antigen in a useful context, involves the major histocompatibility proteins (MHC proteins) on the cell surface.

Most T cells can only recognize antigen if it appears on cell surfaces in association with certain of these MHC proteins. In the first instance this prevents killer T cells, for example, interacting unproductively with free viral or bacterial antigens. A further constraint is also imposed. The identical MHC antigen must be present on both the T cell and the cell with which it interacts. Although one set of MHC antigens – the **class I antigens** (HLA-A, B and C in humans) – is ubiquitous, occurring on virtually all the cells of the body, the other – the **class II antigens** (HLA-D) – occur normally only on T cells, B cells and antigen-presenting cells such as macrophages, although they can appear on other cells during an immune response, possibly allowing these cells to function as antigen-presenting cells.

T cells with different functions recognize antigen in conjunction with *either* class I *or* class II MHC molecules. Cytotoxic T cells, which have to be able to react against any virus-infected cell for example, recognize antigen in combination with class I MHC antigens. Helper T cells, on the other hand, recognize antigen only in association with class II MHC antigens, which effectively restricts their interactions to other T cells, to B cells, and to antigen-presenting cells.

MHC restriction is 'learned' by T cells during their passage through the thymus. T cells recognize antigen in combination with the MHC type they encounter on the non-lymphoid cells of the thymus. Normally this is, of course, the same type as they bear themselves, but the crucial role of the thymus was uncovered by sophisticated experimental manipulation of the immune system by bone marrow transplantation between different MHC strains of mice. Bone marrow cells of one MHC type that had developed into mature T cells in the thymus of another MHC type, when tested in *in vitro* culture, could recognize antigen when presented to them in combination with the host thymus MHC type but not when it was in combination with their own MHC type.

T cells bearing antigen receptors that are complementary in some rather special way to the MHC antigens displayed on the surfaces of the thymus cells are positively selected by some means as yet unknown. Those that fit the MHC molecule completely are presumably selected against, as they would otherwise provoke autoimmune reactions against all body cells.

The way in which the T-cell receptor interacts with an antigen–MHC combination is now under intense investigation and will certainly throw light on which part of the T-cell receptor contacts the MHC molecule and which contacts antigen.

Helper T Cells and Antibody Production

Incoming viral and bacterial antigens are processed and presented to helper T cells in a form they can recognize. They become activated, and start to proliferate and produce various lymphokines (see below) that stimulate other immune system cells. Meanwhile, antigens also become attached directly to B cells bearing complementary antigen receptors (IgM) on their surface.

In some cases, antigen binding itself is sufficient to trigger the B cell to start dividing and differentiating into cells actively producing antibody. Such antigens (the lipopolysaccharides of bacterial cell walls are one example) are termed **T-cell independent antigens**. For many naturally encountered antigens, however, B cell activation requires the cooperation of activated helper T cells.

The details of the cooperation between B and T cells are still somewhat speculative, and it is not yet determined exactly what reacts with

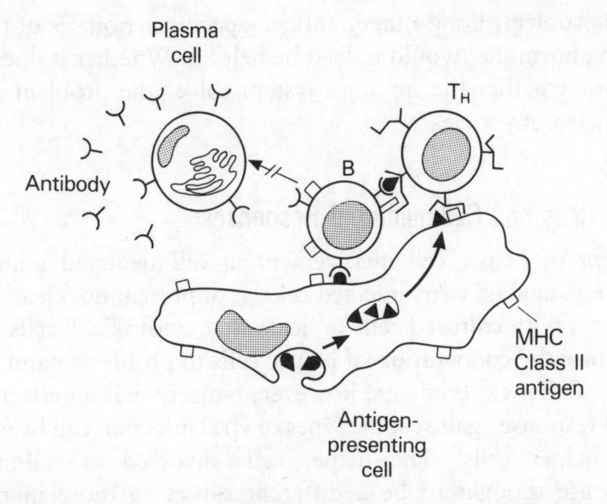

Fig. 12.13 Possible interactions between antigen-presenting cell, helper T cell and B cell.

what at the cellular and molecular level. Some evidence suggests a three-way interaction involving antigen-presenting cell, helper T cell and B cell (Fig. 12.13).

At the molecular level it seems likely that the T cell antigen receptor and the immunoglobulin receptors of the participating B cell recognize different antigenic determinants on the foreign antigen bridging the B and T cells. B cells may also internalize and process antigen bound to their receptors and present antigen fragments on their surface.

Stimulation of the B cell through its antigen receptors and/or through antigen on its surface in conjunction with the action of lymphokines produced by the T cell triggers internal changes that lead to proliferation and antibody production.

It may, however, not be necessary for the T cell always to make direct contact with its target B cell. Helper T cells may release antigen-specific **helper factors** that could act on their target cells at a distance. The nature of these factors and their relevance to the real-life situation is still controversial. They appear to be related in structure to the antigen-specific portion of the T cell's antigen receptor. They could therefore represent a specially synthesized, secreted form of the receptor, or they may be actual receptors shed from the T-cell surface.

Antigen-specific activation of target cells at long range could be useful to the immune system. The numbers of any particular antigen-specific T cell in the virgin lymphocyte population are very small indeed, as are the numbers of its antigen-specific target cell. Bringing them together to initiate an immune response presents a considerable logistical problem. Long-range, antigen-specific action, as of that of a circulating hormone, would indeed be helpful. Whether it does in fact happen, or whether the immune system solves the problem in some other way is not yet clear.

Helper T Cells and Cell-mediated Responses

The extent of helper cell involvement in cell-mediated immune responses (e.g. against virus-infected cells) is at present not clear. *In vitro* experiments with cultured cells indicate that cytotoxic T cells in some cases require the cooperation of helper cells to proliferate and become cytotoxic. Other experiments, however, indicate that an effective cell-mediated response against some types of viral infection can be mounted without helper cells. The helper cells involved in cell-mediated reactions are thought to be a different subset to those involved in antibody production.

In cases where helper cells seem to be involved they appear to produce lymphokines that act on cytotoxic T cells once the latter have themselves encountered antigen. Antigen contact induces the production of receptors for lymphokines emanating from the helper cells. Therefore, although the lymphokines themselves are non-specific, they

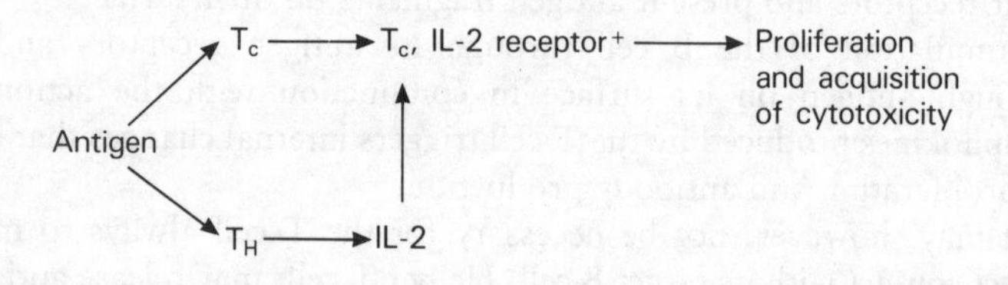

Fig. 12.14 Activation of T_C cells by interleukin 2 (IL-2).

activate only the appropriate antigen-specific cytotoxic cells (Fig. 12.14). The most important lymphokines involved in cytotoxic T-cell activation are thought to be the polypeptides interleukin 2 and gamma-interferon.

Antigen elimination

Opsonization and Phagocytosis

Antigens such as bacterial cells and virus particles complexed with antibody are efficiently cleared from the tissues by phagocytosis. Blood **monocytes** and their **macrophage** counterparts in tissues, and **polymorphonuclear leukocytes** are the chief 'professional' phagocytes in mammals. Microorganisms taken up by phagocytic cells are first killed and then isolated in intracellular **phagolysosomes** where they are digested by hydrolytic enzymes. Some bacteria have evolved ways of evading these lethal mechanisms and set up a chronic infection inside macrophages, making their clearance from the body more difficult. Phagocytes appear to attach to 'foreign' matter such as a bacterial cell surface through non-specific cell-surface glycoproteins.

Complexes of antigen with IgM and most IgG antibodies are more efficiently phagocytosed than antigen alone. Macrophages and polymorphonuclear leukocytes bear specific receptors for the Fc portion of IgM and IgG heavy chains. Antigens coated with these antibodies (**opsonization**) are preferentially bound to the phagocyte surface via these receptors, and receptor stimulation triggers a specific and highly efficient phagocytic response. Antibodies that can promote phagocytosis in this way are termed **opsonins**. Macrophages also bear receptors for one of the complement components (C3) and so also preferentially take up antigen/antibody/complement complexes.

Macrophages, polymorphonuclear leukocytes and eosinophils (a type of leukocyte) are also involved in the various types of inflammatory reactions that usually accompany an immune response (see below).

Complement Action

The presence of some types of bacteria and/or the binding of IgG and IgM antibodies to their corresponding antigens on cell surfaces activates a set of potentially destructive proteins – the **complement system**. The proteins of the complement system are chiefly proteolytic enzymes. They are always present as inactive precursors in fairly large amounts in the blood. Binding of IgG or IgM antibody to an antigen on the surface of a cell uncovers a binding site for the 'first' complement component on the Fc portion of the antibody heavy chain. When it binds it triggers a cascade of enzymatic reactions in which complement components are successively bound and activated by proteolytic cleavage – the process known as **complement fixation** (Fig. 12.15).

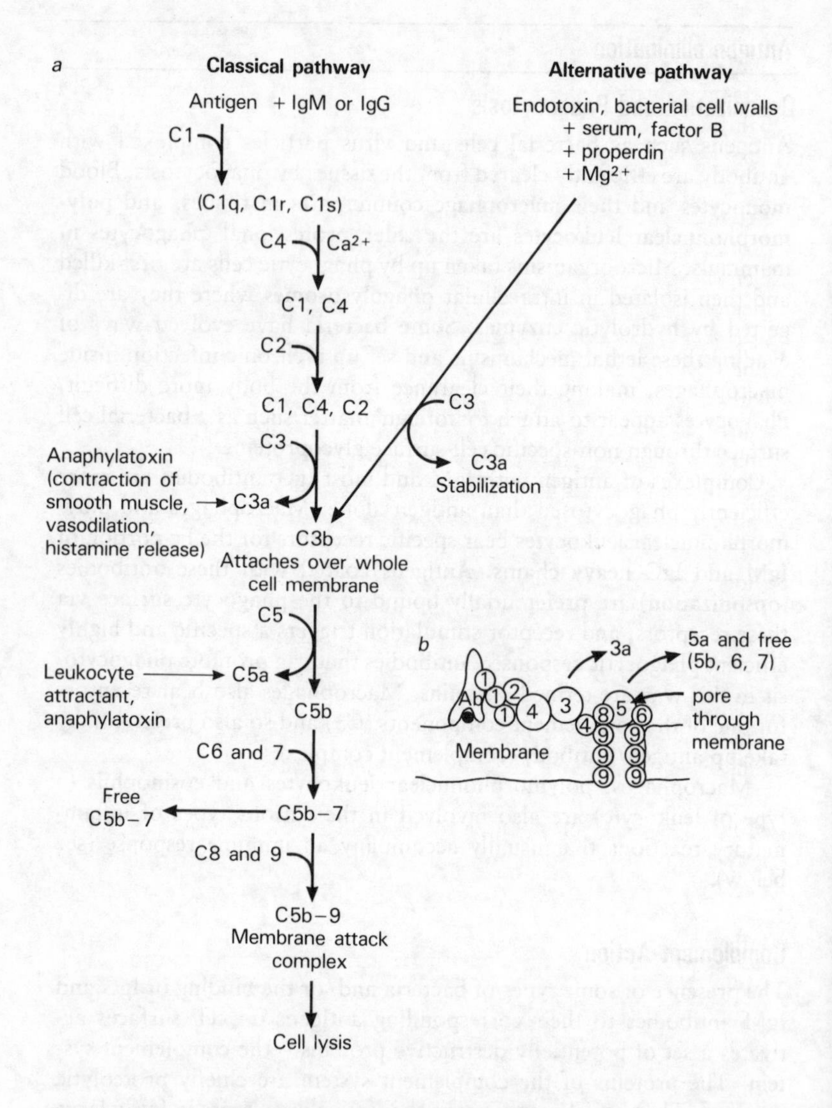

Fig. 12.15 The complement system. *a*, A much-simplified scheme of the two pathways by which the complement system (C1–C9) may be activated. The classical pathway is activated by antigen/antibody binding on the surface of a cell. The alternative pathway is independent of antigen and is triggered by interactions between various serum components and certain bacterial components. *b*, Schematic diagram of the successive binding of complement components to antibody localized on the surface of a bacterial cell, culminating in formation of a pore through the membrane.

Complement components are numbered 1 to 9 (some represent more than one protein). Some of the components and their proteolytic products have distinct physiological properties, all of which are directed to fighting infection by producing an acute inflammatory response (see next section).

The complement cascade also kills antibody-coated cells directly. Activated complement components bind to each other, forming a lethal protein complex on the bacterial cell membrane. When the last complement component – C9 – is bound, the complex forms a hole in the membrane destroying its integrity as an osmotic barrier. The contents of the cell leak out and it disintegrates (**complement-mediated lysis**). The complement pathway can be activated via an alternative route without the intervention of antibody by the lipopolysaccharides typically present on the surface of Gram-negative bacteria that cause intestinal disease. This important component of innate immunity provides an immediate local protective response to intestinal infection before antibody is formed. But if these lipopolysaccharide **endotoxins** get into the bloodstream, they trigger a fatal, systemic, complement-mediated shock reaction caused by massive vasodilation.

Circulating antigen–antibody complexes also bind complement components 1–3.

Regulation of the Immune Response

Antibody-producing plasma cells and activated effector T cells are not long-lived (a matter of hours and days). Once antigen is removed from the system by the mechanisms outlined above and by cell-mediated lysis of infected cells, the immune response dies down. Immune responses may also be regulated in more subtle ways, by suppressive interactions between T cells (see below, Suppressor T cells).

Inflammation

Many immune responses are accompanied by inflammatory reactions. Acute inflammation is a largely non-specific response to tissue damage by microorganisms, which attempts to contain and destroy the invaders. This may be triggered by complement activation, bacterial products or chemicals released by injured cells. It may also be a sequel to an immune response and be induced by local antibody/antigen reactions and consequent complement fixation.

Complement activation or tissue damage releases local chemical mediators that dilate blood capillaries. Their walls become leaky, allowing serum proteins to diffuse into the tissue fluid and blood cells to squeeze out between the cells of the capillary walls. Polymorphonuclear leukocytes (PMNs) and eosinophils in particular are attracted to the site in large numbers by some of the by-products of complement activation and tissue injury. PMNs attempt to ingest the agents that are causing the damage. If they prove indigestible, hydrolytic enzymes and inflammatory lipid derivatives (e.g. prostaglandins, thromboxanes and leukotrienes) released by the injured leukocyte exacerbate the inflammatory reactions and attract more PMNs and eosinophils.

The influx of fluid, cells and plasma proteins causes swelling. Swollen tissue presses on nerve endings, which are also stimulated by some of the chemicals released, resulting in tenderness and pain. If the leukocytes clear the agents responsible the inflammation subsides and tissues are repaired. If not, the site of inflammation develops into a purulent abscess filled with dead and active PMNs, tissue debris and bacteria, and walled off by a ring of actively phagocytic cells and clotted blood.

If a local infection persists a chronic antigen-specific inflammatory response may be initiated. T cells of the helper class infiltrate the tissues. The T cells are stimulated by cell-bound antigen to release various lymphokines that attract macrophages in large numbers (**macrophage chemotactic factor, macrophage migration inhibitory factor, macrophage aggregation factor**), damage cells other than lymphocytes (**lymphotoxin**) and stimulate the division of other lymphocytes (e.g. various **interleukins**).

Hypersensitivity

Immune reactions, complement activation and inflammation are the body's way of protecting itself, but when provoked to excess or in inappropriate circumstances they can have pathological consequences as the result of tissue damage. In certain circumstances an encounter with antigen provokes a **hypersensitive immune response** that can range from a local reaction (the weeping eyes and runny nose of hay-fever) to a systemic and fatal anaphylactic shock caused by massive activation of the complement system and consequent vasodilation.

There are many types of hypersensitive reactions, the immune system components involved and the physiological outcome depending on the type of antigen, the route it enters the body and the individual's previous history of exposure. Most hypersensitive reactions occur on second encounter with an antigen or after prolonged exposure to it. They fall into two main classes, those mediated by antibody – **immediate (or antibody-mediated) hypersensitivity** – and those that involve largely cell-mediated reactions and cause **delayed hypersensitivity** (see Table

12.2). Two examples are given below (see table for other types of hypersensitivity).

Table 12.2 Types of hypersensitivity reaction

Class	*Name*	*Cells and effectors involved*
I	anaphylactic	IgE + mast cells or basophils. Release of histamine, leukotrienes, etc. Vasodilation, smooth muscle contraction, local inflammatory responses.
II	cytotoxic	IgM or IgG + complement. Antibodies directed against a cell-surface component bind and activate complement leading to complement-mediated cell lysis.
III	immune complex	IgM or IgG in immune complexes with antigen and complement. If not cleared by phagocytosis they can settle in blood vessels and tissues where complement activation, platelet aggregation and release of vasoactive amines result in tissue damage.
IV	cell-mediated	T cells (helper type and cytotoxic), macrophages. Various lymphokines released by the activated T cells.

Adapted from L. E. Hood *et al. Immunology*, 2nd edn. Benjamin/Cummings Publishing Company, Menlo Park, CA, 1984.

Allergy and IgE

The familiar **allergies** in which unpleasant symptoms appear within hours of exposure to a particular substance – pollen, house dust, certain foods, etc. – are examples of immediate hypersensitivity reactions.

Around 10 per cent of the population make an allergic response to one or more of these common environmental **allergens**. Why some substances act as allergens and why they produce an allergy in one person and not in another is still far from clear, but is believed to be due at least in part to genetic predisposition, and the history of past encounters with the antigen. Allergic reactions occur at the site of contact with antigen. In this type of hypersensitive reaction sensitized individuals appear to make IgE antibodies rather than circulating IgG or IgM antibodies.

IgE has a particular type of heavy-chain constant region that attaches via a specific receptor to various cells, especially **mast cells** in tissues or **basophil leukocytes** in the blood. Mast cells and basophils contain numerous granules containing several biologically active compounds – chiefly histamine, heparin, and leukotrienes. These have multifarious activities including dilation of blood vessels and stimulation of smooth muscle contraction. Antigen entering the body is not mopped up by circulating IgG but binds to the tissue-fixed IgE.

Antigen-binding stimulates the mast cells to release granules and their contents into the tissue fluids or bloodstream, depending on the site at which antigen is encountered. Contraction of the smooth muscle of the bronchial tubes, for example, causes the asthma attacks that sometimes accompany hay-fever and other allergies involving the respiratory tract. Preventive treatment for common allergies usually consists of giving small repeated doses of antigen to provoke a greater IgG response. When antigen is then encountered in any quantity the IgG blocks it before it can reach the tissues.

The 'natural' role of IgE antibodies seems to be to respond to certain parasitic infections – for example, by schistosomes – where chemical mediators released by IgE-bound mast cells attract eosinophils to the site, which release enzymes that kill the parasite. The pharmacological agents released by mast cells, by causing smooth muscle contraction, may also possibly help to expel the parasite from the organs in which it has settled.

Delayed Hypersensitivity

Delayed hypersensitivity involves T cells. The weals that arise at the test site in response to the tuberculin test in people already sensitized to tuberculin are a classic example of delayed hypersensitivity. Memory helper T cells specific for tuberculin react with incoming antigen to contain it at its site of entry by initiating an inflammatory response (see above). On encountering antigen carried on antigen-presenting Langerhans cells of the skin, they secrete lymphokines that attract macrophages to the site, producing after a day or two the characteristic swollen weals of a delayed skin hypersensitivity response. A similar but more destructive reaction occurs on second contact with a plant such as poison ivy where an exceedingly unpleasant **contact sensitivity** response is initiated by helper T cells and cytotoxic T cells to sensitizing compounds in the plant that become attached to skin cells.

LYMPHOKINES: THE CHEMICAL MESSENGERS OF THE IMMUNE SYSTEM

The cooperative action of immune system cells during an immune response depends to a large extent on a variety of chemical signals, known generally as **lymphokines**, which are produced and released by cells activated by antigen, and act either on the producing cell itself or on other cells of the immune system. Lymphokines are produced chiefly by T cells and by macrophages, and to a lesser extent by B cells.

Their most general effect is to stimulate proliferation of immune system cells, amplifying the initial antigen-specific response. Lymphokines are not themselves antigen specific and will stimulate any susceptible cell. Lymphokine action also provides a link between antigen-specific immune responses and the body's general non-specific defence mechanisms, such as inflammation and increased phagocytic activity.

Lymphokines are polypeptides and proteins, and are present in only minute amounts in blood and tissue, making them difficult to isolate and purify by standard techniques. Most have been isolated and produced in sufficient quantity for analysis only over the past decade, as a result of the introduction of recombinant DNA technology. Previously, they had been largely identified by their activity in various biological assays, which led to a confusing proliferation of names for the same protein. The nomenclature used below is that in current use in the scientific literature on the subject.

There is considerable interest in the pharmaceutical industry in some of these factors and their mode of action, with the aim of producing a new generation of antiviral and antitumour agents that might work largely through stimulating the body's natural defences.

The **interleukins** are a subset of lymphokines produced by leukocytes, chiefly lymphocytes and macrophages and which stimulate the proliferation and maturation of other lymphocytes, leukocytes and and their precursors.

Some important lymphokines and their roles in immune responses are outlined below. The list is by no means complete, and it is also becoming apparent that some lymphokines have effects on cells outside the immune system. The body produces an enormous number of biologically active polypeptides loosely termed 'growth factors', because they were first recognized by their effects on cell proliferation. Lymphokines are part of a functional network of polypeptide activity, whose complexity is only beginning to be perceived.

Lymphokines and interleukins

1. Polypeptide factors produced by activated macrophages

Tumour necrosis factor (TNF, TNF-α; see also lymphotoxin below) (recently identified with **cachectin**)

TNF is produced by activated macrophages and other cells and has a very wide range of apparent biological activities. Its main function in the immune system appears to be to stimulate the macrophages that produce it to produce more of another lymphokine – interleukin 1 (see below). It is part of a complex regulatory circuit that links macrophages, T cells and B cells during an immune response and in triggering inflammation (Fig. 12.16).

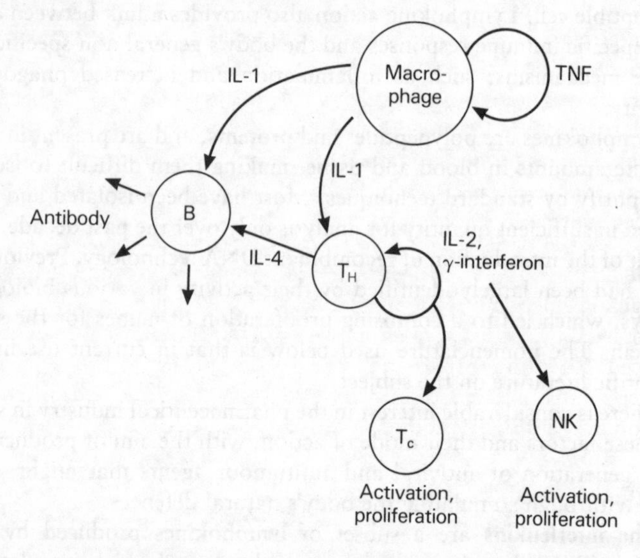

Fig. 12.16 Some lymphokine circuits. IL, interleukin; TNF, tumour necrosis factor; NK, natural killer cells; T_C, cytotoxic T cells; T_H, helper T cells. See above, Lymphokines, for the actions of these factors.

The present pharmaceutical interest in TNF stems from its other demonstrated effects. TNF was first discovered as a 'factor' that can produce haemorrhagic necrosis in tumours in experimental animals by attacking the walls of the blood capillaries supplying them, and that also attacks cancer cells directly both *in vivo* and *in vitro* to a greater extent than normal cells. The reason for its preferential cytotoxic action on cancer cells is so far unknown. TNF has also recently been shown to be capable of inducing the growth of new blood capillaries (angiogenesis). This could give it an important role in wound repair and inflammation but also implies paradoxically, that it could also help provoke the outgrowth of the new capillaries that invade tumours and help nourish them.

The TNF gene has been cloned and TNF is now produced semi-commercially by several biotechnology companies. Antitumour activity in human patients has not yet been established. A variety of other biological effects of TNF have been demonstrated experimentally that may be relevant to the regulation of immune responses *in vivo*. It also has antiviral activity, but since it can also induce ß-interferon production, it is not clear whether TNF acts directly against viruses or indirectly via interferon. Many of its activities are considerably enhanced when it is administered in combination with γ-interferon (see below) which is itself a lymphokine with a wide variety of effects.

TNF was recently identified with a protein factor - **cachectin** – isolated quite independently, and which is involved in the wasting (cachexia) often seen in patients with cancer. Cachectin apparently exerts its effects by inhibiting an enzyme involved in lipid accumulation in fatty tissue.

Interleukin 1 (IL-1) (formerly: endogenous pyrogen, leukocyte endogenous mediator, lymphocyte-activating factor)

Produced by activated macrophages and other cells. It is involved in T-cell activation, stimulating the production by T cells of other interleukins, and in stimulating B-cell proliferation and differentiation into antibody-producing cells.

2. **Lymphokines produced by T cells**

Interleukin 2 (IL-2)

Produced by T cells activated by interleukin 1. It is a key T cell proliferation factor *in vivo*, acting in conjunction with antigen stimulation (see Fig. 12.16). In general, T cells only start synthesizing receptors for interleukin 2 after they have been stimulated by antigen. IL-2 therefore only triggers proliferation of cells specific for the challenging antigen.

Interleukin-3 (IL-3)

Produced by T cells activated by interleukin 1. It acts on bone marrow, apparently inducing a general proliferation of blood cell precursors.

Interleukin 4 (IL-4) (formerly: B-cell growth factor 1 (BCGF-1), BSF-1)
Produced by activated T cells. It acts on B cells, inducing proliferation and differentiation to antibody-producing cells. Various effects have been demonstrated experimentally that depend on the state of differentiation of the target B cell: it promotes proliferation of B cells in response to stimulation of their immunoglobulin receptors and stimulates secretion of IgG in plasma cells. It also stimulates resting lymphocytes to increase expression of class II MHC antigens (those involved in MHC-restricted responses between helper T cells and B cells).

Lymphotoxin (TNF-ß)
A cytotoxin produced by activated T cells that attacks many types of cells but spares lymphocytes. It is also active against cancer cells. Although it is different from TNF (see above) it binds to the same receptors on cell surfaces and many of its biological effects mimic those of TNF.

γ-interferon (IFN-γ) (see also Interferons)
As well as possessing antiviral activity, γ interferon has a wide variety of effects on cells of the immune system. It is produced by T cells stimulated by interleukin 1. It induces natural killer (NK) cells (see below, Immunology and cancer) to proliferate and also stimulates further production of tumour necrosis factor (TNF) (see above) from macrophages. It enhances the effects of TNF, possibly by increasing the number of receptors for TNF on cell surfaces. It also stimulates immature cytotoxic T cells to acquire cytotoxicity.

3. **Lymphokines produced by activated helper T cells at site of antigen stimulation** (**e.g. lymph node, wound, localized infection, etc.**) **and which are involved in the general stimulation of macrophage activity**

Macrophage chemotactic factor
Attracts macrophages to the site of antigen entry.

Macrophage migration inhibition factor (**MIF**)
Activates macrophages, making them more efficient at phagocytosing and digesting microorganisms. It inhibits macrophage movement, presumably preventing their departure from the site of an inflammatory response.

4. Interleukin produced by B cells

Interleukin B (IL-B) (formerly BEF, B-cell-derived enhancing factor, IL-B4)

Produced by unstimulated B cells. It is unlike other lymphokines in both molecular structure and biological activity. It enhances antigen-driven B-cell responses *in vitro*. Its role *in vivo* is not yet known but it has been shown to inhibit the action of suppressor T lymphocytes.

5. Others

Interleukin 6 (IL-6, interferon-ß2, BSFII)

Synthesized by fibroblasts and various tumour cells. It increases immunoglobulin synthesis and secretion from B lymphocytes (amongst other activities; see also Interferons)

TGF-ß

Suppresses T cell activity (see also Table 9.9)

Interferons

The **interferons** are small proteins produced by many animal cells and that have a wide range of biological effects. 'Interferon' was first discovered as an antiviral protein produced and released by animal cells infected with certain viruses. It exerts its antiviral effects not by direct attack on the virus itself, but by acting on uninfected cells to render them resistant to virus attack. Interferon binds to the plasma membrane and this stimulates increased production of three enzymes – an oligonucleotide synthetase, an endonuclease and a protein kinase. If the cell is subsequently attacked by a virus these enzymes are activated and their concerted action effectively shuts down protein synthesis in the cell, which in turn blocks virus multiplication which is dependent on cellular protein synthesis. Enzyme activation is triggered by the viral nucleic acid, double-stranded RNAs being the most effective. The protein kinase inactivates one of the protein synthesis initiation factors by phosphorylating it; the oligonucleotide synthetase synthesizes an adenylate compound which in turn activates an endonuclease that degrades messenger RNA.

Interferons and synthetic interferon inducers are being extensively investigated as possible antiviral agents possessing none of the side-effects associated with many currently used antiviral drugs, which tend to attack both infected and uninfected cells indiscriminately. Interferon has proved effective in trials against ocular herpes, rhinovirus infections (common cold), and in ameliorating the severe shingles (herpes zoster) that occurs after immunosuppressant cancer treatments. It has, however, not proved to be as effective or as universal an antiviral agent as previously hoped. In particular, there are problems of delivering a sufficient dose to the target cells.

Until the end of the 1970s interferon research was hampered by the fact that supplies were very limited and not highly purified. Only human interferon is effective in humans, and for many years it could be extracted in any amount only from the white blood cells that are routinely discarded from blood destined for transfusion.

With the advent of recombinant DNA technology, the problems of interferon supply and purity have been solved. A human interferon was one of the first 'recombinant' proteins to be produced in bacteria, and pure human interferons are now being produced commercially from genetically-engineered recombinant cells. There are three main types which can be distinguished by their structure and by their biological activities – α-interferon (formerly called leukocyte interferon), ß-interferon (formerly called fibroblast interferon) and γ-interferon (immune interferon). γ-interferon enhances the action of the other interferons by ten times or more.

Although interferons were first discovered as a result of their antiviral effects, they are also produced in small amounts by many types of uninfected cell. Some interferons (γ and ß2 in particular) are important lymphokines, regulating the activity of the immune system and stimulating other defensive mechanisms. Interferon also possesses antitumour activity, although interferon therapy has so far proved only sporadically and partially successful.

IMMUNOLOGY AND CANCER

The role of the immune system in preventing and fighting cancer has always been the subject of much debate. The idea that cancerous cells are continually arising and that most are then eliminated by natural *immunosurveillance* mechanisms was first proposed early this century and has remained more-or-less fashionable ever since. Much work was influenced by the theory, and the need for immunosurveillance against cancer cells has been proposed as a possible driving force behind the evolution of a cell-mediated immune response.

Cancer cells are indeed often altered in ways that make them antigenic in their own host and they provoke both cell-mediated and antibody responses. Some successful cancer cells have lost the cell-surface MHC glycoproteins that are crucial for immune responses to be mounted, and might therefore have escaped the attentions of their own immune system. But, intuitively attractive as the theory of immunosurveillance may be, evidence against it as a universal mechanism for preventing cancer has also steadily accumulated over the past 20 years.

There are now many people who have undergone long-term immunosuppression to prevent transplant rejection, and naturally immunodeficient individuals (and animals) have been enabled to live longer by improved treatment. People who are immunosuppressed as a result of disease, congenital deficiency or medical treatment are indeed at higher risk than the general population of developing cancer – but only of certain types. The cancers that typically arise are of two main types. The incidence of virus-induced cancers increases – which can be explained by the lack of normal cell-mediated immunity against virus-infected cells in general. There is also an increase in cancers of lymphoid origin, possibly reflecting the consequences of disturbance to the immune system itself, rather than a breakdown in immunosurveillance.

The general view at present is that a fully cancerous cell may not be so easily generated as was previously thought. Work on oncogenes suggests that only mutations in a relatively small number of genes out of the total complement will cause a cell to become malignant, and that several mutations or other changes, occurring usually over many years in humans, are required for full malignancy. Complex networks of polypeptide factors that regulate the growth and division of cells are now beginning to be uncovered and may have an important role in suppressing the growth of potential cancer cells *in vivo*. Ways of using and enhancing the immune system to fight cancer are nevertheless still provoking much interest, although attention is at present focused on the newly-discovered lymphokines and non-specific killer cells rather than more 'conventional' specific immune responses.

Various cells other than cytotoxic T lymphocytes kill altered cells by direct contact. These include killer (K) cells, natural killer (NK) cells, and natural cytotoxic (NC) cells. In all cases cell killing is not antigen-specific. They all probably play a part in the elimination of virus-infected cells and are the subject of great interest at present because some also appear to have a preference for attacking cancer cells.

Natural killer (NK) cells are large granular lymphocytes related to T cells. They are cytotoxic, and form part of the body's non-specific defences, attacking and destroying a wide range of 'foreign' cells (cf. cytotoxic T cells which are antigen specific in their action). They are particularly active against virus-induced leukaemia cells and some other types of tumour cell. NK cells proliferate during immune responses, the chief stimulus being γ-interferon. This may be one of the reasons for interferon's antitumour activity. NK cells are thought to produce proteins that breach the target cell membrane by plugging into it and forming a pore, destroying its integrity and leading to lysis. Two such proteins have been identified – **perforin** and **cytolysin**. Why they preferentially attack cancer cells is so far unknown. **Natural cytotoxic (NC) cells** are very similar to NK cells in appearance and properties, but differ slightly in the cell-surface marker proteins they bear.

Killer (K) cells are related to macrophages but are non-phagocytic. They bear receptors (Fc receptors) for heavy chain constant regions of antibody molecules, and their killer activity is dependent on the presence of antibodies (the specificity of the antibody is immaterial).

Several of the lymphokines produced by activated immune system cells also preferentially attack or suppress cultured cancer cells *in vitro*. Attempts to use them directly as therapeutic agents have been largely unsuccessful. But killer cells preactivated by lymphokines are now undergoing clinical trials in the USA as a possible cancer therapy (**LAK therapy**).

Another immunological approach is to target drugs selectively to cancer cells by attaching them to **monoclonal antibodies** (see below, Applications) specific for cell-surface molecules present on the cancer cells but not on normal cells. If this approach is successful, it would provide a means of giving much larger doses of drugs than at present possible, killing cancer cells more efficiently without damaging other cells of the body.

CONTROL OF IMMUNE REACTIVITY

Self-tolerance

The immune system does not automatically react to every potential antigen presented to it. The prime natural example is the body's tolerance of its own antigens – **self-tolerance**. A mother also does not mount an antigenic response against the foetus she carries.

Tissues from any individual contain a host of macromolecules, many of which readily provoke an immune response if experimentally injected into another species or even into an unrelated individual of the same species. The immune system is constantly generating potentially self-reactive lymphocytes (see Generation of Diversity). To avoid autoimmune reactions self-reactive lymphocytes must be destroyed or permanently inhibited, presumably at their first encounter with antigen. In particular self-reactive T cells, which are largely responsible for initiating immune responses, must be eliminated. In the absence of their corresponding helper T cells, self-reactive B cells pose no threat; indeed, mature, potentially self-reactive B cells have been found in the adult immune system.

Self-tolerance is developed during embryonic and foetal development. It has been known for many years that any antigen that is present continually during foetal development will not provoke an immune response in later life. The immune system also takes some time to mature after birth and tolerance is much easier to induce in the immediate neonatal period.

Because self-reactive cells are being generated throughout adult life, there must be some active mechanism maintaining the tolerance patterns laid down during the immune system's development. How this is achieved is still far from clear. Part of the answer may lie in a class of T cells that suppress, rather than induce, immune responses to their cognate antigen. Long-lived memory suppressor cells generated at the first encounter with self antigens may be one way of ensuring that self-reactive helper T cells that arise are immediately inhibited. Why suppressor cells should be generated so efficiently to the body's own proteins and not to incoming antigens on pathogenic microoganisms is still a mystery. Clues may come from diseases in which suppressor T cells are routinely generated against certain antigenic determinants on pathogens and are thought to be responsible for the body's non-response to these antigens.

One important and ubiquitous set of self antigens are the MHC antigens (see below, The major histocompatibility antigens). These are being continually presented to the immune system and represent a rather special case of self-tolerance. T cells do not react against self MHC antigens on their own, but during their time in the thymus they 'learn' to recognize and react to a combination of self MHC protein and foreign antigen. The non-lymphoid cells of the thymus, the epithelial and dendritic cells, bear self MHC antigens (as do virtually all body cells). T cells with antigen receptors reactive to MHC antigens alone are presumed to be destroyed on first contact with these antigens in the thymus. The other side of the coin is the positive selection (again, probably as a result of interaction with the MHC-bearing thymus tissue) of developing T cells that bear antigen receptors capable of recognizing antigen *and* self MHC molecules together.

Genetically determined unresponsiveness

When the immune system encounters an antigen it has to choose whether to initiate an immune response or a suppressive response or perhaps, simply ignore it. One reason for ignoring it is a genetically determined inability to 'see' the antigen at all. This is due to possession of a type of MHC protein with which that antigenic determinant cannot combine; it is therefore not recognized by the T cells that initiate most immune responses. Since most naturally occurring antigens have many different antigenic determinants, in practice at least one will usually be recognizable by any given set of MHC proteins and so people are able to mount efficient immune responses to the pathogens they normally encounter.

Induced tolerance

In adults, specific temporary tolerance to certain antigens may be induced if they are administered by a particular route and/or at a certain dose. Not only do they fail to provoke an immune response when first administered, but for some time afterwards the animal (or person) is specifically unresponsive to that antigen, although still able to mount a perfectly good response to other antigens. This implies that the antigen has encountered the immune system, but has in some way led to the inactivation or suppression of the appropriate T cells and/or B cells. Avoiding tolerance is important in immunization procedures, because administration of certain sorts of vaccines in too low (or too high) a dose, or by an inappropriate route, could have exactly the opposite effect to that intended.

The mechanisms of tolerance induction are still largely unknown. Both T cells and B cells may be inactivated, depending on the particular combination of antigen, dose and route of entry. In some cases, suppressor T cells may be generated; in others, there may be a massive blockade of antigen receptors on B cells by excess antigen that paralyses antibody production.

Suppressor T cells

As well as being able to activate cells in response to antigen, the immune system can also prevent their activation. One class of T cells –

suppressor T cells – suppresses immune responses, in an antigen-specific manner, probably by suppressing the activation of the appropriate helper T cells. Active suppression (as opposed to the passive running down of an immune response as antigen is cleared from the system) is believed to operate towards the end of an immune response. Suppressor T cells have also been implicated in the unresponsiveness that occurs to some bacterial and viral antigens in chronic diseases, such as leprosy, where the patient becomes unable to mount an immune response against the pathogen. They may also be involved in maintaining the body's tolerance towards its own antigens.

Less is known about suppressor T cells and their mode of action than about helper or cytotoxic T cells. Their existence was inferred from the ability of cultured lymphocytes taken from an immunized animal to suppress *in vitro* antibody production or cell-mediated immune responses to the original immunizing antigen. Suppressor T cells have since been isolated and can now be cultured as pure cell lines. Some suppressor T cells are known to carry MHC-restricted antigen receptors of the type found on helper and cytotoxic T cells but their co-operative interactions with other cells of the immune system are still highly speculative.

Suppressor T cells are thought to exert their effects chiefly by inhibiting the appropriate helper cells. One general view is that some suppressors may be specific not for the incoming antigen itself but for idiotypes – antigenic determinants borne on T-cell receptors, antibodies and B-cell surface immunoglobulins (see below). They can presumably act at short-range, by cell contact through their specific receptors, but they also release soluble antigen-specific molecules – **suppressor factors** – that may represent the antigen-specific portion of their receptors.

How suppressor T cells actually prevent or turn off the activity of the cells with which they interact is not known.

Contrasuppressor Cells

Yet another functional class of T cells has been discovered recently – **contrasuppressor cells** – cells that suppress suppressors. Their effect is to neutralize the inhibitory effect of suppressor T cells on helper T cells. One of their roles in the immune system may be to allow local responses to antigens in cases where the overall immune response to that antigen is suppressed. This type of response occurs, for example, against some antigens that provoke localized reactions in the gut and oral lymphoid tissue when taken in by mouth, but do not trigger a generalized immune response. This may be the body's way of preventing massive allergic responses to common antigens in food, while still retaining the active defences that are needed against invasion by disease-causing micro-organisms across superficial tissue surfaces (lining of gut, skin, etc.).

Anti-idiotypic networks

The network theory of Niels Jerne, first put forward in the early 1970s, proposes a novel way of thinking about the immune system, and although difficult to test and prove, is influential in immunological research today.

Network theory essentially perceives the pool of antigen-specific cells of the immune system not as totally independent entities, which act out their parts in an immune response solely on the grounds of their specificity for the stimulating antigen, but as part of an elaborate network. The links in this network are forged by recognition by lymphocytes of particular features termed **idiotypic determinants** on antibodies, immunoglobulin receptors and T-cell receptors (Fig. 12.17).

The basic observation behind network theory is that an antibody or T-cell receptor bears features that are recognized as 'foreign' by its own immune system. Each antigen-binding site in particular presents an entirely new molecular configuration to the immune system, one it has not learnt to tolerate during development.

In the total pool of antigen-specific T cells will inevitably be some that recognize these idiotypic determinants and can react to them, either stimulating or suppressing the proliferation of the cells that bear them. These anti-idiotypic cells will in turn be recognized by other cells and so on, forming a network.

Anti-idiotypic antibodies and anti-idiotypic T cells have indeed been found. One of the predicted consequences of the network is that when amounts of a particular antibody (or T-cell receptor as the case may be) increase as a result of an externally provoked immune response, the idiotypic determinants on those antibodies provoke the formation of anti-idiotypic antibodies. The antibodies formed during the first wave of an immune response contain a 'mirror image' or *anti-image* of the original antigen, fragmented between their various antigen-binding sites (Fig. 12.18). These antibodies trigger a set of anti-idiotypic antibodies, which because of the connection between idiotype and antigen-binding site, contain within their binding sites a 'true' image of the original antigen, what Jerne calls the **internal image** of the antigen. This wave of antibodies cannot react with the original antigen. However, they provoke a further wave of anti-idiotypic 'mirror image' antibodies that can react with the original stimulating antigen (pathway *a* in Fig.

12.18). Oscillations in the amount of antigen-specific antibody late in an immune response do indeed occur, and together with the demonstrated presence of various types of anti-idiotypic antibodies predicted by the hypothesis, have led to a general acceptance of at least this aspect of network theory.

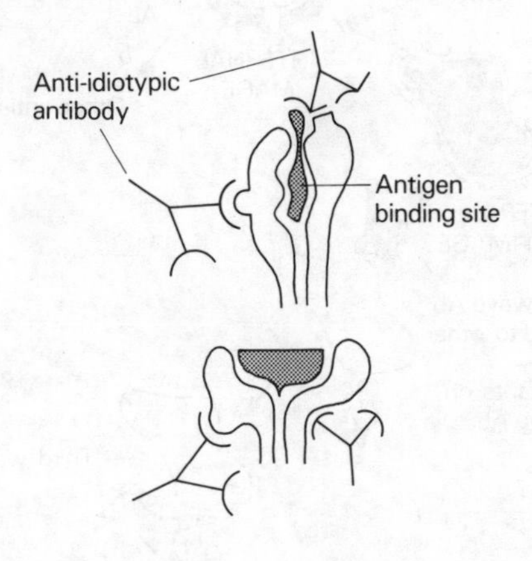

Fig. 12.17 Idiotypic determinants. Individual determinants are called idiotopes, and collectively they form the unique idiotype or signature of that particular antibody. Most, but not all, idiotopes (idiotypic determinants) are found on or around the antigen-binding site. Some (private determinants) are unique to that particular antibody, some (public determinants) are common to a range of different antibodies.

As well as moving along a direct antigen-specific path, anti-idiotypic interactions can also spread outwards to involve cells not remotely connected to the original antigen. Idiotypic determinants other than those associated with the antigen-binding site provoke formation of antibodies unrelated to the original antigen (pathways *b* in Fig. 12.18). How these proposed interactions are prevented from involving the entire immune system is one problem for network theory. The answer may lie with suppressor T cells. Anti-idiotypic suppressor T cells do exist and can suppress the activation of cognate T cells.

Whether the network is simply a consequence of inevitable idiotype–anti-idiotype interactions or whether it has a major role in the development and regulation of immune responses has still to be proven.

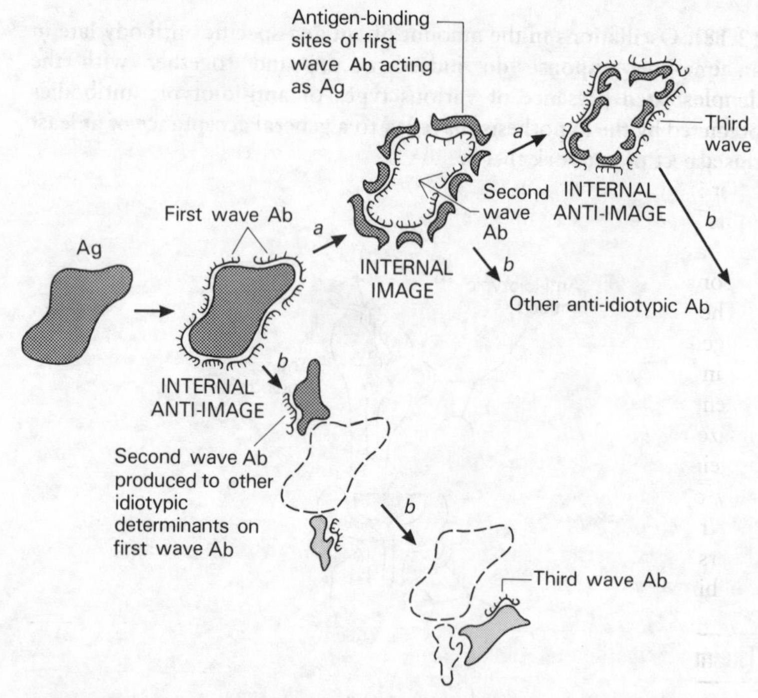

Fig. 12.18 The image and anti-image of an antigen. After Fig. 10-6 in L. E. Hood *et al. Immunology*, Benjamin/Cummings, CA, 1984.

THE MAJOR HISTOCOMPATIBILITY ANTIGENS

(See also pp. 388 and 405)

Tissues transplanted from one individual to a genetically different individual of the same species (an **allograft**) (or of another species, a **xenograft**) are almost always rejected, whereas tissue transplanted between genetically identical individuals, or within the same individual (a **homograft**) invariably heals in successfully. Rejection is due to a massive and rapid cell-mediated immune response directed against the 'foreign' antigens the recipient's immune system perceives on the surfaces of the grafted cells. The cell-surface proteins that provoke rejection are called **transplantation** or **histocompatibility antigens** (*histo* – tissue). The most important are those specified by the genes of the **major histocompatibility complex** (MHC), which lead to rapid rejection of the graft. Each individual possesses a characteristic set of these **MHC antigens** (called the **HLA antigens** in humans) on almost all the cells of their body. This constitutes their **tissue type.**

There are a large number of alternative forms of each MHC antigen in the population as a whole (see below, The major histocompatibility complex). Closely related individuals have some MHC antigens in common as a result of shared inheritance, but two unrelated individuals picked at random from a human or wild mammal population usually differ from each other in one or more of their MHC proteins, and thus grafts between them will be rejected. (There are also **minor transplantation antigens** that provoke weaker immune responses and are responsible for rejection later in transplantation.)

The reason that the MHC antigens out of all the other potential antigens carried on the surface of a cell provoke such a strong response lies in the part they play in normal immune reactions. The immune system is set up, through the structure of the T-cell receptors, to recognize incoming foreign antigens in combination with its own MHC proteins (see above, Antigen recognition). Variant MHC proteins, possibly complexed with fragments of self antigen, or possibly because they resemble the self MHC proteins in some respects but differ in others, present the immune system with new sets of MHC/antigen combinations.

The major histocompatibility complex

The MHC antigens are specified by a cluster of genes termed the **major histocompatibility complex** (**MHC**) (Fig. 12.19). For historical reasons, MHC antigens in humans are called HLA antigens and in the mouse, which has contributed greatly to our understanding of their role in immune responses, they are known as **H-2 antigens**. MHC antigens are glycoproteins distantly related evolutionarily to the immunoglobulins.

Six main HLA proteins have so far been identified in humans – HLA-A, HLA-B, HLA-C, HLA-DR, HLA-DQ and HLA-DP. They are divided into two subgroups, **class I** and **class II**, on the basis of structure (Fig. 12.20).

Class I antigens consist of a single polypeptide chain encoded in the MHC complex to which is bound the small polypeptide, ß-microglobulin, which is encoded elsewhere. Both ß-microglobulin and parts of the MHC polypeptide chains have structural homology with the immunoglobulin 'unit' domain. Class II antigens are heterodimers composed of an α and a ß chain, encoded by separate genes. The MHC-encoded polypeptide chains have a 'constant' region that is very similar between different MHC antigens, and a 'variable' region which carries most of the difference between the many variant forms that exist in the

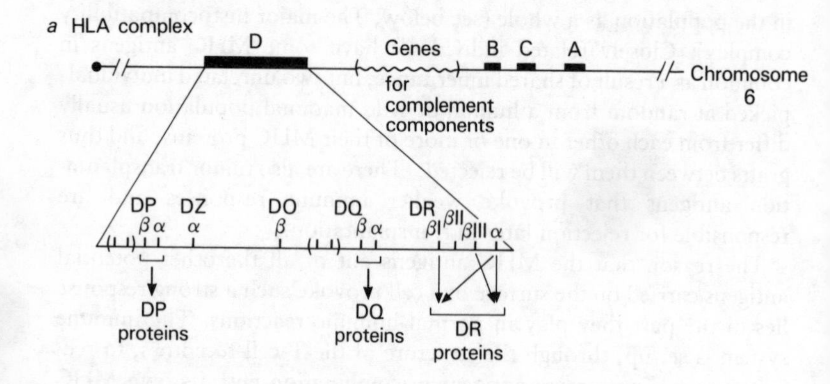

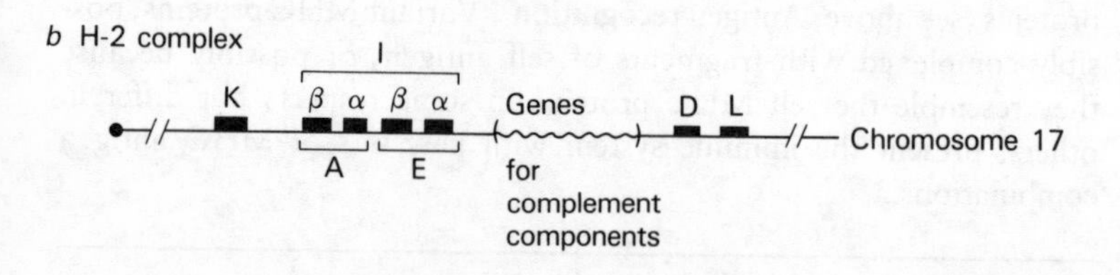

Fig. 12.19 The major histocompatibility complex (MHC) region in humans and mice. Mouse H-2K, H-2D and H-2L correspond to human HLA-A, -B, and -C, and H-2I (I-E and I-A) to the human HLA-D group. At least two proteins are typically produced by the DR genes, being combinations of the relatively invariant α chain with either ßI or ßIII. The MHC complex also includes genes for some complement components. (Map of the D region after J.A. Todd, J.I. Bell and H.O. McDevitt, *Nature*, 329 (1987) 599–604.)

population. Unlike immunoglobulins, however, their genes do not undergo rearrangement.

The two classes have different immunological roles. Class I MHC proteins, human HLA-A, -B and -C, occur in large numbers on most body cells and are the transplantation antigens that are perceived as 'foreign' by cytotoxic T cells, which then attack the graft. In normal immune responses, foreign antigenic determinants produced by virus- or other parasite-infected cells are displayed on the cell surface in association with one or other of the class I antigens and in this form are recognized and attacked by cytotoxic T cells.

The HLA-D antigens – the class II MHC antigens – have a different structure and cellular distribution. They are present mainly on cells of the immune system (lymphocytes, macrophages and other antigen-presenting cells) and play a crucial part in the initial recognition of foreign antigen by helper T cells and in the subsequent cooperative interactions between lymphocytes. Class II MHC antigens in mice are also often called Ia antigens (*i*mmune-*a*ssociated).

In the population as a whole each MHC antigen gene is present in a very large number of different forms. Variants of the same MHC protein have a very similar C-terminal 'constant' region but differ in a 'variable' N-terminal region, which is thought to contain the antigen-binding site. MHC antigens are not antigen specific in the same way as antibodies or T-cell receptors, but appear to be able to bind a wide range of small peptide antigens. A three-dimensional structure for the variable portion of a human class I antigen (HLA-A) has recently been resolved. This will provide immunologists with a much clearer picture of how MHC antigens are able to complex with so many different peptides in their role as antigen presenters to the immune system.

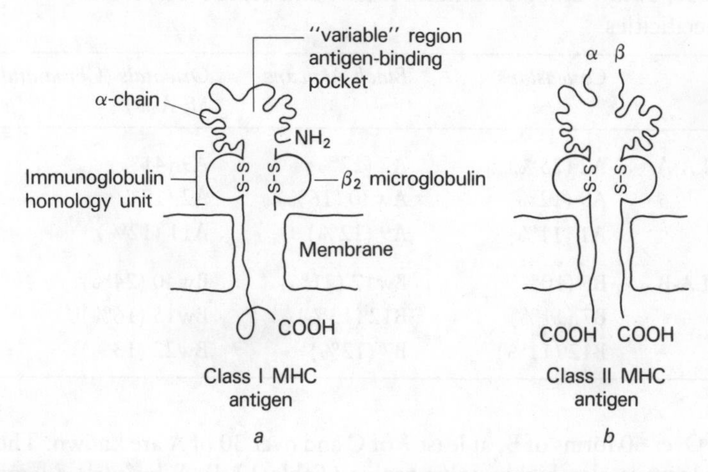

Fig. 12.20 Schematic representation of the structure of *a*, class I, and *b*, class II MHC antigens.

Inheritance

Each person (or animal) inherits a set (a **haplotype**) of MHC specificities from their father and another from their mother (Fig. 12.21). Because the HLA genes are closely linked on the chromosome recombination between them is infrequent and a particular haplotype is often passed on unchanged through several generations. Both parental haplotypes are expressed on the cells of their progeny, and tissue types are now widely used in tracing family relationships, and in forensic work. Tissue types are routinely determined by typing cells with a battery of antibodies specific for the antigenic determinants borne by each variant. Differences detected in this way are termed **serological specificities.**

The great number of different MHC haplotypes that can be detected serologically in the human (or mouse) population results from the exceptional number of different variants (**alleles**) of each MHC gene that exist, more than for any other set of genes discovered so far. In the mouse there are thought to be around 100 different alleles of the K and D genes, and similar variability (**genetic polymorphism**) at the I region. The situation in humans worldwide is probably similar, although certain alleles are commoner than others in different ethnic groups (Table 12.3).

Table 12.3 Some common HLA-A and HLA-B serological specificities

	Caucasians	*Black Africans*	*Orientals (China and SE Asia)*
HLA-A	A2 (25%)	A2 (17%)	A9 (41%)
	A9 (12%)	Aw30 (16%)	A2 (17%)
	A1 (11%)	A9 (12%)	A11 (12%)
HLA-B	B5 (10%)	Bw17 (21%)	Bw40 (24%)
	B7 (11%)	B12 (13%)	Bw15 (16%)
	B12 (11%)	B7 (12%)	Bw22 (13%)

Over 60 forms of B, at least 7 of C and over 30 of A are known. The D locus is also highly polymorphic (Table 12.4). Why such a large number of different alleles of MHC genes are maintained in mammalian populations is at present the subject of much research, impinging as it does on questions of population genetics and evolutionary theory as well as immunology. The immediate answer may be that it is useful for an animal to possess as many different MHC antigens as possible, as it is then more likely to be able to respond to any antigen it may encounter.

Factors other than immunology may be at work in maintaining diversity. There is some evidence that, if given a choice, rats prefer to mate with rats carrying different MHC variants to themselves. How they detect the difference is still unknown, one idea being that fragments of different MHC antigens excreted in urine attach different volatile compounds and the animal detects the resulting spectrum of odours. If this turns out to be a real and widespread phenomenon it would provide another force tending to maintain polymorphism at the MHC locus.

Transplantation

Tissue matching for at least some MHC antigens is essential in preventing graft rejection, but can be made extremely difficult by the large number of variants of each antigen, some of which are extremely rare. As well as tissue-typing with MHC-specific antibodies, compatibility between potential host and recipient is also tested functionally by the **mixed lymphocyte response**, in which cells from one individual are mixed with lymphocytes from another in tissue culture. If the two individuals are incompatible the lymphocytes are stimulated to start proliferating and cytotoxic T lymphocytes are activated, which kill the opposing cells. Brothers and sisters are relatively likely to share an identical set of MHC antigens with each other and HLA-matched bone marrow or kidney transplants between siblings are virtually always successful.

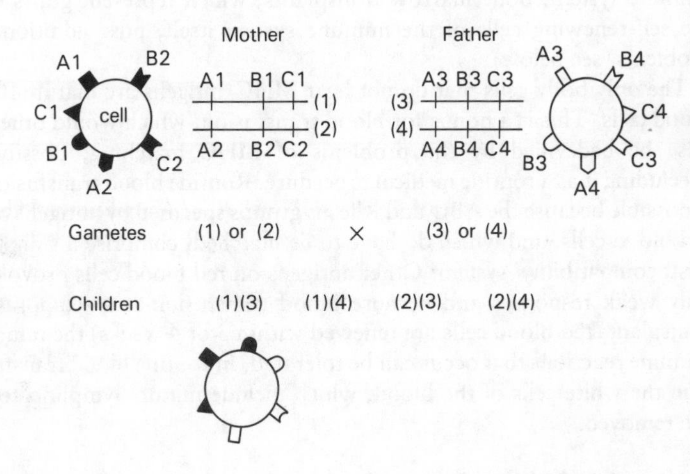

Fig. 12.21 Inheritance of MHC antigens (symbols are purely conventional), assuming no recombination between MHC genes. None of the children will be fully compatible with either parent but there is a one in four chance that a child will be compatible with a brother or sister. The set of alleles on each chromosome is known as the haplotype (represented here by a number in parentheses).

Transplantation across the MHC barrier can only be achieved by severe and continued immunosuppression of the recipient which weakens their defences against infection. Milder immunosuppression can, however, prevent the weaker rejection responses caused by the minor histocompatibility antigens. Transplant surgeons rely both on complete or partial matching for the main MHC antigens and also on general suppression of the body's immune system by irradiation or immunosuppressive drugs during a critical period. New ways of overcoming the problems of transplantation across the MHC barrier, based on a greater understanding of the immune system cells involved in graft rejection are now the subject of intensive research.

One goal is selectively to remove T cells from the recipient that are reactive to the donor's MHC antigens, while leaving the rest of the T cells intact – this might be achieved by the use of suitable monoclonal antibodies, either administered to the patient, or perhaps used to 'clean' the blood outside the body during dialysis. Another approach is to develop specific suppressor T cells and suppressor factors that again would only suppress reactions involving the 'foreign' MHC antigens.

Among tissues routinely transplanted, corneas alone are not subject to immune rejection, probably because the site of grafting in the eye is poorly supplied with blood vessels and is virtually isolated from the immune system. Bone marrow transplants, which represent grafts of the self-renewing cells of the immune system itself, pose additional problems (see below).

The only body cells that do not bear MHC antigens are mature red blood cells. This is a bonus for blood transfusion, which would otherwise be bedevilled by the problems of MHC matching, possibly precluding it as a routine medical procedure. Routine blood transfusion is possible because the ABO and Rhesus groups specified by antigens on red blood cells, and which do have to be matched, comprise a simpler 'histocompatibility' system. Other antigens on red blood cells provoke only weak responses and because blood transfusion is a temporary transplant (red blood cells are renewed within 3 or 4 weeks) the minor immune reactions that occur can be tolerated. In routine blood transfusion the white cells of the blood, which include mature lymphocytes, are removed.

Graft-versus-Host Reactions

As well as the rejection of a transplant by the recipient, some transplants are also subject to the opposite reaction – a **graft-versus-host** reaction. In this, the graft itself attacks and destroys the recipient's tissues. The graft-versus-host reaction is a cell-mediated immune response, carried out by immunocompetent cells in the graft reacting against the recipient's MHC antigens.

Severe graft-versus-host reactions occur in **bone marrow transplantation** in which the cells of the immune system itself are being transferred. Bone marrow contains a small number of mature T cells, presumably derived from the blood, which immediately mount a continuing and fatal attack on the MHC antigens of their new host – in the first instance those of the lymphoid organs, to which the T cells home. Graft-versus-host reactions, not rejection of the transplant itself, are the chief problem in bone marrow transplantation (the recipient's immune system, if functional, is knocked out by immunosuppressive treatment before transplantation). However, if an initial reaction can be avoided, HLA-matched bone marrow transplants are highly successful and long-lasting. One immunological approach that is being tested to avoid graft-versus-host reactions is to 'clean' the marrow to be grafted with monoclonal antibodies specific for T cells, eliminating the T cells that are thought to be primarily responsible for the graft-versus-host reaction. This removes T cells, leaving behind the undifferentiated self-renewing stem cells, which will duly generate new T cells in the host thymus. The 'new' immune system develops tolerance to the self antigens of its host during lymphocyte development.

Bone marrow transplantation, as well as being able to correct some inherited immunodeficiency states and the heritable anaemias, could also alleviate other genetic deficiencies, and has also proved a successful treatment for childhood leukaemia. The transplantation procedure itself is not particularly difficult, as it does not require the delicate surgery of organ transplantation, but it is at present limited in its use by the difficulties of matching donor and recipient. For some genetic deficiencies, it may some day be possible to introduce a good copy of the required gene into a sample of the patient's own bone marrow cells, which can then be replaced, thus avoiding the need for tissue matching (see Chapter 4: Human gene therapy).

The MHC and susceptibility to disease

The possession of certain MHC variants carries an increased predisposition to some diseases. The strongest association discovered so far is between HLA-B27 and ankylosing spondylitis, an inflammatory disease of the spine. HLA-B27 is 100–200 times more common in people suffering from the disease than in the general population. Weaker associations have been found in other diseases. Coeliac disease (intolerance of the protein gluten, found in wheat and other cereals) is associated with B8 and Dw3 (9–10 times more common) and multiple sclerosis with DR2 (5 times more common). Rheumatoid arthritis shows a significant association with DR4 haplotypes, and insulin-dependent diabetes mellitus (an autoimmune disease in which the pancreatic cells that produce insulin are destroyed) with DR4, DR3 and DR1 haplotypes. In the case of insulin-dependent diabetes, the real association has now been found to lie not with the DR antigen itself but with particular DQ alleles (DQw3.2, w2, and w1.1) that are often found in these DR haplotypes (see Table 12.4). Conversely, possession of other DQ alleles (e.g. DQw1.2) makes it less likely that one will develop the condition. MHC type is only part of the answer to why people develop a particular condition, as most of those carrying the requisite alleles stay perfectly healthy.

Table 12.4 Some DR-DQ haplotypes defined serologically in Caucasian populations

Haplotype			*Frequency*
DR1	Dw1	DQw1.1	20%
DR2	Dw2	DQw1.2	26%
DR3	Dw3	DQw2	22%
DR4	Dw4	DQw3.2	9%
DR4	Dw14	DQw3.2	7%
DR5	Dw5	DQw3.1	15%
DRw6/w13	Dw18	DQw18	10%
DR7	Dw7	DQw2	27%

The relationship of serologically-defined specificities to the actual gene variants is complicated for the D group determinants. For example, serological specificities DR1 to Dw14 represent antigenic determinants on the protein specified by the genes DRβ1 and DRα. DQ serological specificities represent determinants on the protein encoded by genes DQβ and DQα. Data from J. A. Todd *et al.*, *Nature*, **329** (1987) 599–604.

The reasons for increased susceptibilities (and resistances) are in general unknown but, given the important role of the MHC in immune responses, they may reflect malfunctions in normal immune reactions.

There are several ways in which MHC proteins could affect susceptibility to disease. Since the immune system recognizes antigens in association with MHC determinants, a combination of some antigens and a particular MHC variant might look sufficiently like one of the body's own components to provoke an autoimmune reaction.

Alternatively, some MHC variants might provide a preferred entry route for a virus or bacterium into a cell (in the way that the T-cell surface antigen CD4 does for the HIV virus). A chronic infection could then lead to massive destruction of host tissue by immune responses designed to get rid of the pathogen (see above, Hypersensitivity).

The MHC genes are now beginning to be cloned and differences between MHC haplotypes at the molecular level are being determined. This will enable a much more precise linking of MHC type with disease susceptibility and identification of the reasons behind the associations.

PRACTICAL APPLICATIONS

The power of a specific antibody to discriminate one antigen from another forms the basis of the **immunoassays** and immunodiagnostic methods used widely in medicine, industry and academic research. With these techniques one can detect, identify, measure or purify almost any substance against which specific antibodies can be raised. Specific antibodies are used, for example, to type blood for transfusion, and tissues and organs for transplantation, to detect and identify disease-causing viruses and bacteria, and to monitor levels of hormones and other physiologically important molecules in blood and urine.

Immunoassays are highly sensitive and in ideal conditions can detect minute amounts of a target substance. Present techniques are sophisticated and varied but they all fundamentally depend on the fact that a specific antibody binds to its corresponding antigen to form an antigen–antibody complex that can then be detected and measured in a variety of ways. Chemical, enzymatic or radioactive tags attached to antibody or antigen (in radioimmunoassays (RIA) and enzyme-linked immunoassays (ELISA)) allow very small amounts indeed of target antigen to be detected.

The enormous quantities of antibodies now needed for the routine immunoassays carried out every day in hospital and research laboratories are produced by the appropriate immunization of animals or, increasingly, by the revolutionary new methods of hybridoma cell culture (see below).

Monoclonal antibodies

A potential complicating factor in all immunological assays and techniques using antibodies raised by immunization in animals is that the **antiserum** obtained by immunization inevitably contains a mixture of antibodies. Even when a highly purified protein is used as antigen, the antisera raised against it will contain antibodies of different specificities that have been induced by the various antigenic determinants borne on the antigen molecule. Even the antibodies specific for the same antigenic determinant are by no means identical in their detailed molecular structure and represent the products of many clones of antibody-producing cells. (A clone is comprised of the identical descendants of a single cell.) The antiserum will cross-react with any other antigen bearing one or more antigenic determinants in common with the original antigen, and it is extremely difficult to distinguish two very similar antigens that differ, say, in a single antigenic determinant. This is not a trivial problem. The characteristic antigens of different strains of virus or bacterium, of different variants of histocompatibility antigens etc. often differ from each other in very subtle ways, involving only a few antigenic determinants.

The situation is particularly difficult when one wishes to raise antibodies against antigens that are not available in pure form, or are not immunogenic in pure form. Many molecules borne on cell surfaces (hormone receptors, histocompatibility antigens, cell type 'markers', etc.) are of great biological and medical interest and antibodies against them have many uses – for example, identifying and isolating different cell types, typing tissues for transplantation and studying hormone action. Most of these cell-surface molecules are difficult to isolate and purify and the easiest way of raising antibodies against them is to use whole cells as the immunogen. However, the resulting antiserum contains a staggering number of different antibodies, many of which will cross-react with, for example, shared antigenic determinants on other variants of the same protein, or antigenic determinants shared by other cell types. Cross-reacting antibodies can be partially removed by laborious treatment with carefully chosen antigens but the resulting antiserum is still a highly variable mixture of antibodies, and its general strength and specificity can differ widely from one immunization to another.

Monoclonal antibodies – antibodies derived from a single clone of cells and therefore all of identical structure and specific for one antigenic determinant only – are the answer to these problems. It is impossible to isolate individual monoclonal antibodies from the mixture produced in a normal immune response, but monoclonal antibodies of a desired specificity can now be produced in virtually unlimited quantities in laboratory cultures of **hybridomas** (artificially constructed hybrid lymphocytes).

The antibody-producing cells of the immune system are the B lymphocytes, which proliferate to form short-lived antibody-producing plasma cells on encounter with antigen. The spleen of an immunized animal contains large numbers of proliferating B cells and antibody-producing plasma cells specific for the immunizing antigen. But lymphocytes, although easily isolated and maintained in culture for short periods, cannot be cultured for long-term antibody production.

Myeloma cells, on the other hand, are cancer cells representing a particular clone of B cells that has proliferated uncontrollably, and, like all cancer cells, will grow and divide indefinitely in culture. Some produce antibodies which, as they are all of a single specificity, were invaluable in determining the structure of antibody molecules. However, the antibodies produced by myeloma cells themselves are of little practical application. Their specificity is usually unknown, and entirely unpredictable, depending entirely on the antigen specificity of the founder cancerous cell. But myeloma cells can be made to produce antibody of desired specificity by introducing into them the antibody genes of an activated B cell from an immunized animal.

This is usually achieved by fusing a myeloma cell with a B cell from the spleen of a recently immunized animal (rat or mouse). The rare stable hybrid cells (**hybridomas**) that result, combine the required characteristics of both parents (Fig. 12.22). They grow and divide indefinitely in culture like myeloma cells, but they produce an antibody typical of their B lymphocyte parent. As each hybridoma is derived from a single B cell it produces antibody of a single specificity. From each sample of spleen lymphocytes, therefore, a range of hybridomas producing different monoclonal antibodies can be derived. Those of the required specificity can then be selected and cultured indefinitely. The most difficult step in the procedure is in practice the final selection of clones producing the desired antibody.

The first hybridomas were produced by Cesar Milstein and Georges Kohler in 1975 using rat myeloma cells and spleen lymphocytes, as an aid to basic research into antibody diversity and mode of synthesis. Researchers in other fields soon realized the usefulness of monoclonal antibodies, and by the 1980s large-scale commercial production was underway. Monoclonal antibody production (chiefly murine antibodies) is now a worldwide, multimillion pound industry, developing and producing tailor-made antibodies for research and clinical diagnosis.

As well as their greater discriminatory power, monoclonal antibodies have the overwhelming advantage over conventionally produced antibodies of being readily mass-produced without the need for immunizing large numbers of animals. Each sample is of a standard strength, and because all the antibodies it contains are of a single specificity it is

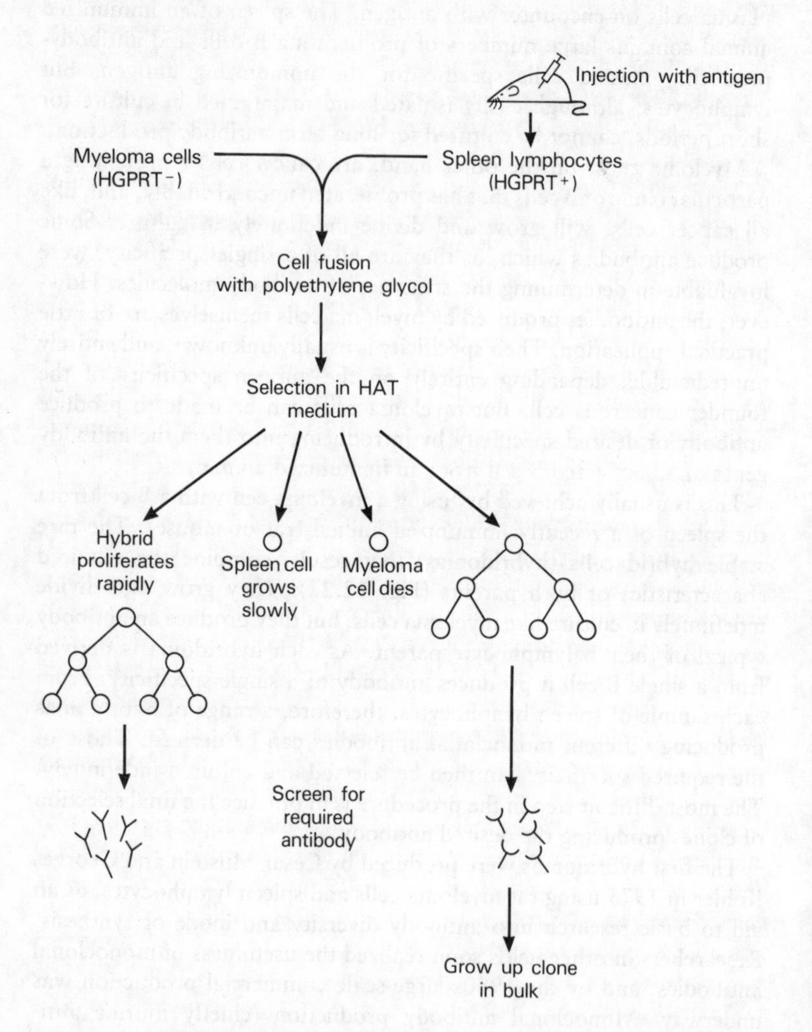

Fig. 12.22 Hybridoma production. Spleen cells from a mouse recently immunized against the desired antigen are separated and fused with myeloma cells that do not synthesize antibody themselves and are genetically deficient in the enzyme hypoxanthine:guanine phosphoribosyltransferase (HGPRT). This enables hybridoma cells to be selected on HAT medium (containing hypoxanthine, aminopterin and thymine), which cannot support the growth of HGPRT$^-$ cells. The spleen cells are HGPRT$^+$ but grow only slowly in HAT medium. Hybridoma cells expressing HGPRT from their spleen cell parent grow rapidly. Individual hybridomas represent the progeny of a fusion between a single spleen cell (B lymphocyte) and myeloma cell.

much more sensitive than a conventional antiserum, needing less test antigen to produce a detectable reaction. Monoclonals have now supplanted conventionally produced antibodies for routine blood and tissue typing, to provide earlier diagnosis of pregnancy or ovulation (by monitoring the level of various sex hormones in the blood), and for more rapid identification of disease-causing bacteria and viruses. They have proved invaluable in many biological research fields, to identify and isolate different cell types and track their development and differentiation by recognizing typical 'marker' cell-surface proteins.

Antibody Engineering

Because of their exquisite specificity and purity and availability in quantity, monoclonal antibodies have also provided a new impetus to the search for immunologically based therapies (see below). For any therapy requiring repeated administration of antibodies, human antibodies are desirable, as antibodies from other species usually soon provoke an immune response against the 'foreign' antigenic determinants on the immunoglobulin molecules themselves. This neutralizes their activity.

It has proved difficult to produce human monoclonals to order using the methods outlined above. An alternative approach to large-scale production of human antibody relies on the new art of 'protein engineering'. Using cloned antibody genes, the hypervariable regions (which determine the antigen-binding site) from the genes encoding a useful rat or mouse monoclonal antibody can be stitched into a framework provided by human immunoglobulin genes. The resulting mosaic gene is then reintroduced into a hybridoma and a permanent cell line secreting that antibody established. This has the antigen specificity of the original rodent antibody, but should be tolerated much better by humans as the species-specific antigenic determinants that provoke the neutralizing response lie for the most part in the constant and variable framework regions.

Therapeutic Uses

One approach being actively pursued at present is the targetting of anticancer drugs and other toxins selectively to cancerous, virus-infected or otherwise diseased cells by linking them with a monoclonal antibody that recognizes some characteristic unique to those cells. These complexes of antibody and toxin are known as **immunotoxins**. IgG antibodies in particular can also trigger complement-mediated lysis of cells to which they attach and an antibody-dependent form of cell-mediated cytotoxicity in which macrophage-type killer cells recognize the Fc region of the IgG heavy chain and are activated to kill the cell to which the antibody is bound.

Both cancerous and virus-infected cells express 'novel' proteins on their surfaces not found in normal cells, and to which an antibody could be targetted. Immunotherapeutic approaches are being pursued especially in cancer chemotherapy, where the powerful drugs currently used kill healthy as well as cancerous cells, producing serious side-effects and limiting the doses that can be given. Many anti-viral drugs suffer from the same limitations and would be safer and more effective if they could be delivered only to virus-infected cells.

Monoclonal antibodies against the specific cell-surface markers on T cells are also being investigated with a view to using them to remove T cells involved in unwanted immune responses such as those that occur in autoimmune diseases, transplant rejection and the unwanted formation of antibodies against proteins (such as antibodies, protein hormones) given as long-term therapy. Monoclonal antibodies specific for T cells generally are already on trial to 'clean' bone marrow before transplantation to prevent graft-versus-host reactions and to disable T cells in transplant recipients. The ultimate goal is to disable only the unwanted antigen-specific T cells while keeping the remainder of the immune system and its responses intact. Present methods of immunosuppression including wholesale T cell removal affect total immune function leaving the patient temporarily defenceless against infection.

Vaccine development

Preventive vaccination against infectious disease has a long history predating that of immunology as a scientific discipline. Despite the great advances in understanding the immune system since Jenner carried out the first vaccinations against smallpox 200 years ago, there is still a considerable element of trial and error in vaccine development. But over the past few years, advances in immunology have made possible some new approaches to vaccine design. Recombinant DNA technology also offers the prospect of novel types of vaccine, and easier and safer production. When the immune system encounters a pathogenic (disease-causing) microorganism for the first time it takes several days to mount an immune response against it. The infection has time to take hold and may overwhelm or escape the subsequent response. Preventive immunization primes the immune system's memory by introducing it to the viral or bacterial antigens that provoke a protective specific response, but in a form that does not cause overt disease. The response it makes against these antigens sets up specific immunity and any future encounter with the pathogen itself will provoke a strong immediate **secondary response** that neutralizes and limits the infectious agent before it can do any damage.

Protection is conferred by vaccines containing inactivated virus or killed bacteria, or inactive toxoids derived from the potent protein toxins that cause the symptoms of, for example, tetanus. Live vaccines contain 'attenuated' strains of pathogen that induce protective immunity but cannot develop the full disease (e.g. the oral polio vaccine). The first reliable vaccination procedure, developed by Jenner at the end of the eighteenth century against smallpox, used live vaccinia (cowpox), a virus that is now known to carry antigens in common with smallpox. As well as the familiar vaccines against diphtheria, whooping cough, tetanus, typhoid and tuberculosis (all bacterial diseases) and against the viruses of measles, german measles, polio and influenza, many others have been developed (e.g. against yellow fever, typhus and cholera).

Vaccine research today is directed primarily at developing vaccines against viral and parasitic diseases, especially those for which drug treatment is non-existent, ineffective or expensive. The body's natural immune response to some viruses, parasites and intracellular bacterial parasites is weak, and as well as developing conventional prophylactic vaccines, there is also the hope that it might be possible to stimulate a better immune response by appropriate vaccination even after infection.

The choice of inactivated or live vaccine varies from pathogen to pathogen. Can the microorganism be made inactive without destroying its capacity to provoke the required response? Are there naturally occurring, non-pathogenic strains of a virus or its relatives that can be developed as 'attenuated' vaccines? Can a harmless virus be constructed by the deliberate deletion of genes that make it virulent? Recombinant DNA technology now offers an additional option – vaccines based not on the microorganism itself, but on the isolated antigens that are responsible for the protective response.

Subunit Vaccines

Viral or parasite protein antigens that might induce a protective immune response can now be isolated and produced in quantity from recombinant bacteria or yeast carrying transplanted genes (see Chapter 4). Vaccines made from these proteins alone, rather than containing live or killed microorganisms, are inherently very safe, both in use and to produce. Virus proteins on their own, freed from the viral nucleic acids, cannot of course direct the formation of new virus, and unlike some bacterial protein products, which are potent poisons, do not themselves appear to be harmful.

An effective vaccine of this type has recently been developed against viral hepatitis B, using as antigen the viral HBAg protein produced either in recombinant yeast or mammalian cells, or isolated from blood plasma from hepatitis B carriers and convalescents. Its use to combat hepatitis B in areas where the virus is endemic and is the cause of widespread cirrhosis of the liver and of liver cancer, is now being planned, but mass immunization programmes will depend on the vaccine becoming available much more cheaply than it is at present. A recombinant hepatitis b vaccine recently approved for use, for example, costs around $15 per dose whereas mass immunizations in endemic regions need vaccine at no more than $1 per dose.

There are several problems to overcome before a successful subunit vaccine is developed. It is not always obvious which of all the antigens displayed by the virus or parasite will be most effective at eliciting a protective response. An educated guess depends on a thorough knowledge of the natural progress of the infection and the body's response to it, and the stages at which the pathogen may be most vulnerable. Antigens are needed that are recognized both by B cells (antibody-producing cells) and by their helper T cells and ideally also provoke the development of cellular immunity mediated by cytotoxic T cells. Recent discoveries that help define the types of antigenic determinants recognized by B cells and by T cells may prove useful in this respect. How to present a purified protein antigen to the immune system in such a way that it provokes cell-mediated immunity is still largely a matter of trial and error, but again, recent advances in understanding how cytotoxic T cells recognize antigenic determinants on virus-infected cells may lead to more rational approaches to the problem.

Antigens to which the immune system makes a response during a natural infection, however apparently ineffectual, are usually the first choice. People infected with the AIDS virus (HIV), for example, make antibodies against the 'env' proteins of its outer envelope (the presence of these antibodies is used as the diagnostic test to detect HIV infection). Various types of vaccine based on env proteins are now being tested (see below).

The various vaccines being developed against falciparum malaria incorporate several approaches. The malaria parasite goes through various developmental stages in the body – a sporozoite stage in which it is transmitted from mosquito to human, and two different stages inside erythrocytes (red blood cells). At each stage it carries different cell-surface antigens, and parasite-specific antigens are displayed on infected red blood cells. The vaccines now being developed chiefly aim to incorporate antigens against more than one of these forms. Malaria presents a difficult problem. In endemic malarial areas natural immunity is not developed after a single attack, but, eventually, if someone survives repeated attacks, a degree of immunity is built up. Immune individuals carry antibodies to various parasite proteins but it is not known whether antibodies or cell-mediated responses are more important in defence against the pathogen. The aim of a vaccine, however, must be to provide immunity immediately, especially to very young children in whom mortality is high. Although protective immunity has been achieved in adult volunteers with some test vaccines there are still many problems to be solved before they are ready for use.

Immunogenicity

Another problem with purified proteins as vaccines is that although they are antigenic (i.e. they can provoke the formation of specific antibodies), they are often not very *immunogenic* – they do not provoke a sufficiently strong response from the immune system to develop effective immunity.

Their lack of immunogenicity can be overcome by mixing the proteins with **adjuvants**, materials that are non-immunogenic themselves but that boost the immune response (probably mainly by preventing dispersion and immediate destruction of antigen at its site of entry). The most effective adjuvants discovered in the past, however, cause an unacceptably severe inflammation at the site of injection, precluding their use in human or veterinary medicine. Several new adjuvants are now being used and the search for means of boosting the immune response has also led to the development of **iscom** vaccines – immune stimulatory **complexes** – in which the relevant protein antigens are carried in a matrix material. Success has been reported with an iscom vaccine against feline leukaemia virus.

Ways are also being developed of targetting antigens directly to the antigen-presenting cells of the immune system, thus ensuring that they encounter the immune system in as efficient a way as possible.

Recombinant Vaccinia Vaccines

Another way of introducing an antigenic protein to the immune system is to use the well-established vaccine strains of vaccinia virus (used as smallpox vaccine for many years) as 'carriers'. **Recombinant vaccinia virus** containing a foreign gene encoding the required protein antigen can multiply in infected cells and express the foreign protein, mimicking in many ways a natural viral infection. Live recombinant vaccinia may therefore be better at provoking the cell-mediated immunity that is important in responses to viruses and parasites. Live vaccinia strains containing genes from the viruses that cause rabies, AIDS, hepatitis B, influenza and herpes simplex are amongst those now being developed. Several live recombinant vaccinia vaccines aimed at preventing rabies in animals have already been successfully tested.

Live vaccinia vaccines containing the *env* protein gene of the human immunodeficiency virus, HIV (the virus responsible for AIDS) are at present undergoing preliminary small-scale trials in healthy human volunteers to determine how good they are at provoking an immune response and to detect any general adverse side-effects. Results so far suggest that it is possible, using certain immunization schedules, to produce a reasonably high level of 'immunity' to the env protein itself. (The immunization schedule that gave the best results in one report arose from a researcher accidentally injecting himself with a suspension of human cells infected *in vitro* with the virus. Fortunately the cells had been killed and the virus inactivated.) Whether immunity against the env protein would in practice prevent HIV infection or the development of AIDS is of course unknown. Trials in chimpanzees are not encouraging on this point. Using human volunteers at such an early stage in vaccine development is unusual, but in the case of HIV there is no readily available animal model for AIDS. Only the chimpanzee (which is now a rare and endangered species) shows symptoms resembling AIDS after HIV infection. Other types of AIDS vaccine are also being tested in humans.

Live vaccinia itself can provoke serious (but rare) complications, which although acceptable when the alternative is a severe and life-threatening disease, might limit its use. To avoid this problem inactivated recombinant vaccinia vaccines are also being developed.

FURTHER READING

See general reading list (Alberts et al.; Hood et al.; Roitt; Sell)

M.M. Davies and P.B. Bjorkman, T-cell antigen receptor genes and T-cell recognition, *Nature*, 334 (1988) 395–402.

C.A. Janeway, jr., Frontiers of the immune system, *Nature*, 333 (1988) 804–6, and references therein. A brief review of recent work on the γ/δ T-cell receptor.

P.J. BJORKMAN et al., Structure of the human class I histocompatibility antigen, HLA-A2, *Nature*, 329 (1987) 506–12. The foreign antigen binding site and T cell recognition regions of class I histocompatibility antigens, 329 (1987) 512–18; and accompanying commentary by A. Townsend and A. McMichael (p. 482, same issue).

M. ROBERTSON, Tolerance, restriction and the Mls enigma, *Nature*, 332 (1988) 18–19, and references therein; A.L. DEFRANCO, Fate of self-reactive B cells, *Nature*, 334 (1988) 652–3, and references therein. Commentaries on recent work on self-tolerance relating to T cells and B cells respectively.

C. MILSTEIN, Monoclonal antibodies, *Scientific American*, 243 (1980) 66–74.

Science, 238 (1987) 1065–98. Frontiers in biology: immunology. Articles on progress towards a vaccine for schistosomiasis; the T-cell receptor; development of the antibody repertoire; B cells; and designing anti-tumour agents.

L. MILLER, Malaria: an effective vaccine for humans?, *Nature*, 332 (1988) 109–10; F.E.G. COX, Malaria vaccines: the shape of things to come, *Nature*, 333 (1988) 702. Short commentaries on the development of malaria vaccines.

L. REICHMANN et al., Reshaping human antibodies for therapy, *Nature* 332 (1988) 323–7.

AIDS, *Scientific American*, 259 (4) (1988). An issue devoted to "what science knows about AIDS".

CHAPTER 13

THE NERVOUS SYSTEM

A nervous system allows an animal to sense and respond to external stimuli and communicate between different parts of its body virtually instantaneously, far more rapidly than would be possible if communication within the body were restricted to chemical diffusion between and within cells. Without a nervous system animals such as the sponges lead a sedentary, plant-like existence. With the development of a nervous system linked to muscular tissue and sensory organs, multicellular animals became mobile and could actively search for food and escape from predators. Even in the simplest organisms this entails the rapid processing and integration of information coming from many sources, another fundamental property of nervous systems, which reaches its greatest development in the complexities of the human brain.

The nervous system conveys information in the form of impulses of electrical activity transmitted along a cable network formed of electrically excitable cells – nerve cells or **neurones**. Electrical impulses can travel along neurones at the rate of up to about 100 metres per second in the fastest axons. Neurones come in many shapes and sizes but all have the same functional organization (Fig. 13.1). They receive input via their **dendrites** and **cell body** and transmit a signal along the single **axon**, which becomes more or less highly branched after it leaves the cell body. In large animals, the axons of individual nerve cells making up nerves that lead to and from the extremities of the body may be several metres long; in the brain and spinal cord on the other hand, thousands of millions of smaller neurones with highly branched dendrites and axons are packed together, each making many precise connections with other neurones to form an ordered information-processing network of great power.

The vertebrate nervous system is functionally and anatomically differentiated into an information-processing **central nervous system** (CNS) comprising the brain and spinal cord, and a **peripheral nervous system** comprising **sensory nerves** that convey information from sense organs to the central nervous system for interpretation and **motor nerves** that convey executive signals from the central nervous system to muscles. The axons of peripheral neurones (**nerve fibres**) serving a particular organ or region of the body are bundled together to form **nerves** providing a continuous pathway for fast long-distance signalling. The mammalian peripheral nervous system is functionally differentiated into an **autonomic** nervous system and a **voluntary nervous system**. The former mediates involuntary, automatic physiological responses such as control of heartbeat, breathing, the contraction and relaxation of the smooth muscle linings of gut and airways, hormone secretion from glands under direct neural control and so on; the voluntary nervous system consists of the nerves supplying skeletal muscles that are under 'conscious' control.

Neurones convey information along their axons in the form of electrical impulses or **action potentials**, which are generated in the cell body in response to signals arriving from other neurones or from sensory receptors (e.g. photoreceptors in the retina, stretch receptors in muscle, touch receptors in skin and the proprioceptors in joints that monitor the body's position). Electrical activity in neurones and other electrically active cells is the result of the special properties of their cell membranes.

The individual electrical impulses are always of a fixed size whatever the type or strength of stimulus that elicits them. Information is encoded in the **frequency** of the electrical impulses (see below), which changes with the strength of the stimulus, and is interpreted in the light of the ordered connections each neurone makes with another. The high degree of organization within the nervous system means that the stereotyped signals conveyed by an individual neurone can be recognized and interpreted as a particular type of information, coming from a specific point in the body, and the required response relayed to other neurones, muscles or glands.

Nothing in neuronal biochemistry and the ways in which one neurone communicates with another suggests that neurones possess any properties different in kind from any other living cell. They carry out their special tasks at the cellular level using biochemical machinery found in many other cell types. Neurones from an earthworm use much the same biochemical means of transmitting and conveying impulses as do those from mouse or humans and the electrical signals they convey are indistinguishable. It is the number of neurones and the complexity and specific organization of their interconnections that gives the brains of vertebrates, and especially those of humans, their remarkable capabilities.

Neurones make connections with each other and with other target cells such as muscle or gland at **synapses** (see Fig. 13.12). Although information is conveyed along an axon in the form of an electrical impulse, communication across the majority of synapses in the mammalian nervous system is by *chemical* messengers (**neurotransmitters**). At a synapse, the incoming electrical signal is converted into a chemical signal that diffuses across the tiny gap between the two neuronal membranes to be reconverted into an electrical signal at the opposing membrane. The mechanism of synaptic transmission and its control at peripheral synapses such as those between motor nerves and skeletal muscle are now well understood, and have been dissected in molecular detail. Transmission within the mammalian central nervous system is far more difficult to study and the roles of many of the potential signalling molecules that have been found there are not yet certain.

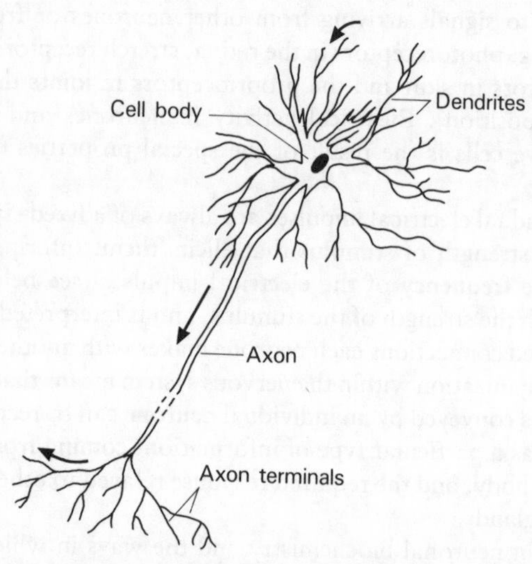

Fig. 13.1 Structure of a generalized neurone. Arrows indicate direction of input and output.

A typical neurone in a mammalian nervous system receives input through thousands of synapses on its dendrites and cell body, and even on the axons, made by the axon terminals of other neurones. In operational terms there are only two sorts of input – **excitatory** and **inhibitory**. Excitatory inputs tend to make the neurone fire, inhibitory inputs tend to suppress firing. Each neurone is continually summing up and interpreting the inputs it is receiving and responding by either transmitting impulses at a particular frequency or remaining silent.

As a general rule, an individual neurone (and therefore the synapses it makes on other neurones) supplies either inhibitory or excitatory input because the same transmitter is released at all synapses – Dale's principle. It is specified as inhibitory or excitatory as the nervous system develops, and remains so throughout the animal's life, forming the basis for orderly function. (As there are some instances where the same neurone makes both excitatory and inhibitory synapses on other neurones, it is more accurate to say that *synapses* are specified as excitatory or inhibitory.)

Much of the basic hard-wired circuitry of inhibitory and excitatory connections is laid down during embryonic development. The central nervous system in particular, however, continues to mature and develop new functions after birth. In mammals at least, the pattern of connections involved in visual, auditory and tactile perception is crucially influenced by sensory experience during a period immediately after birth. Animals temporarily deprived of the sight of one eye during this time cannot develop correct binocular vision, and unless the sight of that eye is restored within a limited period will never do so. Rats whose whiskers are trimmed for the first few weeks after birth do not develop the corresponding tactile signal-processing capacity in the brain. Even after the whiskers regrow the rats cannot interpret the stimuli they normally receive this way. In the same way children born blind or deaf do not develop some types of visual or auditory processing capability in the brain. There is also a considerable degree of operational **plasticity** in the fully developed brain. This is most evident in the ability of animals to learn – to acquire new information about the world – and to retain it permanently in the form of a memory.

The sections below focus chiefly on the basic communication mechanisms in the nervous system (see The generation of electrical signals; Synaptic transmission; Chemical signals in the nervous system). Much neurochemical research, for example, is aimed at trying to uncover and correct the deficiences and disturbances in particular types of chemical transmission that cause the symptoms of conditions such as Parkinson's disease, schizophrenia, epilepsy and Alzheimer's disease with the aim of developing better drugs with less harmful side-effects. Novel techniques for isolating and identifying human genes (see Chapter 4: Genetic disease) are also being applied to try to uncover the underlying defects in some neurological diseases.

Understanding the information-processing ability of the brain in terms of its cellular organization and neuronal function is far more difficult. At a descriptive level electrophysiological and anatomical approaches have identified general regions and anatomical pathways involved in various aspects of brain function. The study of brain-damaged patients has identified remarkable and specific disturbances in language, memory and perception that can be traced to damage in particular areas. Considerable progress has also been made in understanding the way in which the brain breaks down and analyses sensory information as it travels towards the higher centres of perception (see the reading list at the end of this chapter)

But the almost unimaginable complexity of the human brain with its millions of interconnected neurones has so far defeated attempts to analyse the 'higher functions' of the cerebral cortex – such as perception, language, consciousness, sensation and memory – in even the most general organizational terms. These higher-level functions are currently largely the material of the 'top-down' approaches of cognitive psychology and the systems approaches of artificial intelligence (see reading list). However, one area in which neurophysiology and, more recently, cell and molecular biology make an important contribution is in understanding the nature of the changes that take place during learning and that underlie the acquisition of memory (see below, Learning and memory).

The ability to learn implies a permanent alteration in the patterns of information flow through the system, and is thought to be mediated at the cellular level by modifications to the connections between neurones – the synapses – which alter their transmitting properties. Despite the fact that learning and memory in higher animals, and most of all in humans, cannot yet be firmly grounded in neuroanatomy and neurophysiology, the biochemical changes that may underlie them are beginning to be elucidated with much help from simple invertebrate nervous systems, where the cellular and biochemical basis for processes akin to learning and memory has been worked out in considerable detail.

ORGANIZATION OF THE VERTEBRATE NERVOUS SYSTEM

Information about the outside world and the internal state of the body is collected by **sensory receptors** in eyes, ears, nose, mouth, muscles, joints, the body surface and in internal organs, and channelled to the central nervous system by sensory nerves. Sensory information from limbs and the trunk is conveyed first to the spinal cord where it is transmitted to the brain through well-defined **ascending pathways** of nerve fibres. Sensory information from eyes, ears, nose, mouth and the surface and muscles of the face enters the brain directly through cranial nerves. In the brain, the mass of sensory input is analysed and interpreted, and results in outgoing instructions to muscles and glands. Commands leave the brain via cranial motor nerves (such as those that supply the muscles of the face and the eyes) and through **descending** pathways in the spinal cord from which motor nerves leave at intervals to innervate skeletal muscles, internal organs and the skin (Fig. 13.2).

Output concerned with the automatic regulation of physiological processes is conveyed by the **autonomic** nervous system, which has **sympathetic** and **parasympathetic** branches. Autonomic nerve fibres leaving the spinal cord (**preganglionic** fibres) terminate in groups of nerve endings (**ganglia**) where they synapse with **postganglionic** fibres that serve the effector organs. Some internal organs are innervated by both sympathetic and parasympathetic fibres, which often have opposing effects. The autonomic nervous system innervates the ubiquitous smooth muscle (which makes up the walls of the gut and large blood vessels and is present in most internal organs), cardiac muscle, and some glandular tissue (e.g. sweat glands).

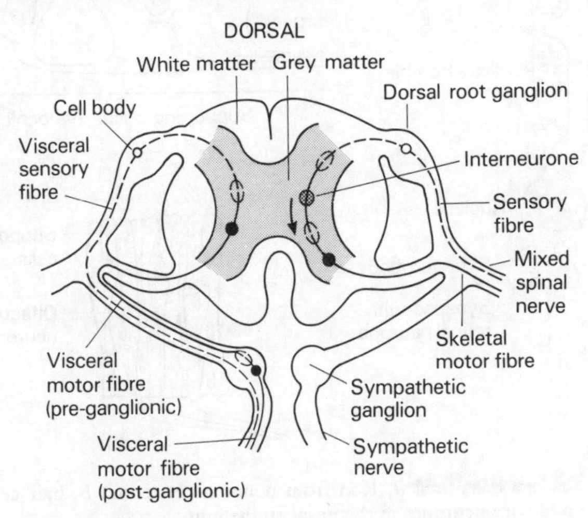

Fig. 13.2 Schematic cross-section through the spinal cord showing sympathetic and skeletal muscle sensory–motor circuits. For convenience, only one pathway is shown on either side. Solid lines indicate outgoing signals from the central nervous system, dashed lines indicate incoming signals from organs and muscles.

Commands to skeletal muscles travel via the **voluntary nervous system.** Skeletal motor nerve fibres leave the spinal cord and go directly to their target muscle, each muscle fibre being innervated by a single axon terminal in mature mammalian muscle.

Sensory receptors (Fig. 13.3) are modified neurones with specialized endings that respond to a stimulus by producing an electrical signal for onward transmission. The rods and cones of the retina are photoreceptors, containing light-sensitive protein pigments. The hair cells lining the cochlea of the inner ear are exquisitely tuned to vibrations in the cochlear fluid caused by sound waves. In the nose, olfactory cells bear cell surface receptors for odorant molecules. The stretch receptors that gauge the degree of extension of skeletal muscle are modified terminals of sensory neurones and are sensitive to mechanical deformation.

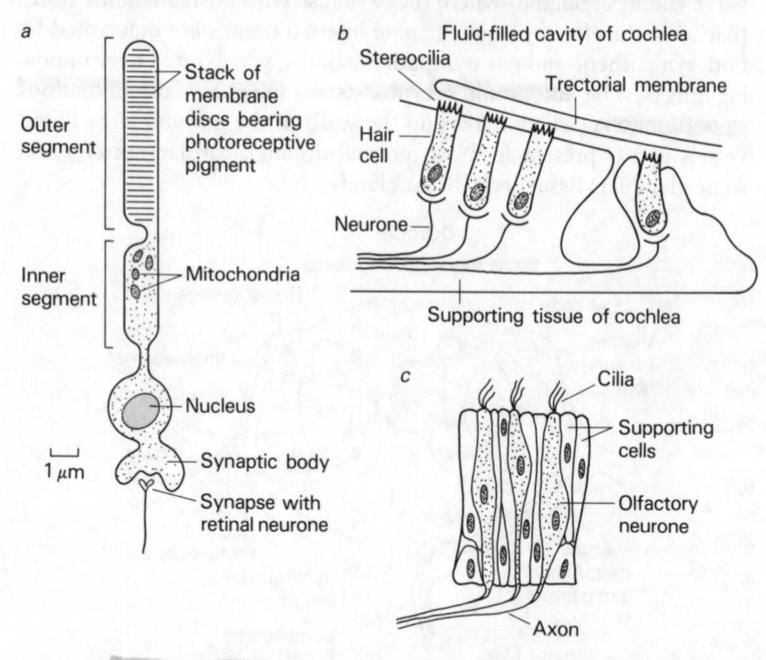

Fig. 13.3 Sensory receptors. *a*, Rod from mammalian retina; *b*, hair cells in the cochlea; *c*, olfactory neurones in the nasal epithelium.

Within the spinal cord, input from the periphery is channelled upwards through several tracts of nerve fibres that deliver the signals to appropriate relay stations within the brain. The orderly arrangement of neurones within these tracts, and the connections they make at the relay stations maintains a spatial 'map' of the source of the information that is more-or-less retained intact as the information is conveyed to higher levels in the brain. There is continual feedback from one level to another and between motor and sensory systems, mediated by spinal interneurones. There are some operations that are mediated entirely by interactions between motor and sensory systems at the level of the spinal cord (such as the knee-jerk reflex and other automatic muscular adjustments) without any brain involvement. In most cases, however, even for 'unconscious' movements, input from both brain and spinal circuits is involved.

The central nervous system

The brain (Fig. 13.4) and spinal cord are entirely surrounded and isolated from the rest of the body by three membranous layers, the tough dura mater overlying the arachnoid membrane and the delicate pia mater. Within the brain are several fluid-filled cavities – the **ventricles**. The cerebrospinal fluid (CSF) that fills these cavities and bathes the whole central nervous system is produced at the choroid plexus – an area of intermingled nervous and vascular tissue. The CSF distributes nutrients, etc. within the central nervous system. Nutrients and other essential materials are delivered to the brain by cerebral blood vessels that run over its surface. There is a functional **blood–brain barrier** formed by tight seals between the endothelial cells lining blood vessels. This prevents the passive entry of many substances from blood into the brain. The brain is, for example, normally isolated from the immune system, as immune system cells, antibodies, etc. cannot pass the blood–brain barrier.

As well as thousands of millions of nerve cells, the brain contains massive numbers of **neuroglial cells**. Unlike neurones these are non-excitable. They surround the neurones and are probably involved in maintaining appropriate concentrations of ions at the neuronal surface and in general nourishment and support. Some glial cells – the **oligodendrocytes** – surround the axons of individual CNS neurones with a fatty insulating covering, which corresponds to the myelin sheaths produced around peripheral nerve fibres by Schwann cells. Brain and spinal cord tissue is often described in terms of **grey** and **white matter**. The grey matter corrresponds to aggregations of cell bodies – as in the cortex and the various brain **nuclei** – and the white matter to tracts of myelinated fibres.

The mammalian brain is conventionally divided into five main divisions (see Fig. 13.4).

1. The **telencephalon** (the **cerebral hemispheres**) is the most obvious feature of the brains of humans and other primates but is less well developed in other mammals. The highly convoluted outer few

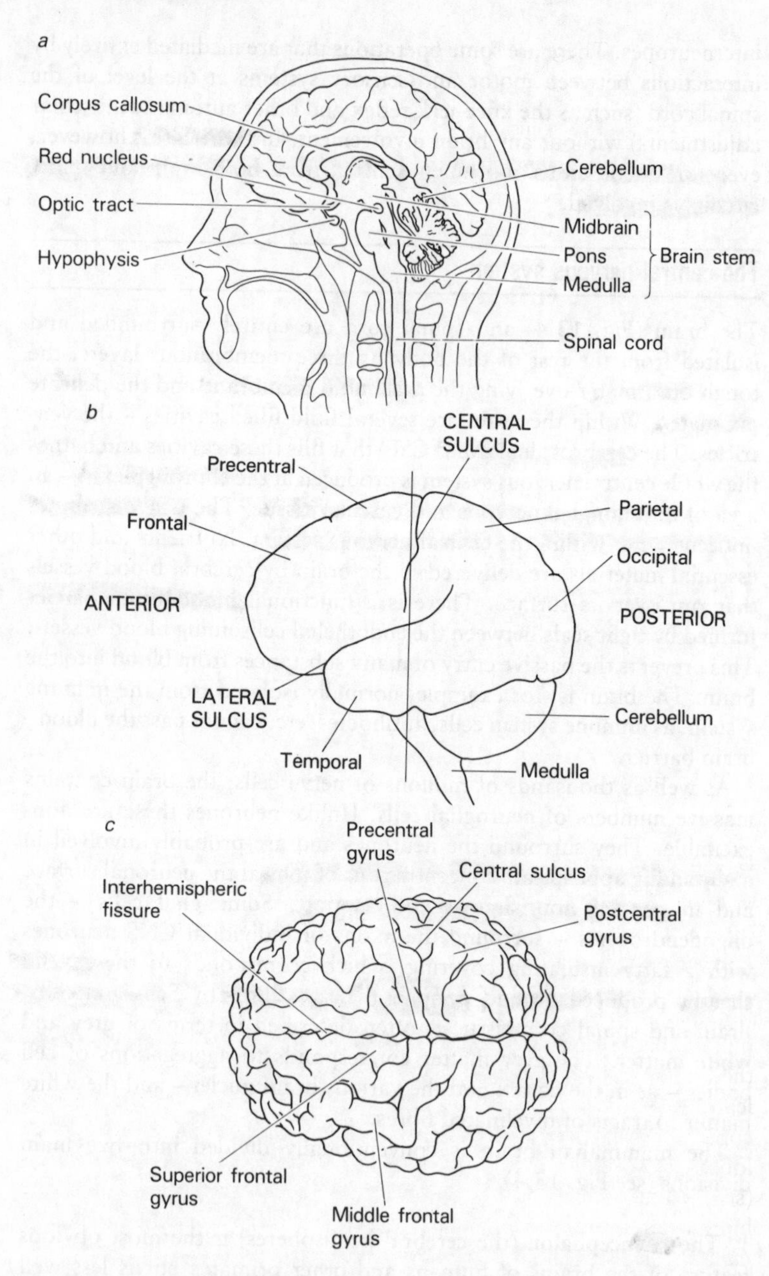

Fig. 13.4 The human brain. *a*, Longitudinal section *in situ*; *b*, lateral view with main regions; *c*, dorsal view; *d*, sagittal section (through midline between hemispheres); *e*, cross-section along line indicated in *b*.

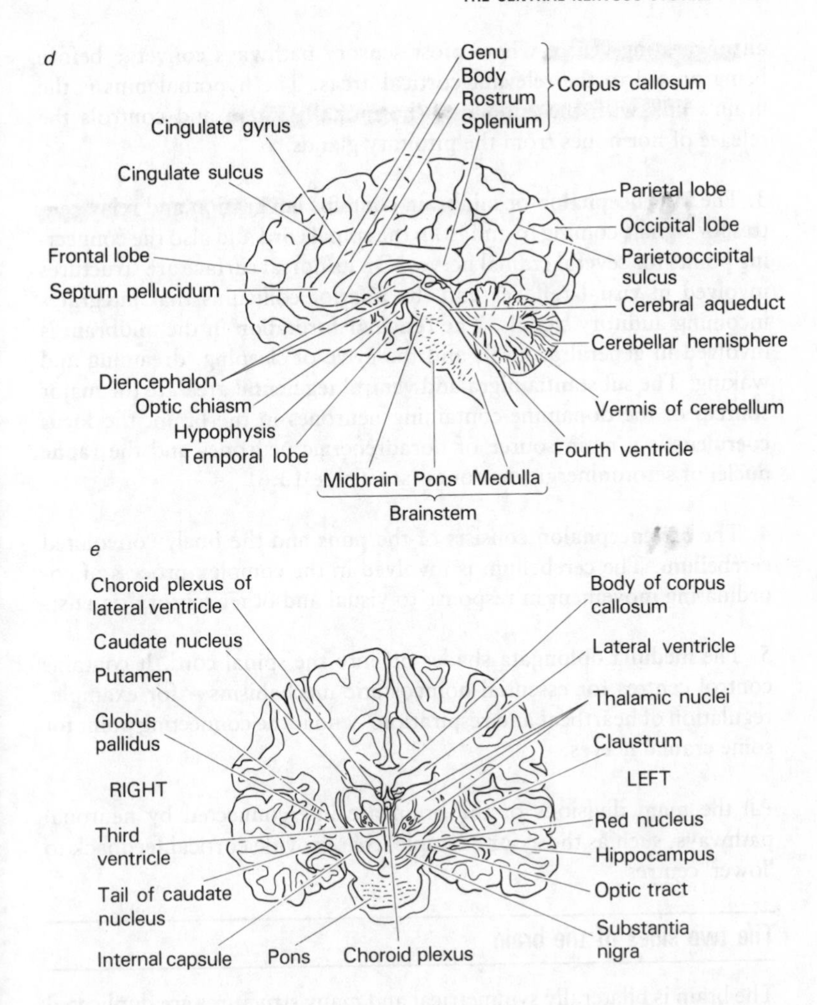

millimetres of cerebral tissue is the **cerebral cortex**, in which are located the highest processing centres for sensory perception, motor control, learning and memory, and, in humans, the functions of language and conscious sensation. Each cerebral hemisphere is divided into several lobes (e.g. occipital, temporal, frontal) delimited by major infoldings (sulci, sing. sulcus). Other structures of the hemispheres include the **hippocampus** and the **amygdala**, which are important in learning and memory, and the **basal ganglia**, which are involved in motor control.

2. The **diencephalon** underlying the cerebral hemispheres contains the important regions of the **thalamus** and **hypothalamus**. The thalamus is

an integrating centre where most sensory pathways converge before being passed to the relevant cortical areas. The hypothalamus is the brain's link with the endocrine (hormonal) system and controls the release of hormones from the pituitary glands.

3. The **mesencephalon** or **midbrain** contains integration and relay centres for signals coming to and from the spinal cord and also the connecting points for several cranial nerves. On its dorsal surface are structures involved in visual reflexes and the inferior colliculus that integrates incoming auditory signals. The **reticular formation** in the midbrain is involved in general arousal, and the cycle of sleeping, dreaming and waking. The **substantia nigra** and **ventral tegmental area** are the major sources of the dopamine-containing neurones in the brain, the **locus coeruleus** is a main source of noradrenergic neurones and the **raphe nuclei** of serotoninergic neurones (see Table 13.6).

4. The **metencephalon** consists of the **pons** and the finely convoluted **cerebellum**. The cerebellum is involved in the complex process of co-ordinating movement in response to visual and other sensory signals.

5. The **medulla oblongata** shades off into the spinal cord. It contains control centres for essential homeostatic mechanisms – for example, regulation of heartbeat and respiration – and is the connecting point for some cranial nerves.

All the main divisions of the brain are interconnected by neuronal pathways, such as those for example, that provide cortical feedback to 'lower' centres.

The two sides of the brain

The brain is bilaterally symmetrical and many structures are duplicated on both sides, receiving information from one side of the body (but see below). The cerebral hemispheres are divided by a deep fissure and are connected only by a broad band of nerve fibres – the **corpus callosum**. Split-brain studies in humans, where the corpus callosum has been severed (e.g. to prevent epileptic attacks spreading from one hemisphere to the other) show that the two hemispheres are not equivalent in respect to subtle manifestations of higher functions such as language, with some aspects being localized to one hemisphere (Fig. 13.5). Normally, information is transferred between the two hemispheres through the corpus callosum.

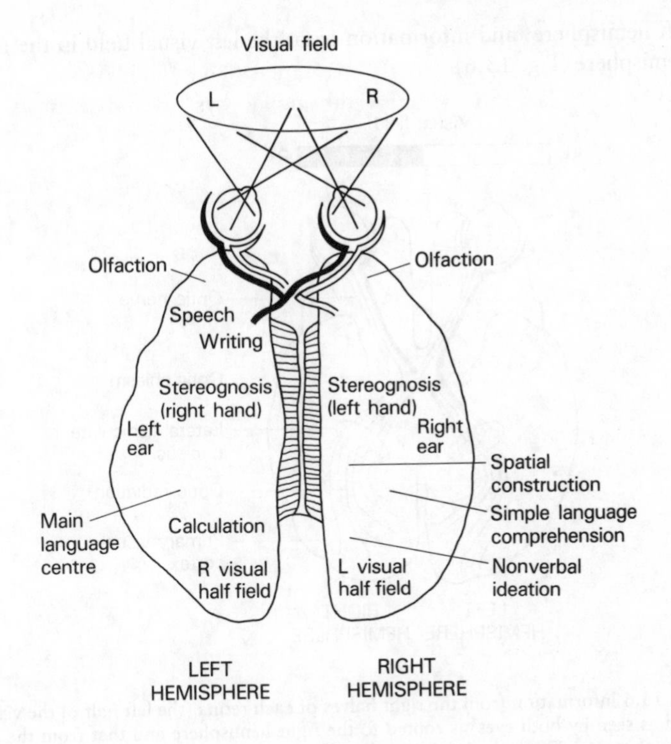

Fig. 13.5 Location of functions in the right and left hemispheres of the human brain.

The separation of some functions can be illustrated by the case of a right-handed man with a severed corpus callosum who cannot verbally name an object put into his left hand, although he can show that he recognizes what it is by using it. The inability to name it arises from the location of the language centre in the left hemisphere, whereas information from the left hand is routed to the right hemisphere. If the corpus callosum is cut and there is no connection between the two hemispheres this information is inaccessible to the language centre. If the object is put into the right hand, the man can name it.

In many sensory pathways, there is a major crossover that sends some of the information from one side of the body to the other side of the brain for processing, and that serves to integrate the signals received by each side. In the visual system, for example, there is a partial crossover at the optic chiasm, so that information from the left halves of both retinas (which record the right visual field) is all processed in the left hemisphere, and information from the left visual field in the right hemisphere (Fig. 13.6).

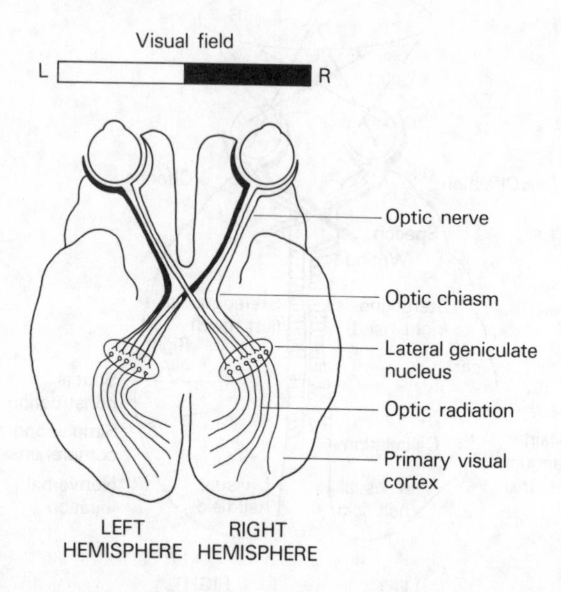

Fig. 13.6 Information from the right halves of each retina (the left half of the visual field as seen by both eyes) is routed to the right hemisphere and that from the left sides of each retina to the left hemisphere.

Localization of brain functions

Extensive studies over the past 100 years in animals and in humans with brain damage as a result of injury or disease, has shown that many aspects of brain function are localized to particular cortical areas. This is true not only for the senses but for certain 'higher' functions such as language and speech (Fig. 13.7).

Humans and other primates, who are highly visual creatures, devote a very large area of the cortex to visual processing. There are primary visual areas which receive the first input from the eyes and analyse it further and relay the information to secondary visual areas which interpret different aspects of the total picture. For example, separate areas seem to be involved in recognizing what an object is, and where it is, using in the one case information on colour, size and shape, on the other position and movement. This information appears to be already separated out and encoded in parallel neuronal channels at the retina.

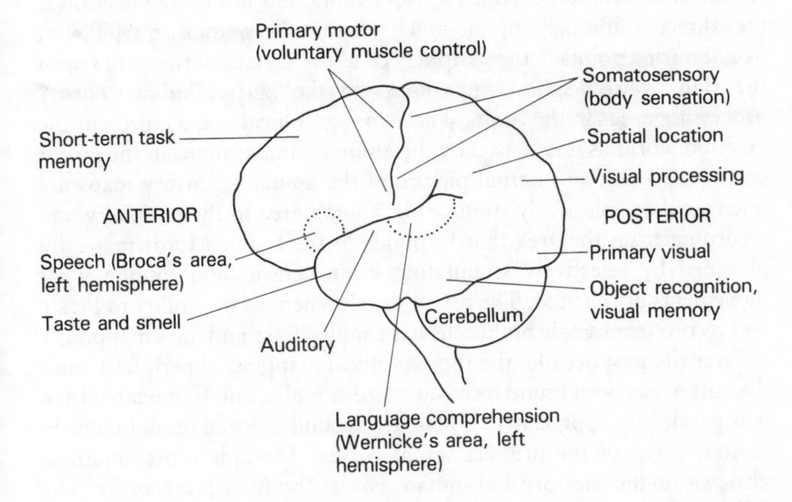

Fig. 13.7 Location of some functions in the human cortex. The posterior part of the cortex (parietal, occipital and temporal lobes) is concerned largely with the reception, analysis and storage of incoming sensory information; the anterior part of the cortex is concerned with generating actions. The frontal lobes are concerned with the organization of complex goal-directed activity (e.g. purposive movement, carrying out a complicated sequence of actions). The appropriate motor commands are generated in the primary motor cortex.

Other areas are involved in combining information from different senses and in coordinating sensory and motor output.

How an abstracted representation is put together again to form a perception is still not known. A currently fashionable metaphor is the brain as a parallel distributed processor (see reading list), with parallel independently processed channels of abstracted features being fed into a distributed network. In this model a final perception would entail the simultaneous activity of a circuit of cells possibly distributed throughout different processing areas. Other models envisage pathways converging onto cells that successively perceive a more complete representation. There are neurones in the areas of the primate cortex involved in object recognition, for example, that are attuned specifically to respond to views of faces, or stimuli representing certain aspects of a face, or of particular types of body movement.

Topographical maps

A crucial correlation between sensory input and brain organization is the direct (although often much distorted) spatial mapping of neighbouring points in the periphery (e.g. the surface of the retina or of the skin) onto neighbouring points in the corresponding sensory processing areas in the brain. These **topographical maps** represent the external world as it is sensed by the animal. Similar maps in the motor cortex represent an internal picture of the animal. (Sensory maps are uncovered by selectively stimulating a small area in the periphery and recording from the area that responds in the brain. Motor maps are obtained by selectively stimulating brain centres and finding what movements are elicited. The recent development of the ability to locate and record from single brain cells has enabled finer and finer mapping.)

Over the past decade, the topographical mapping of periphery onto the cortex has been found to be far more complex and fragmented than was previously appreciated. There are around a dozen retinal maps in various areas of the primate visual cortex. Multiple representations also exist in the auditory and somatosensory (body surface) cortex. The full significance of this organization to the way information is processed is not yet entirely clear, but it is thought that each is devoted to processing different aspects of the sensory input.

Structural organization

The neuroanatomical work started in the nineteenth century, some of which still remains unsurpassed, reveals an underlying orderliness of structural organization in the brain. This may embody and eventually reveal important clues to the brain's internal logic. Although there are an estimated 10^{11} neurones in the human brain there are only a relatively small number of anatomically distinguishable types. The cerebellum, for example, has a layered structure and only five main types of neurone. Each type is restricted to a particular layer and connects with the others in a precise and regular way. A very great deal has already been achieved in tracing the architecture of neuronal connections, the pathways of information through the brain and the connections between different areas.

Under the light microscope, brain tissue shows various structural motifs that are repeated in many areas and brain formations. Many brain structures (such as cortex and cerebellum) are made up of successive distinct layers (laminae) of cell bodies. Their layered construction appears to be directly related to the way they process information, different types of input being segregated into each layer. At right angles to the layers, an electrophysiologically detectable, functional organization into vertical 'columns' of neurones, has also been found in some areas of the cortex. Each column (in reality a narrow vertical slab) is concerned with a tiny area of the visual or sensory field and contains neurones that 'see' a particular abstracted feature – such as a line of a certain length, orientation and direction of movement (Fig. 13.8). Similar columns are detectable in the somatosensory cortex where neurones responding to light pressure (touch) are segregated from those responding to deep pressure from joints, etc, and in the motor cortex where neurones are segregated in respect of the muscles they serve.

Neuroanatomy confirms in many respects the columnar organization revealed by electrophysiology. Sections through the cortex show extensive connections running vertically with in general only short-range horizontal (lateral) connections. Long-range lateral connections are chiefly made between different areas by bundles of fibres that leave the cortex and run beneath it to resurface at another point. Much attention is now being given to the exact paths of these connections and the groups of cells they connect, both within and between cortical areas.

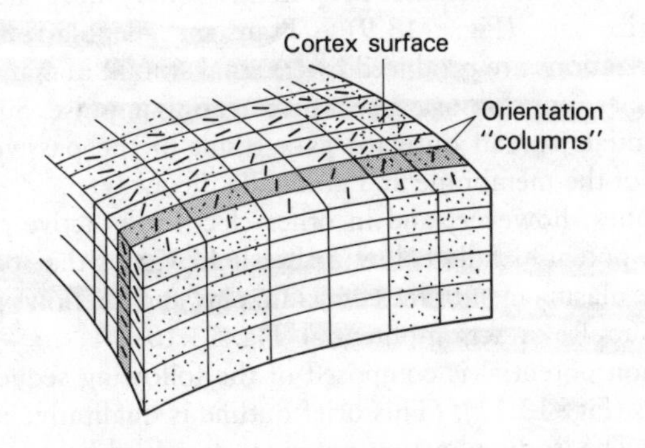

Fig. 13.8 Schematic view of the organization of orientation columns in the primate visual cortex. Section perpendicular to the cortical surface. A single 'column' is shown shaded. Each column is specifically responsive to a bar of light presented in a particular orientation. Similar groupings of cells responsive to other stimuli – for example, bars moving in a certain direction or different wavelengths of light (i.e. colour-coded columns) – have also been uncovered. Stippled areas represent cells driven by input from one eye, unshaded areas from the other eye (ocular dominance columns).

THE GENERATION OF ELECTRICAL IMPULSES IN NERVE CELLS

The action potential

When a nerve cell is excited by an incoming impulse (see below, Synaptic transmission) it generates an electrical impulse (or a series of impulses) which is propagated down its axon at speeds of up to 100 metres per second. The ability of a neurone to generate a propagating electrical signal is due to the special properties of its cell membrane, properties shared only with muscle cells and some specialized gland cells.

All living cells maintain an electrical potential difference across their cell membrane – the so-called **membrane potential** (see Chapter 6), which is the result of the inherent permeability of the cell membrane to certain ions and the active pumping in and out of ions against concentration gradients to maintain the required concentrations within the cell. As a result an electrochemical gradient is set up across the membrane (with the inside of the cell negative with respect to the exterior), which stabilizes at a particular level depending on the type of cell. In neurones the **resting potential** is around −70 to −80 mV. A decrease in the membrane potential towards zero is said to **depolarize** the membrane, an increase in potential (i.e. more negative) to **hyperpolarize** it (Fig. 13.9). Transient depolarizations and hyperpolarizations are produced by external stimuli in many cells, not only neurones, but do not cause a propagating impulse. Such **electrotonic** potentials spread outward as a result of the passive electrical properties of the membrane and gradually die away.

In neurones, however, and in other electrically active cells, if the membrane potential falls below a **threshold** value, the special properties of the plasma membrane come into play and ion flows produce an electrical impulse or **action potential** (Fig. 13.10).

The action potential is composed of the following sequence of ion movements (Fig. 13.11). (This brief outline is qualitative and greatly simplified. The 'sodium' action potential described below is found in many vertebrate and invertebrate neurones, and was the first to be described in terms of ion movement in the classic electrophysiological work on the giant axon of the squid (*Loligo*) in the 1950s and 1960s.)

The axonal membrane possesses many thousands of **voltage-sensitive Na^+ channels** that open transiently when the membrane potential reaches threshold in their vicinity (see Chapter 6: Ion channels). This results in the rapid inflow of sodium down its concentration gradient into the cell (Na^+ concentration within a cell is actively maintained at a much lower value than that in the surrounding medium). The rapid influx of positive charge leads to a further fall in membrane potential, more sodium enters and there is a sudden depolarization until membrane potential spikes at a peak value (around +55 mV, interior positive), as a result of opposing ion movements (see below).

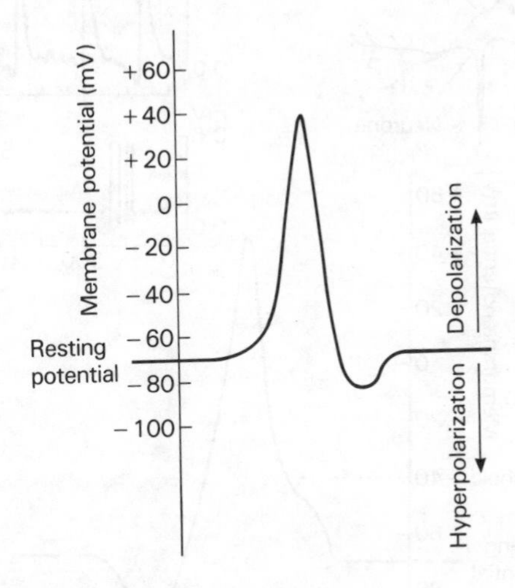

Fig. 13.9 Hyperpolarization and depolarization of the nerve cell membrane.

At the same time a more slowly acting *inactivating* gating mechanism has also been stimulated by depolarization and the sodium channels now begin to close, and to stay closed for a period. The membrane is now once again relatively impermeable to sodium. Depolarization has also caused a more slowly activated set of voltage-gated **K^+ channels** to open, which stay open as long as the membrane remains depolarized. Potassium ions (representing positive charge) therefore start to flow out of the cell down their electrochemical gradient (potassium concentration is normally maintained at a higher level inside the cell with respect to the external medium). Potassium outflow initially takes the membrane potential to below its resting level and ion flows then gradually restore it to the resting level. Until resting potential is restored, the neurone cannot fire again, however strongly it is stimulated; this is known as the **refractory period**. A continuing depolarizing stimulus after the refractory period will initiate another action potential, and so on – prolonged stimulation may produce a **train** of equally spaced action potentials.

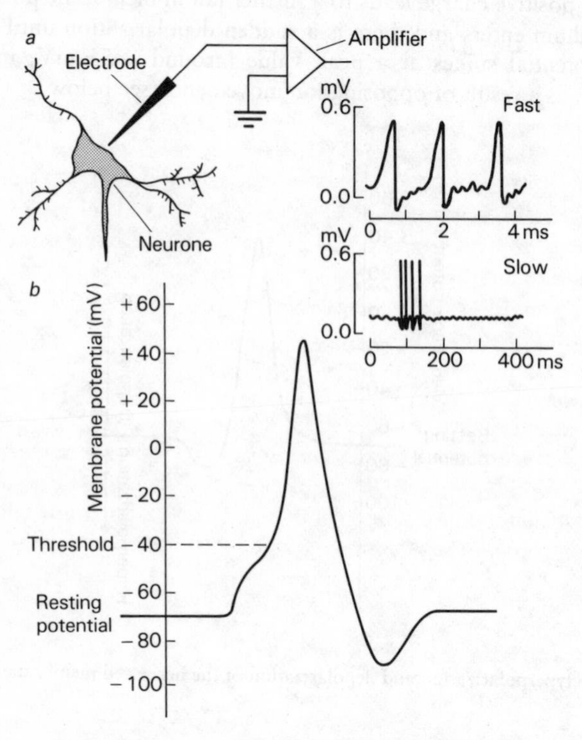

Fig. 13.10 *a*, Examples of patterns of action potentials recorded by an extracellular electrode from a firing neurone. *b*, The course of the action potential in terms of changes in membrane potential.

Because the actual numbers of ions required to discharge and recharge the membrane are minute, intracellular concentrations of potassium and sodium do not change significantly from 'normal' during the process. Thousands of action potentials can be discharged before metabolically dependent active transport by the sodium–potassium pump (which is powered by ATP hydrolysis) is required to reform the sodium and potassium gradients across the membrane (see Chapter 6).

An action potential is initially generated in the **axon hillock**, where the axon meets the cell body. It is propagated unchanged along the axon as waves of voltage-gated Na^+ channels open before it.

Action potentials in some mammalian neurones, in vertebrate

cardiac muscle and in some non-neural endocrine cells are generated not by the inflow of sodium but by calcium. Such cells possess plentiful voltage-gated Ca^{2+} channels instead of, or as well as, gated Na^+ channels. A striking example is the Purkinje neurone of the cerebellum, in which calcium action potentials are generated in the dendrites and sodium action potentials at the cell body.

The speed of electrical conduction along a nerve fibre is influenced by its passive cable properties. Large diameter fibres, for example, conduct impulses more rapidly than small diameter fibres. In the peripheral nervous system, large nerve fibres are surrounded by a fatty insulating sheath rich in **myelin**, composed of flattened **Schwann cells** wrapped round and round the axon, and which is interrupted at intervals at the **nodes of Ranvier** where patches of membrane are exposed.

Ion movement in and out of the axon is restricted largely to the nodes, and impulses 'jump' between nodes. Myelinated fibres also exist in the central nervous system. Myelinated nerves not only conduct signals more rapidly, but are more efficient, requiring less activity by the sodium–potassium pumps. They are involved in the transmission of 'urgent ' information, concerned, for example, with position, balance and movement; unmyelinated fibres conduct information concerned with, for example, regulation of blood flow and control of glandular secretion.

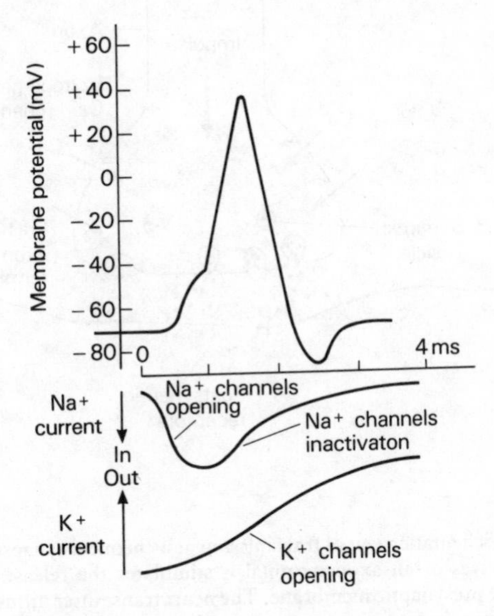

Fig. 13.11 The ion movements that generate an action potential.

SYNAPTIC TRANSMISSION

Neurones make connections with each other and with their target cells at restricted sites known as **synapses** at which information is passed from cell to cell. **Electrical synapses** occur when two cell membranes are closely apposed and current can flow between them through low-resistance junctions in the membrane – **gap junctions** (see Chapter 9). Most information flow in the nervous system, however, occurs through **chemical synapses**. These are sites at which the two opposing cells are separated by a narrow fluid-filled cleft that blocks direct electrical transmission (Fig. 13.12). Instead, the electrical signal (the action potential) is converted into a chemical signal – molecules of **neurotransmitter**. These diffuse across the cleft to interact with specific receptors on the opposing membrane. This interaction generates a new electrical signal in the receiving neurone.

Chemical synapses are formed between axon terminals of one neurone and the dendrites, cell bodies and even the axon terminals of other neurones. Signal transmission from nerve to muscle and other target cells is also through chemical synapses.

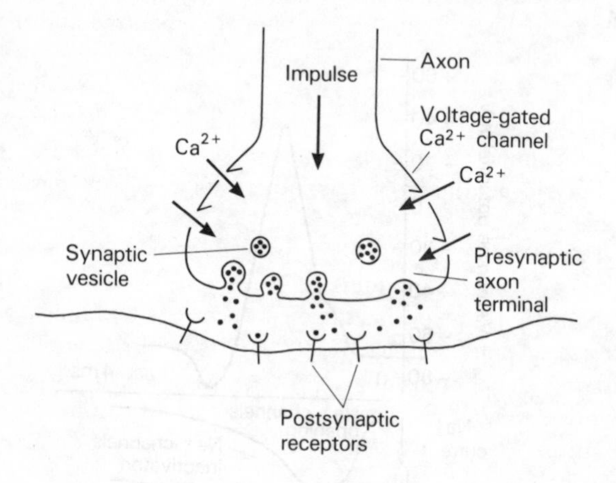

Fig. 13.12 Schematic view of transmission at a chemical synapse. When an electrical impulse arrives at an axon terminal it stimulates the release of neurotransmitter through the presynaptic membrane. The neurotransmitter diffuses across the narrow space between the two membranes to impinge on the postsynaptic membrane of the other, where it binds to specific receptors on the membrane surface.

Neurotransmitters

The first two neurotransmitters to be discovered were acetylcholine (ACh) and noradrenaline (NA or norepinephrine (NE)) which give their names to classes of **cholinergic** and **noradrenergic** neurones (i.e. those secreting, respectively acetylcholine or noradrenaline from their nerve endings), and are important transmitters in both the peripheral nervous system (where their actions were first demonstrated) and in the central nervous system.

Central nervous system transmitters are more difficult to identify, but more than 40 substances are now known to be produced and apparently have effects within the central nervous system (see below) and within peripheral neurones. The exact significance and roles of many of these are not yet certain.

Only a minority have been unequivocally identified as neurotransmitters. Although many of them have profound effects when administered experimentally, a true neurotransmitter has also to satisfy the following criteria: it must be shown to be synthesized and released by an axon terminal in response to stimulation, and in sufficient quantities to have an effect on the postsynaptic cell. It must also be rapidly removed after action. All this is technically difficult to demonstrate, requiring the analysis of minute amounts of compound and recording from single cells. Therefore, there is still considerable uncertainty about what the neurotransmitter may be, even at well-studied synapses. Acetylcholine, noradrenaline, serotonin (5-hydroxytryptamine (5-HT), dopamine, adrenaline, glutamate, glycine and GABA (γ-aminobutyric acid) and a few others have so far passed these tests (Table 13.1).

As well as carrying a signal across a well-defined synapse, some of these compounds (e.g. dopamine) are believed to act more diffusely, like circulating hormones, affecting neurones some distance from their site of release. Some neurochemicals appear to have modulatory effects, altering signal transmission across synapses in various ways, either by preventing or increasing the release of neurotransmitter from the transmitting cell, or by altering the properties of the postsynaptic receiving cell.

At any given synapse, a particular neurotransmitter will be either excitatory or inhibitory, depending on the properties of the postsynaptic receptors (see below) but never both. However, the same neurotransmitter can be inhibitory or excitatory at different sites. Acetylcholine, for example, excites the contraction of skeletal muscle, but inhibits the contraction of cardiac muscle (slowing the heartbeat)

Table 13.1 Chemicals with identified transmitter action in the mammalian nervous system*

Acetylcholine (ACh)

$$CH_3-\overset{\overset{O}{\|}}{C}-OCH_2CH_2N(CH_3)_3$$

An excitatory neurotransmitter at skeletal neuromuscular junctions where its effects are mediated by the nicotinic acetylcholine (ACh) receptor. Has inhibitory and excitatory effects in the autonomic peripheral nervous system and in the CNS, where it acts via muscarinic ACh receptors. In the CNS, it is thought to act chiefly by modulating the actions of other neurotransmitters. A marked decline in brain cholinergic function has been implicated in the development of Alzheimer's disease (senile dementia).

L-*glutamate*

$$HOOC-CH_2-CH_2-\underset{}{\overset{\overset{COOH}{|}}{CH}}-NH_2$$

A widespread, chiefly excitatory neurotransmitter in the central nervous sytem. Acts at both ion channel and second messenger-type receptors.

GABA (γ-aminobutyric acid)

$$HOOC-CH_2-CH_2-CH_2-NH_2$$

Major inhibitory neurotransmitter in the central nervous sytem, acting at both ion channel and 'second messenger'-type receptors. The benzodiazepine tranquillizers seem to act by facilitating the action of GABA at one of its receptors.

Glycine

$$HOOC-\underset{H}{\overset{H}{C}}-NH_2$$

Table 13.1 contd.

Inhibitory neurotransmitter in the CNS. Can also regulate the action of glutamate at its own receptor.

Serotonin (5-hydroxytryptamine, 5-HT)

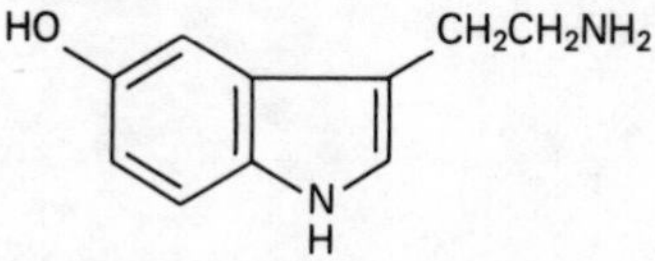

HO $CH_2CH_2NH_2$

N
H

Mediates profound and various effects throughout the brain, especially on the level of wakefulness and on enhancing pain sensation. Produced by relatively few neurones whose terminals branch widely. Acts at 'second messenger'-type receptors, of which several have been identified.

Histamine (also produced by mast cells in peripheral tissues where it is involved in inflammatory reactions, causing blood vessels to dilate and become leaky)

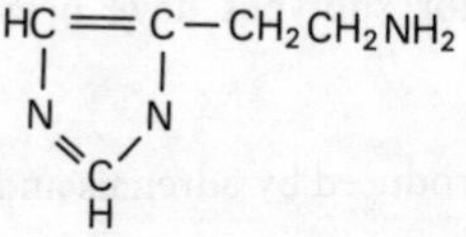

HC = C – $CH_2CH_2NH_2$

N N

C
H

Has neurotransmitter activity in the central nervous system. The action of antihistamines in relieving motion sickness seems to be due to their inhibiting histamine action in the region of the brain concerned with the routine sensing of position/movement. Three histamine receptor subtypes have been identified in the nervous system.

Noradrenaline (NA) (known as norepinephrine (NE) in the USA) (also produced in small amounts by adrenals)

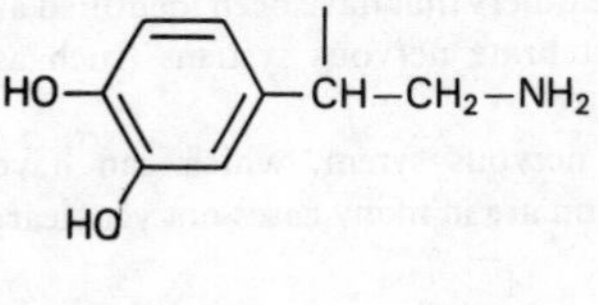

OH

HO– –CH–CH_2–NH_2

HO

Table 13.1 contd.

The main identified neurotransmitter in the peripheral sympathetic nervous system. Has excitatory and inhibitory effects in both peripheral and central nervous system. In the brain, its manufacture is limited to particular small groups of neurones whose terminals branch widely. Involved in many aspects of brain function.

Dopamine (DA) (also produced by adrenals)

HO–(benzene ring, with HO)–$CH_2-CH_2-NH_2$

Present in many regions of the brain. Like noradrenaline and serotonin, it is released only by relatively few neurones. A deficit of dopamine has been identified as a main cause of the deficient motor control symptoms of Parkinson's disease, which can be controlled to a certain extent by the drug L-dopa, a dopamine precursor. Appears to be involved in many other brain functions. The major tranquillizers such as chlorpromazine, which are sometimes effective in treating schizophrenia, appear to act by blocking dopamine action at its receptors. Several receptor subtypes have been identified.

Adrenaline (termed epinephrine in the USA; also produced by adrenals and acts as a major circulating hormone)

HO–(benzene ring, with HO)–$CH(OH)-CH_2-NH-CH_3$

Acts as a neurotransmitter in the sympathetic nervous system.

In addition to the substances listed above there are others that have been identified as neurotransmitters in other vertebrate and invertebrate nervous systems (such as aspartic acid, taurine, and some peptides).

*The many peptides produced in the central nervous sytem, which can have profound effects but whose role and mode of action are in many cases not yet clear, are dealt with in the next section.

and within the rest of the nervous system acts both as an excitatory and inhibitory neurotransmitter by a variety of different mechanisms.

Individual neurones become programmed during development to manufacture at most a very few types of neurotransmitter, and this restriction, together with the precise pattern of interconnections laid down during development, provides much of the functional order within the nervous system. In many cases, neurones have been found to synthesize a neuropeptide as well as a 'classical' neurotransmitter.

Neurotransmitter Release

Neurotransmitter is manufactured within the neurone and stored ready for use at the axon terminal packaged in membranous vesicles – **synaptic vesicles**. The arrival of an electrical impulse down the axon causes depolarization of the membrane. At the terminal this causes voltage-gated Ca^{2+} channels in the axon membrane to open, allowing Ca^{2+} to flow in down its concentration gradient (the internal concentration of free Ca^{2+} ions is always low compared with the extracellular fluid). The change in the internal Ca^{2+} concentration triggers the process of exocytosis by which neurotransmitter is released.

Neurotransmitter is released in packets or **quanta** each containing several thousand molecules. The amount of neurotransmitter released depends on how long the membrane remains depolarized, so each stereotyped impulse releases a set number of quanta. As soon as it has been released into the synaptic cleft, neurotransmitter begins to be removed, either by enzymatic destruction (by enzymes on the outer membrane faces of the synaptic cleft) or by re-uptake into its releasing neurone by the process of endocytosis. Rapid removal of neurotransmitter is essential to produce a sharp signal proportionate to the stimulus. A train of impulses arriving in rapid succession will keep up the concentration of neurotransmitter in the synaptic cleft, providing a larger signal to the receiving cell. This provides a way of grading the signal that is passed on with respect to the frequency of electrical impulses reaching the axon terminal, and thus the strength of the stimulus that is evoking them.

Postsynaptic receptors

The postsynaptic receptors for neurotransmitters fall into two broad classes. Some are ion channels, which open when the neurotransmitter binds, causing changes in the membrane potential that excite or inhibit the firing of the target cell. Of these, the best characterized so far are the ACh receptor in skeletal muscle, which is a cation channel and mediates the excitatory effect of acetylcholine on muscle cells (see Fig. 13.13), and the GABA receptor from CNS neurones, which is an anion channel and mediates the inhibitory effects of GABA.

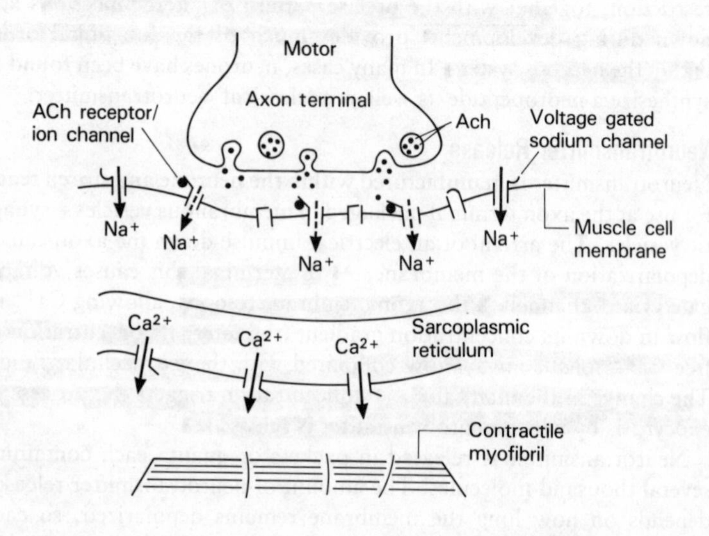

Fig. 13.13 Action of acetylcholine on muscle cells. Acetylcholine opens cation channels in the muscle cell membrane. Na^+ enters down its electrochemical gradient leading to depolarization of the membrane. At threshold voltage, voltage-gated Na^+ channels open leading to a massive inflow of positive current. This leads to an action potential that is propagated throughout the muscle cell membrane. By some means this signal is transmitted via the T tubules (see Fig. 9.2) to the sarcoplasmic reticulum of the muscle fibre, where it results in the opening of Ca^{2+} channels in the membrane. The release of Ca^{2+} onto the contractile myofibrils stimulates muscle contraction (see Chapter 8: Skeletal muscle).

The other class of receptors act in a more indirect way, transducing the signal across the membrane to stimulate production of 'second messengers' inside the cell that act either directly or indirectly to open or close the appropriate ion channels. The α-adrenergic receptors at which noradrenaline acts and the receptors for ACh in the central nervous system are of this type. They make use of intracellular signalling machinery that is common to a wide range of receptors (e.g. for hormones and growth factors) on many types of cells, not only neurones and muscle.

How neurones integrate incoming signals

A neurotransmitter can evoke one of two responses when it binds to the postsynaptic membrane – it can tend to *excite* the neurone, inducing it to fire or to accelerate the firing rate, or it can tend to *inhibit* or suppress firing. Neurotransmitter binding to receptors leads directly or indirectly to ion channels opening or closing in the postsynaptic membrane. This results in a change in membrane potential and the generation of a localized non-propagating electrical signal – the **synaptic potential** (see below).

As a broad generalization transmission at an excitatory synapse produces a membrane depolarization in the postsynaptic target cell – the **excitatory postsynaptic potential (e.p.s.p.)**, also called an **end-plate potential** in skeletal muscle. Inhibitory input on the other hand tends to keep membrane potential negative or even to hyperpolarize the membrane, which can be recorded as an **inhibitory postsynaptic potential (i.p.s.p.)**.

In the simple case of muscle, in which a muscle cell is innervated by a single axon, it contracts if succeeding e.p.s.p.'s are generated sufficiently rapidly by a train of action potentials in the motor neurone to push membrane depolarization below threshold, at which an action potential is produced in the muscle membrane initiating the sequence of events leading to contraction.

The motor neurone innervating that muscle cell, and most other neurones, has to integrate input from many different sources. There are thousands of synapses on its dendrites and cell body that lie within the spinal cord. Input will include information on the position of the body, and the state of contraction of other muscles, as well as information about the external environment relayed via the brain from the eyes and other sense organs. Some of this input is inhibitory, some excitatory, and at each synapse it produces a local synaptic potential – a change in the state of polarization of the postsynaptic membrane.

The dendrites and cell body of most neurones do not contain many voltage-gated Na^+ channels. Synaptic potentials therefore do not propagate, and die away within about a millimetre as they travel outwards from their point of origin. The neuronal membrane is able automatically to sum this input. The grand total (the **grand postsynaptic potential**) is determined both by the strength and duration of the signal at each synapse, which determines the timescale of membrane potential changes, and by the position of the synapse in terms of its distance from the central processing area of the membrane – the axon hillock – where the grand postsynaptic potential is registered and leads either to the neurone firing a train of action potentials at a certain frequency, or remaining silent.

In addition to the sodium and potassium channels mentioned earlier that generate the action potential, the axon hillock contains other types of potassium channels and a calcium channel whose coordinated action enables the cell to encode the magnitude of the grand postsynaptic potential as frequency of firing.

Neurotransmitters can also cause long-term biochemical changes in nerve cells, leading to the modification of the transmitting or receiving properties of the cell. As well as a means of transiently influencing the flow of information along a neuronal pathway, such changes may be involved in the longer-term alterations that underlie learning and memory (see below).

Presynaptic regulation

An additional common form of regulation is termed **presynaptic inhibition** or **presynaptic facilitation** (inhibitory or excitatory, respectively). This is mediated through synapses formed on the axon terminals themselves (Fig. 13.14). Input at these synapses modifies the strength of the action potentials as they arrive at the axon terminal and thus the amount of neurotransmitter that will be released. Inhibitory input at presynaptic terminals leads to less neurotransmitter being secreted whereas facilitatory input increases the amount of neuro-

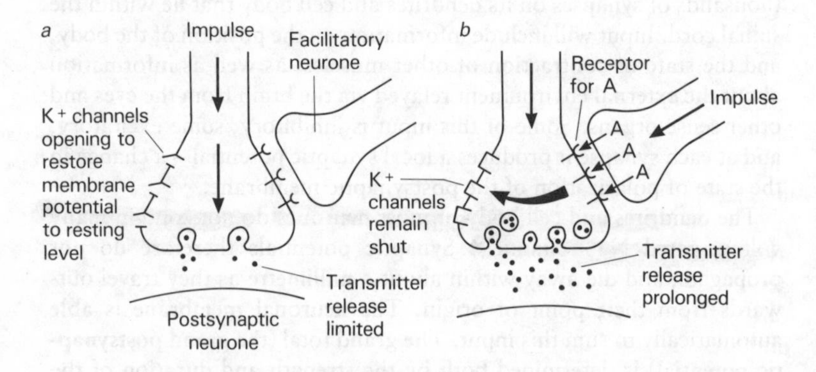

Fig. 13.14 *a*, Synaptic transmission in the absence of presynaptic regulation. *b*, Presynaptic facilitation. Stimulation of the facilitatory neurone prevents K^+ channels opening in the presynaptic nerve terminal and thus prolongs membrane depolarization and release of neurotransmitter.

transmitter by, for example, broadening the action potential, so that the membrane remains depolarized and transmitter release continues for longer. This type of **modulation** of the primary signal represented by the action potential has proved of particular interest in the study of 'learning' in the very simple nervous systems of molluscs.

CHEMICAL SIGNALS IN THE NERVOUS SYSTEM

Around 40 neuroactive substances have so far been found within the nervous systems of vertebrates and invertebrates. They include amino acids, catecholamines, other amines, nucleosides (adenosine), prostaglandins and a host of peptides ranging from the tripeptide thyrotropin-releasing hormone produced by the hypothalamus and the pentapeptide enkephalins, to the large polypeptide insulin. Only a few have yet been unequivocally identified as neurotransmitters (see Table 13.1). Others are strong contenders, whereas others may have more generalized 'neurohormonal' effects on nerve cells.

A notable feature of some brain chemicals is that they are made by only relatively few neurones (a few thousands out of the many millions of neurones in the brain) whose cell bodies cluster in particular areas but whose axon terminals (from which the transmitter is liberated) project to many regions. One theory is that the substances released by these neurones (e.g. noradrenaline, dopamine and serotonin) are acting by a 'garden sprinkler' mechanism and that the specificity of their action, like that of hormones elsewhere, is chiefly the result of specific receptors on appropriate neurones rather than strict one-to-one synaptic connections.

Neurochemistry and its close relative neuropharmacology have made great advances during the past 20 years. Past treatment of psychiatric disorders has been based on drugs discovered serendipitously, such as the benzodiazepine minor tranquillizers (e.g. Valium), and the much more powerful neuroleptics, such as the phenothiazines (e.g. chlorpromazine) used to treat schizophrenia. Some psychoactive drugs such as opium have been known for centuries. Knowledge of their site of action has led to development of new drugs and treatments on a more rational basis. The rational design of new drugs based on a better understanding of the brain's chemistry can be a long process as the difficulties in developing drugs based on the enkephalins (discovered more than 10 years ago) show (see below, Enkephalins and endorphins).

Like other chemical signals neurochemicals act chiefly at cell-surface receptors, and the identification and study of their receptors forms a large part of modern neurochemistry. The existence of different types of receptor for the same neurotransmitter has been known for many years. In some cases different receptors have been found to activate quite different intracellular pathways, or mediate the effects of the chemical on different types of cells. Until very recently the various receptor subtypes have been distinguishable only by their differential binding of pharmacological **agonists** and **antagonists**. These are substances that respectively mimic or inhibit the action of the natural ligand. Some produce these effects by interacting directly with the receptor in a highly specific manner (others interfere with re-uptake of transmitter or transmitter synthesis) and this type of compound has enabled receptor action to be studied and receptors isolated. In the early days of neurology, natural neurotoxins such as nicotine and muscarine were used to uncover the actions of acetylcholine at its 'nicotinic' and 'muscarinic' receptors. Nowadays many natural and synthetic compounds are used to distinguish different receptor subtypes, and to study the effects of neuroactive compounds in the brain.

Nervous system peptides

One of the growth areas in neurophysiology today is concerned with disentangling the complex effects within the nervous system and the rest of the body of the many peptides and polypeptides that are produced both by the nervous system and in non-neural tissues such as gut, heart, pancreas and placenta. During the 1970s, peptide hormones long familiar in other parts of the body started to be found in the brain. Many biologically active peptides are released by the pituitary, gut, heart, pancreas, placenta and other glands and tissues, and have a wide variety of physiological effects (see Table 9.6). Many of these (Table 13.2) and other peptides have now been found in the central nervous system, where the role of most of them is still not clear. It is not even known which act as specific neurotransmitters and which as more generally acting 'neurohormones'. In most cases, they occur in nerve terminals along with a 'classical' neurotransmitter such as noradrenaline, serotonin, dopamine or acetylcholine.

In some cases they may be involved in the central regulation of the same general effects as their counterparts in the rest of the body. Experimental stimulation of angiotensin receptors in the central nervous system of rats, for example, leads to a release of vasopressin from the pituitary, a rise in blood pressure and increase in drinking. Atrial natriuretic factor (atriopeptin, ANF, ANP) has recently been found in

Table 13.2 Some peptides found in both nervous and non-neural tissues

Peptides present in both gut (and associated tissues) and brain
Vasoactive intestinal peptide
Cholecystokinin
Substance P
Enkephalins
Neurotensin
Somatostatin
Gastrin
Insulin
Glucagon
Pituitary hormones also present in neurones and in other tissues
ACTH (pituitary/brain/placenta)
β-endorphin (pituitary/brain/placenta)
Hypothalamic releasing factors present also in non-neural tissue and elsewhere in nervous system
Corticotropin-releasing factor (CRF)
Thyrotropin-releasing factor (TRF)
Others
Atrial natriuretic factor (atriopeptin, ANF, ANP) (heart/brain)
Angiotensin II (precursor) (brain/liver) (angiotensin cleaved by renin and then converting enzyme/brain)
Peptides related to bombesin (gut/brain)
Peptides related to mast cell degranulating peptide from bee venom (gut/brain)

brain areas known to be involved in the central regulation of the cardiovascular system, and may also control the release of vasopressin from the pituitary. A few, such as substance P (see below), are probably straightforward neurotransmitters.

One group of brain peptides, the so-called endogenous opioids – the enkephalins and endorphins – has excited particular interest.

Enkephalins and Endorphins

Enkephalins (Met- and Leu-enkephalin, Tyr–Gly–Gly–Phe–Met (or Leu)) are produced at nerve terminals in the central nervous system, and also in the adrenals and other non-neural tissue. One of their roles in the central nervous system may be to inhibit pain perception and influence mood. They were discovered in 1975 as the natural ligands of the brain's opiate receptors, at which the narcotic opiate drugs act to produce analgesia and euphoria. They are often called the brain's natural painkillers and ever since their discovery they have held out the hope that analgesic drugs as effective as the opium derivative morphine, but without its addictive effects, could be developed. The analgesic and euphoric effects of opium and its derivatives have been known for thousands of years but the receptors at which they act in the brain were only revealed in the mid-1970s, to be followed closely by the discovery of the enkephalins. ß-Endorphin, a longer peptide of 31 amino acids that contains the sequence of enkephalin at its amino terminus, was subsequently discovered in the brain, the pituitary, and later also in the placenta. Its role in the body is not yet clear. If injected experimentally into the cerebral ventricles it produces a profound analgesia and paralysis in rats. It is apparently released from the pituitary in response to stress and may act at opiate receptors elsewhere in the body (e.g. in the gut).

Investigations into the natural role of enkephalins and their possible application as painkillers have been hampered by the fact that they are exceptionally short-lived in the brain, being destroyed almost as soon as they are released. Their short life makes them good candidates for 'classical' neurotransmitters, possibly acting in neuronal pathways that inhibit the 'pain' pathways mediated by another suspected peptide neurotransmitter – substance P. Attempts to develop analgesics based on enkephalins or endorphins have not been successful so far. In animal studies they show many side-effects, paralysis and possible addictiveness. There is also the problem of administration. If taken by mouth, enkephalins and ß-endorphin, like many other peptides, are destroyed by proteolytic enzymes in the gut. Intravenous injection is ineffective as a route for getting ß-endorphin into the brain as it cannot pass the blood–brain barrier. At present, direct infusion or injection into the brain is required to gain any effect. One avenue of approach to increasing the effects of enkephalins is to find ways of suppressing the degradative enzymes that naturally break them down in the brain, and the effects of various selective enzyme inhibitors are now being studied.

The enkephalins have effects other than analgesia. They appear also to be released at the hypothalamus in response to stress, where they stimulate the release of corticotropin-releasing factor from hypothalamic secretory neurones, and therefore the release of adrenocorticotropin (ACTH) from the anterior pituitary.

ß-Endorphin is synthesized as part of a much longer precursor protein – pro-opiomelanocorticotropin – which is subsequently split into several different peptide hormones, including ACTH and ß-lipotropin as well as ß-endorphin. Synthesis from a larger precursor is a fairly general mechanism for producing small peptide hormones. Some are produced from precursor proteins containing several copies of the same peptide sequence (e.g. enkephalin and epidermal growth factor). The precursor to ß-endorphin is particularly versatile (see Fig. 3.1*d* for details).

Substance P and Pain Perception

Information about pain is collected by special **nociceptive** sensory receptors in skin and internal tissues that are distinct from the receptors sensitive to pressure and touch. It is conveyed to the brain through well-defined 'pain pathways', which are relayed via the thalamus to wide-spread areas of the cortex. A transmitter in the pain pathways is thought to be the peptide **substance P** (first isolated in 1931), which is found in brain, in small fibres of peripheral sensory nerves and in the spinal cord. It is also produced in the gut where it causes smooth muscle contraction.

Substance P and two other structurally related neuropeptides are known collectively as the **tachykinins**. The receptor for one of them – substance K – has recently been isolated and its amino acid sequence determined, the first neuropeptide receptor for which this has been done.

The neuroendocrine system

The brain influences the body not only by direct innervation of tissues and organs but also by controlling the release of hormones from the pituitary (or **hypophysis**). Many of these then act on the glands of the endocrine system (adrenals, thyroid, gonads, etc.) to stimulate the manufacture and release of circulating hormones such as the adrenal steroids, the steroid sex hormones and thyroid hormone. In many cases the circulating hormones have feedback effects on the brain, adding a further level of control on pituitary release.

The nervous system is therefore subject not only to the chemical signals it generates within itself in response to direct sensory stimuli, but to hormones such as the oestrogens, androgens and others that direct, for example, reproductive development and the mammalian reproductive cycle by a complex web of interactions between brain and other target tissues. The already highly intricate hormonal relationships between the brain and the rest of the body are being revealed in ever greater complexity as more and more factors that influence the neuroendocrine system are being discovered.

Hypothalamic and Pituitary Hormones

A region of the brain located immediately above the pituitary – the **hypothalamus** – is the main regulator of pituitary secretion. It contains specialized **neurosecretory** neurones that secrete various small peptides (often known generally as **releasing factors**) (see Table 9.2) into the local hypothalamic circulation in response to signals from other brain regions. The hypothalamic circulation links directly with the **anterior pituitary** (the **adenohypophysis**) which acts as a storage depot for various peptide hormones made in the hypophysis.

Hypothalamic releasing peptides stimulate or inhibit the release of hormones from the anterior pituitary (see Table 9.3 for the main effects of these hormones).

The connection between hypothalamus and posterior pituitary (the **neurohypophysis**) is more direct. It receives a rich supply of neurosecretory nerve endings from the hypothalamus, which release the peptides **vasopressin** and **oxytocin** directly into the general circulation.

Activity at the hypothalamus is regulated by routine feedback of physiological signals and also by pain, trauma and emotion. Many brain chemicals have been implicated in generating moods and emotional states and their action at the hypothalamus is of particular interest.

A wide variety of physiological signals feed back to the brain, sometimes directly to the hypothalamus. Circulating androgens and oestrogens act at the hypothalamus to stimulate release of follicle-stimulating hormone (FSH) and luteinizing hormone (LH) from the pituitary, which in turn inhibits the release of prolactin. In contrast, milk production as a result of prolactin release generates stimuli that inhibit FSH and LH release. Some of the anterior pituitary hormones are routinely controlled by feedback from blood levels of the circulating hormones whose production they regulate. The adrenal steroids, for example, act at the hypothalamus to inhibit further release of ACTH from the pituitary. Vasopressin (a potent vasoconstrictor and anti-diuretic) is regulated to a large extent in accordance with the osmotic pressure of the blood. Haemorrhage is the most powerful stimulant yet found to increase vasopressin output from the pituitary.

Recent research has implicated some newly discovered brain peptides in the control of hypothalamic secretion, and the interaction between the action of certain peptides in the brain and in the rest of the body is likely to be exceedingly complex.

LEARNING AND MEMORY

The ability to learn, in the widest sense of the word, is shared to a greater or lesser degree by all but the very simplest animals. The process of learning involves a permanent alteration in the patterns of information flow through the nervous system which implies an equally permanent alteration in the functional architecture of its components – the neurones and their myriad interconnections. The general ability of the brain to alter in this way is known as **plasticity**. How this is achieved, and how information is stored and retrieved with an efficiency that, for any reasonably well-developed brain far outstrips that of the most powerful computer yet devised, is a central problem in neurobiology.

Neurones in the brain are linked in a network, each cell making and receiving many connections. It is on these connections – the synapses – that research into learning and memory is focused at the cellular level. Computer modelling at a systems level of 'neuronal networks' reveals that densely interconnected networks can store a vast amount of information, since a change in the strength of relatively few connections can produce significant changes in the output of the network as a whole.

Present research is based on the idea that the changes that occur during learning involve permanent restructuring of selected synapses so that they preferentially transmit when the neurone is subsequently stimulated. This would channel signals along a preset path through a 'memory circuit', activating a particular pattern of neurones representing a particular memory or element of a memory. As with all higher functions, no-one yet has any idea of what a 'memory circuit' might consist of, or even what a 'memory' at this level might be, but the general hypothesis has led to several fruitful lines of research in cellular and molecular neurobiology, as well as to the 'connectionist' way of looking at brain function as a whole (see reading list).

Neurobiologists are now beginning to identify the various ways in which one neurone can temporarily or permanently change the information-processing properties of another. By clearly identifying the possible mechanisms for change this work will provide a sound biochemical and cellular basis for theories of learning and memory at a higher level of organization.

Animal experiments and human experience distinguish at least two distinct types of memory – short-term and long-term. Short-term applies to memories that last (in humans) a few seconds, minutes or hours, like remembering a telephone number long enough to dial it. Long-term memory, on the other hand, lasts for months and years – it underlies the permanent acquisition of physical skills and learning to use our senses to negotiate the environment as well as more conscious learning processes such as the acquisition of language. Short-term memory is not thought to entail any permanent alterations, whereas long-term memory does. (The situation is undoubtedly more complex in real life, but this simple distinction has proved helpful in analysing memory at the cellular level.)

Short-term memory theories are built on the idea of a temporary modification changing the transmitting properties of a synapse. This could occur in many ways. Each of the individual components of the synaptic transmission machinery – such as ion channels, postsynaptic receptors and cytoskeletal elements involved in neurotransmitter release – are all sensitive to biochemical modifications that can dramatically alter their properties. Enzymes such as protein kinases (p.255) that are activated when a neurone is stimulated, add phosphate groups to ion channels, receptors and cytoskeletal proteins. Phosphorylation can alter the properties of ion channels so that, for example, they open less often, or remain open longer, or remain permanently open or shut when the neurone is stimulated; it can make a receptor unresponsive (or more responsive) to a neurotransmitter; and it can induce some cytoskeletal elements to disassemble. Steps in the pathways through which receptors transduce the signals represented by neurotransmitters and modulators can also be affected by phosphorylation of the proteins involved. Any or all of these changes, which might occur on both sides of the synapse, would have profound effects on synaptic transmission. A common means of temporarily changing the transmitting properties of a synapse is by presynaptic regulation, which alters the signal arriving at the axon terminal and therefore the amount of neurotransmitter released.

Studies on long- and short-term memory in experimental animals suggest that short-term memory does not require the synthesis of any new proteins (and by inference does not require any new gene expression) whereas long-term memory probably does. There is no evidence yet of what that gene expression might entail. It could result in the synthesis of regulatory proteins that change the properties of pre-existing neurotransmitter receptors and ion channels at synapses, or the synthesis of new receptors or ion channels, all of which could permanently alter the receiving and transmitting properties of the neurone.

Aplysia

Progress in investigating the relatively simple neural circuits of lower animals may, however, reveal some general principles that could be applicable in higher animals.

The exhaustive study of the very simple nervous systems of animals such as the sea-hare *Aplysia californica* (a gastropod mollusc), carried out over the past 20 years, has enabled a few instances of 'learning' and 'memory' to be described in great detail in terms of the individual neuronal circuits involved and the biochemical changes within them. In this case, short-term 'memory' involves biochemical modification of ion channels in particular synapses so that they are locked shut, changing the transmitting properties of the synapse.

If threatened by a gentle jet of water on its mantle (the outer, protective fold of epidermal tissue), *Aplysia* withdraws its siphon – this is the basic 'siphon withdrawal reflex'. It is mediated by a set of sensory neurones in the siphon that pick up the stimulus and interact with the motor neurones serving the muscles that withdraw the siphon and gill. (The circuits involved have been mapped in great detail.) A brief series of electric shocks to the tail of the animal **sensitizes** this response. This means that if *Aplysia* is threatened, siphon withdrawal (powered by muscle contraction) is now more rapid and vigorous, and this effect persists for an hour or so. This type of sensitization is taken as a very simple model of short-term memory.

Intensive biochemical and neurophysiological investigations in *Aplysia* have now pieced together much of the picture. The electric shocks activate another set of sensory neurones. These release the chemical serotonin (5-hydroxytryptamine) onto synapses at which mantle sensory neurones are contacting the motor neurones serving the siphon (Fig. 13.15).

Serotonin acts presynaptically on the sensory axon terminal to **modulate** the release of the neurotransmitter that is carrying the excitatory signal between sensory and motor neurone.

Initially, it increases the synthesis of the 'second messenger' cyclic AMP within the sensory neurone terminal. The increased level of cyclic AMP activates a protein kinase that closes a crucial K^+ channel in the presynaptic terminal membrane by phosphorylating it (or an associated regulatory protein). This change lasts for an hour or so, even after the neuromodulatory signal has ceased. It has the effect of prolonging the depolarization of the axon membrane that occurs whenever the sensory neurone is stimulated (by the original water jet stimulus), as the K^+ channel normally helps to restore membrane potential to normal. This leads to more neurotransmitter being released from the sensory neurone, therefore increasing the strength of the signal transmitted to the motor neurone. This in turn increases the rapidity and vigour of the

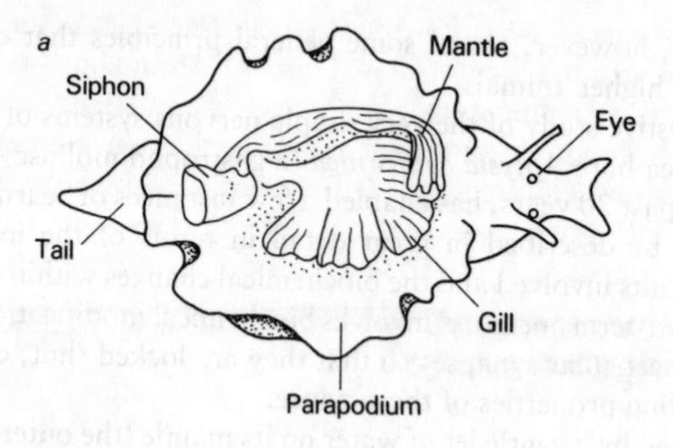

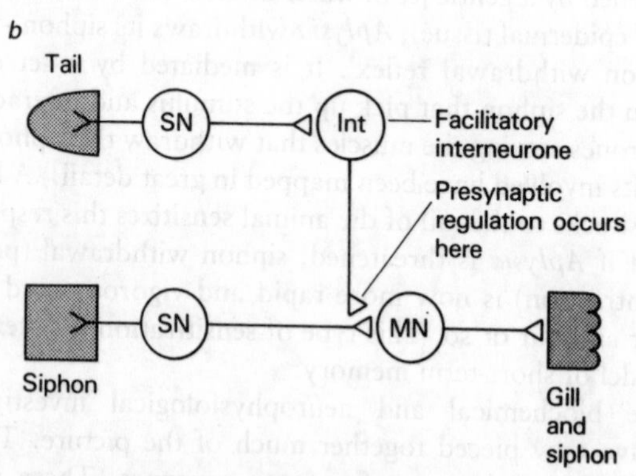

Fig. 13.15 The sea hare *Aplysia californica*. The overlying parapodia and the protective mantle are shown drawn back to reveal the gill. *b*, Simplified representation of the neuronal circuits involved in the siphon withdrawal reflex and its sensitization (see text). SN, sensory neurone; MN, motor neurone.

muscle contraction that withdraws the siphon.

A series of closely spaced tail shocks over a few hours each day for several days produces sensitization that lasts weeks, and has been taken as a model of long-term memory. It seems to involve the same synapses and the same K^+ channel but requires protein synthesis. The nature of this long-term change has not yet been uncovered, but several promising lines of approach are being pursued.

Long-term potentiation in the hippocampus

Progress in uncovering the biochemical and cellular bases for learning and memory in the vastly more complex brains of mammals has naturally progressed more slowly. Evidence from brain-damaged patients seems to show that conscious memories themselves are not localized to any particular area but are distributed diffusely throughout the cortex. One region has been pinpointed, however, in animals and humans, as being crucially important in the process of laying down memory, and this is where much biochemical and neurophysiological research on learning and memory is focused at present.

The **hippocampus** (see Fig. 13.4*c*) is a region of the cortex that has been implicated for many years in learning. A phenomenon known as long-term potentiation (LTP) can be induced experimentally in the hippocampus and is studied as a model of a memory-like process that may have some relevance to 'real' learning. A brief high-frequency stimulation of the hippocampus in rats causes a facilitation of synaptic transmission through hippocampal pathways for a few hours afterwards, changing the pattern of firing.

The biochemical basis of LTP is now beginning to be explored, in particular the involvement of a particular receptor – the so-called NMDA receptor – for the neurotransmitter glutamate. Unusually, NMDA receptors in the hippocampus have been found that only become responsive when the neuronal membrane becomes depolarized, that is, if the cell is being stimulated. This makes them a good candidate for a means by which a cell's properties can be changed by an incoming stimulus. Blocking this receptor by specific drugs prevents LTP being established although ordinary synaptic transmission seems to be unaffected. *In vivo* treatment with the same anti-receptor drug also blocks particular types of learning in rats.

REPAIR AND REGENERATION IN THE NERVOUS SYSTEM

Most tissues can within certain limits repair or renew themselves if damaged, but the mammalian central nervous system, unlike those of invertebrates and lower vertebrates, has lost most of its capacity for repair. Neurones of any sort lose the power to divide once mature, but peripheral neurones can at least regrow axons and other nerve processes if damaged, remaking previous connections with some precision. Mammalian central nervous system (CNS) neurones seem to be able to sprout in a limited way but if they are damaged they usually die, releasing toxic compounds that also lead to the death of surrounding neurones. Severe spinal and brain injuries are therefore devastating and irreparable, whereas a severed peripheral nerve can often be saved by prompt surgery, with little subsequent loss of function.

During the past decade, however, the efforts of neurologists to persuade mammalian CNS neurones to regrow has begun to show the first signs of success. They have found that CNS neurones, like peripheral neurones, will regrow if placed in a more favourable environment – in the peripheral nervous system, for example. It is therefore the environment provided by the brain and spinal cord, and not some inherent property of the neurones themselves, that inhibits regrowth of damaged CNS neurones. There are very good biological reasons for this inhibition. In everyday circumstances, the death of a few damaged neurones is far less disruptive to the complex mammalian brain, with its multiple parallel pathways, than the possibility that they may regrow and form wrong connections. Only when large numbers of brain cells die as a result of a stroke (where their death is caused by lack of oxygen), physical injury or disease do the effects become apparent.

Neurologists are now looking for means of overriding the inhibitory influences exerted by the CNS. One approach is to look at what stimulates nerve regrowth in the peripheral nervous system. An important influence there appears to be the Schwann cells, non-neural cells that form the fatty myelin insulating sheath surrounding each nerve fibre in many peripheral nerves. When peripheral nerve fibres are damaged, the terminal portion beyond the damaged area degenerates, the myelin sheath is lost, but the Schwann cells themselves remain, forming a tube down which the regrowing axon is guided to its correct destination. The Schwann cells may not only provide mechanical guidance but also growth factors or other molecular signals that could be used in the CNS as well, and such factors are now being looked for.

One substance known to be required for the maintenance of some peripheral neurones is **nerve growth factor**, a protein produced by smooth muscle and other non-neural tissue, and which stimulates the differentiation and extension of processes from cultured neurones of the sympathetic nervous system. Its exact role *in vivo* is not yet known, but it is secreted from target cells once they have become innervated and is believed to be involved in maintaining innervation. In recent experiments nerve growth factor injected into the brains of 'aged' rats has been claimed to prevent the atrophy of brain cells and to halt the normal deterioration in the ability to learn particular tasks that rats (like people) display as they age.

Embryonic neurones are another possible source of molecules that might override inhibition of regrowth. Embryonic neurones are the only nerve cells that appear to retain their ability to grow in the surroundings provided by the mature mammalian CNS. Nerve cells from rat embryos have been transplanted into the experimentally damaged brains of adult rats, and have been found to regrow and even form connections with host nerve cells. As well as a source of molecular signals therefore, the possibility of replacing damaged CNS neurones with transplanted embryonic cells is now being actively explored in animals. However, the formidable problem of ensuring that transplanted cells make the *correct* connections is still unresolved. Any extension of the work to humans is also likely to involve difficult ethical decisions, involving as it would, the use of embryonic cells taken from aborted foetuses.

Transplantation of foetal neurones into the human brain has already been undertaken in the UK and Sweden on a small number of patients suffering from Parkinson's disease. Parkinson's disease is the result of the death of groups of brain cells that produce dopamine. Since the effects of dopamine in the brain appear not to rely on a strict one-to-one connection between producing and target cells, function might therefore be restored by introducing a depot of any dopamine-producing neurones into the brain. The foetal neurones in this case are simply introduced as individual sources of dopamine and do not make any contribution to the information communication network. Transplantation of a complete foetal brain is expressly forbidden in the voluntary guidelines in force in the UK. These operations were performed only recently, and the long-term results are not yet available.

FURTHER READING

See general reading list (Kandel and Schwarz; Kuffler et al.; Alberts et al.), and also references in Chapter 8.

P.S. CHURCHLAND, *Neurophilosophy*, MIT Press, Cambridge, MA, 1986. This provides in its first part an excellent general introduction to the brain for the beginner, and in the second explores various ideas on the mind–brain problem and models of brain function.

D. HUBEL, *Eye, brain and vision*, W.H. Freeman, San Francisco, 1988. A very readable account, for a wide audience, of his own and others' work on information processing in the visual cortex, by one of the discoverers of feature detection columns.

D.E. RUMELHART and J.L. MCCLELLAND, *Parallel distributed processing: explorations in the microstructure of cognition. Vol. 1: Foundations*, MIT Press, Cambridge, MA, 1986; and J.L. MCCLELLAND and D.E. RUMELHART, *Parallel distributed processing: explorations in the microstructure of cognition. Vol. 2: Applications*, MIT Press, 1986. The theoretical background to the influential 'connectionist' theories of brain function and cognition.

P. Johnson-Laird, Connections and controversy, *Nature*, 330 (1987) 12, and references therein.

D. Zipser and R.A. Anderson, A back-propagation programmed network that stimulates response properties of a subset of posterior parietal neurons, *Nature*, 331 (1988), 679–84.

M. Livingstone and D. Hubel, Segregation of form, colour, movement and depth: anatomy, physiology and perception, *Science*, 240 (1988) 740–9. A review of recent work on the segregation of different aspects of visual information in the visual system.

D.I. Perret, A.J. Mistlin and A.J. Chitty, Visual neurones responsive to faces, *Trends in Neuroscience*, 10 (1987) 359.

P. Goldman-Rakic, Topography of cognition: parallel distributed networks in primate association cortex, *Annual Review of Neuroscience*, 11 (1988) 137.

S.T. Mason, *Catecholamines and behaviour*, Cambridge University Press, Cambridge, 1984.

D.R. Lynch and S. Snyder, Neuropeptides: multiple molecular forms, metabolic pathways and receptors, *Annual Review of Biochemistry*, 55 (1986) 773–99.

M. Mishkin and T. Appenzeller, The anatomy of memory, *Scientific American*, 256 (1987), 62.

Trends in Neuroscience 11 (4) (1988). A special issue devoted to memory.

T.J. Carew and C.L. Sahley, Invertebrate learning: from behaviour to molecules, *Annual Review of Neuroscience*, 9 (1986) 435. A survey of various invertebrates used to study learning, from bees to *Aplysia*.

Trends in Neuroscience, 10 (4) (1987). A special issue on excitatory amino acids and the NMDA receptor.

G.L. Collingridge The role of NMDA receptors in learning and memory, *Nature*, 330 (1987) 604–5.

A. Fine, Transplantation in the nervous system, *Scientific American*, 255 (1986) 42.

SELECTED READING LIST

The books listed below are a small selection of undergraduate textbooks and other introductory texts. The majority include extensive references to the scientific literature.

General

P.H. RAVEN, R.F. EVERT and H. CURTIS, *Biology of plants*, Worth, New York, 1981.

A general and wide-ranging introduction to the plant kingdom, including chapters on algae and fungi.

R. MCNEILL ALEXANDER, *The invertebrates*, Cambridge University Press, Cambridge, 1979.

R. MCNEILL ALEXANDER, *The chordates*, 2nd edn, Cambridge University Press, Cambridge, 1981.

The world of the molecular and cell biologist contains rather few organisms; these two books provide a good introduction to the marvellous diversity of form and function in the animal kingdom.

SCIENCE, 240 (1988) 1427–75. A series of articles on some of the organisms most used in experimental biological research, and why. Includes articles on retroviruses, bacteria, yeast, *Xenopus*, *Caenorhabditis*, *Drosophila*, plants, transgenic animals and primates.

Biochemistry

L. STRYER, *Biochemistry*, 3rd edn, W.H. Freeman, San Francisco, CA, 1988.

A. LEHNINGER, *Principles of biochemistry*, Worth Publishers, Inc., New York, 1982.

General cell biology

B. ALBERTS, D. BRAY, J. LEWIS, M. RAFF, K. ROBERTS and J.D. WATSON, *The molecular biology of the cell*, 2nd edn, Garland Publishing, New York & London, 1989.

J. DARNELL, H. LODISH and D. BALTIMORE, *Molecular cell biology*, W.H. Freeman, San Francisco, CA, 1986.

Comprehensive cell biology texts requiring little prior knowledge. Coverage is mainly of animal cells, but Alberts et al. has a section on plants. As well as covering the basic principles of cell biology, including basic genetics, they also deal with selected aspects of, for example, development, nervous system function, immunology and the cancerous cell.

R.V. KRSTIC, *Ultrastructure of the mammalian cell: an atlas*, Springer-Verlag, Berlin/Heidelberg/New York, 1979. Beautiful three-dimensional drawings of organelles and cell structure reconstructed from electron micrographs.

Genetics

J.D. WATSON, M.H. HOPKINS, J.W.ROBERTS, J.A. STEITZ and A.N. WEINER, *Molecular biology of the gene*, 4th edn, Vols I and II, Benjamin/Cummings, Menlo Park, CA, 1987. Volume I covers basic principles and facts in prokaryotic and eukaryotic molecular genetics; Volume II deals with the molecular genetics of eukaryotic viruses, certain aspects of development and immunology, oncogenes, and the origin of life and molecular evolution.

M.W. STRICKBERGER, *Genetics*, 3rd edn, Macmillan Publishing Co., New York, 1985.

P.J. RUSSELL, *Essential genetics*, Blackwell Scientific, Oxford, 1987.

N. MCLEAN, S.P. GREGORY and R.A. FLAVELL, *Eukaryotic genes: structure, activity and regulation*, Butterworths, London, 1983.

D.A. HARTL, *Human genetics*, Harper & Row, New York, 1983. A very readable introduction to human genetics.

Evolution

S. LURIA, S.J. GOULD and S. SINGER, *A view of life*, Benjamin/Cummings, Menlo Park, CA, 1981.

J. MAYNARD SMITH, *The theory of evolution*, Penguin, Harmondsworth, 1975.

Semi-popular introductions to evolution and evolutionary theory.

Th. DOBZHANSKY, F.J. AYALA, G.L. STEBBINS and J.W. VALENTINE, *Evolution*, W.H. Freeman, San Francisco, 1977. A comprehensive undergraduate-level textbook.

Developmental biology

L. BROWDER, *Developmental biology*, Saunders, Philadelphia, 1980.

N.J. BERRILL and G. KARP, *Development*, 2nd edn, McGraw-Hill, New York, 1981.

Undergraduate-level textbooks providing a broad introduction to developmental biology.

Immunology

L.E. HOOD, I.L. WEISSMAN, W.B. WOOD and J.H. WILSON, *Immunology*, 2nd edn, Benjamin/Cummings, Menlo Park, CA, 1984. An undergraduate-level textbook dealing comprehensively with cellular and molecular immunology.

I.M. ROITT, *Essential immunology*, 6th edn, Blackwell Scientific, Oxford, 1988.

S. SELL, *Immunology, immunopathology and immunity*, Elsevier, North-Holland, Amsterdam/London/New York, 1987.

General textbooks, somewhat more medically oriented than Hood et al.

Neurobiology

E.R. KANDEL and J.H. SCHWARTZ, *Principles of neural science*, 2nd edn, Elsevier, Amsterdam/Oxford/New York, 1985. A wide-ranging introductory textbook.

S.W. KUFFLER, J.G. NICHOLS and A.R. MARTIN, *From neuron to brain: a cellular approach to the function of the nervous system*, 2nd edn, Sinauer, Sunderland, MA, 1984.

Magazines, journals, serials

Scientific American and its compilations of articles on particular themes, e.g. evolution, neurobiology.

New Scientist (in the United Kingdom) and *Science News* (in the United States) for news about science and short reports and reviews of current research, written for a wide audience.

Nature, *Science*, *The Proceedings of the National Academy of Sciences*, and *Cell* publish much key original research in areas covered by this book, especially the molecular genetical, biochemical and cell biological aspects of immunology, endocrinology, cancer biology, development and neurobiology, and work on the application of recombinant DNA to the study and diagnosis of disease. *Nature* and *Science* also carry general news coverage about science, and up-to-date reviews and explanatory commentaries on current research. *Cell* has regular short reviews.

The 'Trends in' series, published monthly by Elsevier, carries short reviews of current work. It now covers biochemistry, genetics, neurosciences and pharmacology. *Immunology Today*, also published by Elsevier, does the same for immunology. *Science Progress*, published four times a year by Blackwells, is also a useful source of review articles written with the non-specialist in mind. The *British Medical Bulletin*, published annually, is a compilation of articles on a different topic each year.

The New England Journal of Medicine, *The Lancet* and the *British Medical Journal* often carry non-specialist reviews and original reports of applications of the new biology in medicine.

Bio/technology and *Biotechnology* carry news of medical, industrial and agricultural developments in genetic engineering.

International Review of Cell Biology for reviews on many aspects of cell biology. *International Journal of Cytology*, *Journal of Cell Biology* and *Journal of Cell Science* also carry regular review articles.

Development, *Developmental Biology*, *Immunology* and *Journal of Immunology*, also regularly carry review articles.

The 'Annual Reviews' series (e.g. Biochemistry, Genetics, Cell Biology, Neuroscience). Somewhat technical reviews, often assuming considerable background knowledge.

Cold Spring Harbor Symposium on Quantitative Biology. Collected papers from the annual conferences on molecular biology held at the Cold Spring Harbor Laboratory. They provide up-to-date coverage of important developments in selected fields but papers are often highly technical.

INDEX

A page number followed by T refers to a table; and by F, to a figure.